W. Brockmann, P. L. Geiß,
J. Klingen, B. Schröder

Adhesive Bonding

Further Reading

G. Habenicht

Applied Adhesive Bonding

A Practical Guide for Flawless Results

2009
ISBN: 978-3-527-32014-1

Walter Brockmann, Paul Ludwig Geiß,
Jürgen Klingen, Bernhard Schröder

Adhesive Bonding

Materials, Applications and Technology

Translated by Bettina Mikhail

WILEY-VCH Verlag GmbH & Co. KGaA

The Authors

Prof. Dr.-Ing. Walter Brockmann
Brunnenstrasse 71
67661 Kaiserslautern
Germany

Prof. Dr.-Ing. Paul Ludwig Geiß
Technical University (TU) Kaiserslautern
AWOK
P.O. Box 3049
67663 Kaiserslautern
Germany

Dr. Jürgen Klingen
3M Deutschland GmbH
Corporate Research Laboratory
Carl-Schurz-Strasse 1
41453 Neuss
Germany

Dr. Bernhard Schröder
NMI Natural and Medical Sciences Institute
at the University of Tübingen
Markwiesenstrasse 55
72770 Reutlingen
Germany

The Translator

Bettina Mikhail
Pariser Strasse 7
67245 Lambsheim
Germany

Revised and updated translation of the German book "Klebtechnik – Klebstoffe, Anwendungen und Verfahren", ISBN 978-3-527-31091-3, published by Wiley-VCH GmbH & Co. KGaA, Weinheim © 2005

■ All books published by Wiley-VCH are carefully produced. Nevertheless, authors, editors, and publisher do not warrant the information contained in these books, including this book, to be free of errors. Readers are advised to keep in mind that statements, data, illustrations, procedural details or other items may inadvertently be inaccurate.

Library of Congress Card No.: applied for

British Library Cataloguing-in-Publication Data
A catalogue record for this book is available from the British Library.

Bibliographic information published by the Deutsche Nationalbibliothek
The Deutsche Nationalbibliothek lists this publication in the Deutsche Nationalbibliografie; detailed bibliographic data are available on the Internet at http://dnb.d-nb.de.

© 2009 WILEY-VCH Verlag GmbH & Co. KGaA, Weinheim

All rights reserved (including those of translation into other languages). No part of this book may be reproduced in any form – by photoprinting, microfilm, or any other means – nor transmitted or translated into a machine language without written permission from the publishers. Registered names, trademarks, etc. used in this book, even when not specifically marked as such, are not to be considered unprotected by law.

Printed in the Federal Republic of Germany
Printed on acid-free paper

Cover Design Adam Design, Weinheim
Typesetting Thomson Digital, Noida, India
Printing betz-druck GmbH, Darmstadt
Bookbinding Litges & Dopf Buchbinderei GmbH, Heppenheim

ISBN: 978-3-527-31898-8

Contents

Preface *XV*

List of Contributors *XVII*

1	**Adhesive Bonding as a Joining Technique** *1*	
2	**The Historical Development of Adhesive Bonding** *5*	
3	**Adhesion** *11*	
3.1	Introduction *11*	
3.2	Classical Adhesion Theories *14*	
3.2.1	Polarization Theory *17*	
3.2.2	Diffusion Theory *20*	
3.2.3	Chemical Reactions *21*	
3.3	Adhesion in Real Systems *24*	
3.3.1	Gold *25*	
3.3.2	Polyethylene and Polypropylene *25*	
3.3.3	Glass *26*	
3.4	New Concepts in the Field of Adhesion *26*	
3.5	Conclusions *28*	
4	**Survey and Classification of Adhesives and Primers** *29*	
4.1	Noncuring (Pressure-Sensitive) Adhesives *29*	
4.2	Physically Setting Adhesives *33*	
4.2.1	Contact Adhesives *33*	
4.2.2	Plastisol Adhesives *34*	
4.2.3	Hot-Melt Adhesives *34*	
4.3	Chemically Setting Adhesives *35*	
4.3.1	Two-Part Adhesives *35*	
4.3.2	Hot-Setting, One-Part Adhesives *36*	
4.3.3	Cold-Setting, One-Part Adhesives *36*	

4.3.4	Microencapsulated Adhesives 37
4.4	Primers 37
4.5	General Handling Instructions 38

5 Chemistry and Properties of Adhesives and Primers 39

5.1	Pressure-Sensitive Adhesives (PSAs) 40
5.1.1	Introduction 40
5.1.2	Chemistry of Pressure-Sensitive Adhesives 40
5.1.3	Physical Properties of Pressure-Sensitive Adhesives 43
5.1.4	Properties of Pressure-Sensitive Adhesives as a Function of Temperature 44
5.1.5	Tack 45
5.1.6	Peel Resistance 45
5.1.7	Creep 45
5.1.8	Formulations of PSAs 46
5.1.8.1	Typical Formulation of Natural Rubber-Based PSAs 46
5.1.8.2	Typical Formulation of Block Copolymer PSAs 46
5.1.8.3	Typical Formulation of Acrylate-Based PSAs 47
5.2	Contact Adhesives 47
5.2.1	Composition of Contact Adhesives 47
5.2.2	Properties and Fields of Application of Contact Adhesives 48
5.3	Hot Melts 49
5.3.1	Thermoplastic Hot Melts 50
5.3.2	Hot-Seal Adhesives 51
5.3.3	Plastisols 51
5.3.4	Self-Bonding Varnishes 52
5.3.5	Polyurethane-Based Reactive Hot Melts 52
5.3.6	Epoxy Resin-Based Reactive Hot Melts 52
5.3.7	Trends in Hot-Melt Technology 53
5.4	Phenolic Resin Adhesives 53
5.4.1	Chemistry of Phenolic Resins 53
5.4.2	Formulation of Phenolic Resin Adhesives 56
5.4.3	Behavior and Applications of Phenolic Resin Adhesives 57
5.5	Epoxy Resin Adhesives 58
5.5.1	Chemistry of Epoxy Resin Adhesives 58
5.5.2	Reactions of Epoxy Resins 59
5.5.3	Properties of Epoxy Resin Adhesives 60
5.5.4	Formulations of Epoxy Resin Adhesives 63
5.5.4.1	Epoxy Resins 63
5.5.4.2	Crosslinking Agents 64
5.5.4.3	Hardeners 64
5.5.4.4	Flexibilizers and Additives That Improve Impact Toughness 64
5.5.4.5	Fillers and Thixotroping Agents 65
5.5.4.6	Further Additives 65
5.5.4.7	Typical Epoxy-Resin Adhesive Formulations 65

5.6	Polyurethane Adhesives	66
5.6.1	Chemistry of Polyurethanes	66
5.6.2	Raw Materials	67
5.6.2.1	Isocyanates	67
5.6.2.2	Polyols	68
5.6.2.3	Catalysts	69
5.6.3	Structure and Properties of Polyurethane Adhesives	69
5.6.3.1	One-Part Polyurethane Adhesives	69
5.6.3.2	Two-Part Polyurethane Adhesives	70
5.6.4	Formulations of Polyurethane Adhesives	70
5.7	Acrylate Adhesives	71
5.7.1	Physically Setting Acrylates	71
5.7.1.1	Solvent-Containing Acrylates	71
5.7.1.2	Dispersion Acrylates	71
5.7.1.3	Acrylate-Based Contact Adhesives	71
5.7.2	Chemically Setting Acrylates	71
5.7.2.1	Monomers	73
5.7.2.2	Two-Part Acrylates/Methacrylates	74
5.7.2.3	One-Part Acrylates/Methacrylates	75
5.7.2.4	Examples of Application	75
5.7.2.5	Anaerobic Acrylates	76
5.7.2.6	UV-Setting Acrylates	77
5.7.2.7	Cyanoacrylate Adhesives	83
5.7.3	Formulations of Acrylate Adhesives	84
5.7.3.1	Formulation of a Two-Part Acrylate 'No Mix' Adhesive	84
5.7.3.2	Formulations of UV-Setting Acrylate Adhesives	84
5.7.3.3	Formulations of Cyanoacrylate Adhesives	85
5.8	Silicones	85
5.8.1	One-Part Silicone Adhesives	85
5.8.2	Two-Part Silicone Adhesives	86
5.8.2.1	Condensation Crosslinking	86
5.8.2.2	Addition Crosslinking	87
5.9	Natural Adhesives	88
5.9.1	Introduction	88
5.9.2	Natural Raw Materials	89
5.9.3	Modern Adhesives Based on Natural Raw Materials	91
5.9.3.1	Glutine Glue	92
5.9.3.2	Casein Glues	92
5.9.3.3	Dextrin Glues	93
5.9.3.4	Natural Material-Based Pressure-Sensitive Adhesives	93
5.9.4	Future Trends	94
5.10	Chemistry of Primers and Adhesion Promoters	94
5.10.1	Primers	95
5.10.2	Adhesion Promoters	96
5.11	Fillers	99

6 Design, Production and Quality Assurance of Adhesive Bonded Joints *101*

- 6.1 Design and Dimensioning *101*
- 6.1.1 Interaction of Polymer Behavior and State of Stress *101*
- 6.1.2 Design of Adhesive Bonded Joints *103*
- 6.1.3 Dimensioning of Adhesive Bonded Joints *104*
- 6.2 Surface Preparation *108*
- 6.2.1 Mechanical Surface Preparation Methods *109*
- 6.2.1.1 Grit Blasting *109*
- 6.2.1.2 Grinding *110*
- 6.2.1.3 Brushing *110*
- 6.2.2 Chemical Surface Preparation Techniques *111*
- 6.2.3 Physical Surface Preparation Techniques *111*
- 6.2.3.1 Flame Treatment *111*
- 6.2.3.2 Microplasma Treatment *111*
- 6.2.3.3 Atmospheric Plasma Treatment *112*
- 6.2.3.4 Low-Pressure Plasma Treatment *112*
- 6.2.3.5 Corona Treatment *112*
- 6.2.4 The Preconditions of Adhesion *112*
- 6.2.5 Protection of Prepared Surfaces *114*
- 6.3 Application of the Adhesive *115*
- 6.3.1 Processing of Adhesives *115*
- 6.3.1.1 Processing of Two-Part Adhesive Systems *115*
- 6.3.1.2 Processing of One-Part Adhesive Systems *119*
- 6.3.1.3 Hot-Melt Adhesives *121*
- 6.3.1.4 Film Adhesives *121*
- 6.3.1.5 Hybrid Bonding *122*
- 6.3.2 Quality Assurance *123*

7 Adhesives and Adhesive Joints: Test Methods and Properties *125*

- 7.1 Adhesive Bulk Properties *126*
- 7.1.1 Stress–Strain Analysis of Bulk Specimens *126*
- 7.1.2 Determination of T_g by Dynamic Mechanical Analysis or Differential Scanning Calorimetry *127*
- 7.1.3 Coefficient of Thermal Expansion (CTE) *128*
- 7.1.4 Temperature Resistance by Thermogravimetry *128*
- 7.2 Test Categories by Type of Mechanical Stress *128*
- 7.2.1 Tensile Stress (e.g. Butt Joints) *128*
- 7.2.2 Shear Stress (e.g. Single Lap Shear Specimen) *130*
- 7.2.3 Cleavage (e.g. Wedge Test) *132*
- 7.2.4 Fracture Resistance *133*
- 7.2.5 Wedge Test *133*
- 7.2.6 Peel Test *134*
- 7.3 Test Categories by Duration and Load Rate *137*
- 7.3.1 Long-Term Static Tests (e.g. Creep Experiments Under Dead Load) *137*

7.3.2	Impact Tests (e.g. Impact Wedge Test to Evaluate Crash Performance)	138
7.3.3	Cyclic Loading (e.g. Fatigue Tests)	139
7.4	Assessment of Durability and Service Life of Adhesive Joints	139
7.5	Tests Related to Storage and Handling of Adhesives	143
7.5.1	Determination of Words and Phrases Relating to the Product Life	143
7.5.2	Open Time and Working Life	143
7.5.3	Extrudability	144
7.5.4	Sagging	144
7.5.5	Volatile Organic Compound (VOC) Emissions	144
7.6	Nondestructive Testing Methods	145
7.6.1	Visual Techniques	146
7.6.2	Radiographic Techniques	146
7.6.3	Magnetic, Electrical and Chemical Techniques	146
7.6.4	Mechanical Vibration	147
7.6.4.1	Impulse-Induced Resonance	147
7.6.4.2	Pulse Echo	148
7.6.4.3	Through-Transmission Sound Testing	149
7.6.5	Thermal Methods	149
7.6.5.1	Optical and Ultrasonic Lockin Thermography	149
7.6.5.2	Ultrasound Burst Phase Thermography	151
7.6.5.3	Excitation Mechanisms for Active Thermography	153
7.6.6	Assessment of Nondestructive Testing Methods	155
7.7	Applications of Test Methods in Structural Adhesive Bonding	156
7.7.1	Tensile Shear Strength Under Short-Term Loading	157
7.7.2	Tensile Shear Strength Under Long-Term Static and Alternating Loading	162
7.7.3	Peel Strength	164
7.7.4	Impact Behavior	165
7.7.5	Durability of Bonded Joints	166
7.7.6	Adhesive Bonding of Aluminum	168
7.7.7	Adhesive Bonding of Steel	183
7.7.8	Adhesive Bonding of Glass	188
7.7.9	Adhesive Bonding of Polymer Materials	191
7.8	Experiences and Learning from Damage Cases	191
7.8.1	Delamination of Bonded Rotor Blade Pockets	193
7.8.2	Total Separation of a Rotor Blade Profile	194
7.8.3	Bond-Line Corrosion in Wing Honeycombs	196
7.8.4	Delamination of the Outer Cover Sheet of Tail Units of Aircraft	197
7.8.5	Temperature-Induced Damage on a Tail Rotor Blade	199
7.8.6	Temperature-Induced Delamination of a Bonded Glass Fiber Patch	200
7.8.7	Incorrect Design of Bonded Filter Elements	201
7.8.8	Deficiencies of the Peel Ply Treatment	202

8	**Adhesive Bonding Technology: Fields of Application** 205	
8.1	Introduction 205	
8.2	Transportation 206	
8.2.1	Aircraft Manufacture 207	
8.2.1.1	Historical Development 209	
8.2.1.2	Advantages of Bonding in Aircraft Manufacture 213	
8.2.1.3	Adhesives Used in Aircraft Manufacture 216	
8.2.1.4	Surface Preparation 217	
8.2.1.5	Bonding Techniques 218	
8.2.1.6	Quality Control of Bonding in Aircraft Manufacture 218	
8.2.1.7	Conclusions 219	
8.2.2	The Automotive Industry 220	
8.2.2.1	Bonding in Passenger Car Bodywork Construction 223	
8.2.3	Bonding in Rail Vehicle Manufacture 234	
8.2.4	Shipbuilding 235	
8.3	Building Industry 237	
8.3.1	Adhesives for the Bonding of Surfaces 237	
8.3.1.1	Adhesive Mortars for Tiling 237	
8.3.1.2	Floor-Covering Adhesives 238	
8.3.2	Adhesive Anchors in the Building and Construction Industry 238	
8.3.2.1	Adhesive Anchor Systems 238	
8.3.2.2	Adhesives 240	
8.3.3	Building Reinforcement by Means of Bonded Surface Reinforcement 241	
8.3.4	Structural Bonding in Structural Steel and Facade Engineering 241	
8.4	Wooden Constructions 243	
8.4.1	Introduction 243	
8.4.2	The Structure of Wood 245	
8.4.2.1	The Molecular Structure of Wood 245	
8.4.2.2	The Macroscopic Structure of Wood 246	
8.4.3	Consequences 246	
8.4.3.1	Strength, Swelling and Shrinking 246	
8.4.4	Adhesive–Wood Interphase 247	
8.4.5	Adhesive Systems 247	
8.4.6	Adhesives Based on Renewable Resources 248	
8.4.7	Synthetic Wood Adhesives 249	
8.4.7.1	Physically Setting Systems 249	
8.4.7.2	Chemically Setting Adhesives 249	
8.4.7.3	Adhesion Promoters (Primers) 251	
8.4.8	Wood Composites 251	
8.4.9	The Supreme Discipline: Laminated Wood Constructions (Civil Engineering Using Wood as Building Material) 253	
8.4.9.1	Development 253	
8.4.9.2	Laminated Wood Constructions Today 253	
8.4.10	Prefabricated Elements 255	

8.4.11	Prospects of Wood Constructions 257
8.5	Paper and Packaging Industry 257
8.5.1	Introduction 257
8.5.2	Manufacture of Packaging Materials 258
8.5.2.1	Corrugated Board and Paperboard Containers 258
8.5.2.2	Lamination 259
8.5.2.3	Folder Gluer 259
8.5.2.4	Folding Cartons 260
8.5.2.5	Flexible Packages 260
8.5.2.6	Bags, Pouches and Sacks 263
8.5.3	The Setting-Up and Closing of Trays and Boxes 264
8.5.3.1	Setting-Up a Tray 264
8.5.3.2	Carton Closing 264
8.5.4	Pallet Securing 267
8.5.5	Labeling 269
8.5.5.1	Pressure-Sensitive Labels 269
8.5.5.2	Inline Labeling 270
8.5.5.3	Tamper-Evident Bottle Closures 273
8.5.5.4	Membrane Bonding 274
8.5.6	Cigarette Production 274
8.5.7	Manufacture of Tissue Paper Products 274
8.5.7.1	Core Winding 275
8.5.7.2	Lamination 276
8.5.7.3	Pick-Up Adhesives 277
8.5.7.4	Tail Tie Adhesives 277
8.5.8	Graphic Products 278
8.5.8.1	Unsewn Binding 278
8.5.8.2	Unsewn Binding with Dispersion Adhesives 279
8.5.8.3	Unsewn Binding with Hot Melts 279
8.5.8.4	Unsewn Binding with Reactive Polyurethane Hot Melts 280
8.5.8.5	Multilayer Unsewn Binding 282
8.5.8.6	Side Gluing 283
8.5.8.7	Backgluing of Thread-Sewn Book Blocks 284
8.5.8.8	Book Manufacture 284
8.5.8.9	Making the Case 285
8.5.8.10	Casing-In 285
8.5.8.11	Fold Gluing 286
8.5.9	Envelopes and Advertising Mail 286
8.5.10	Reuse and Recycling of Paper and Packages 287
8.5.10.1	Reuse 287
8.5.10.2	Recycling 288
8.5.10.3	Recycling of Paper Products 288
8.5.10.4	Recycling of Plastic Packaging 291
8.5.10.5	Compostable Products 291
8.5.11	Regulations Stipulated by Law 292

8.5.12	Perspectives *292*
8.6	Small-Scale Industries and Handicraft Trades *293*
8.6.1	Joining *293*
8.6.1.1	Structural Adhesives *294*
8.6.1.2	Cyanoacrylate Adhesives *295*
8.6.1.3	Postcuring Hot Melts *296*
8.6.1.4	Hot-Melt Adhesive Systems *296*
8.6.1.5	Aerosol Adhesives *297*
8.6.1.6	Dispersion Adhesives *297*
8.6.1.7	Solvent-Based Adhesives *298*
8.6.1.8	High-Performance Acrylic PSA Tapes *298*
8.6.1.9	Reclosable Fasteners *300*
8.6.1.10	Screen-Printable Adhesives *301*
8.6.2	Protecting, Splicing and Masking *301*
8.6.3	Damping *304*
8.6.4	Labeling, Marking and Coding *305*
8.7	Electronics Industry *307*
8.7.1	Thermally Conductive Adhesives *307*
8.7.2	Electrically Conductive Adhesives *308*
8.7.3	Underfill Materials *310*
8.7.4	Functional PSA Tapes *310*
8.7.5	Spacers *311*
8.7.6	Labeling *311*
8.8	Optical Industry *312*
8.8.1	General Requirements for Adhesives in Optical Industry *312*
8.8.2	Optical Adhesives for the Temporary Fixture of Optical Elements *313*
8.8.3	Optical Cementing *314*
8.8.4	Bonding Optical Elements to Mounts *315*
8.9	Mechanical Engineering and Process Equipment Construction *316*
8.9.1	Mechanical Engineering *316*
8.9.2	Domestic Appliance Industry *317*
8.9.3	Manufacture of Internal Combustion Engines and Transmission Manufacturing *317*
8.9.4	Stainless Steel Structures *319*
8.10	Textile Industry *320*
8.10.1	The Early History of Fabric-Reinforced Rubber Materials *320*
8.10.2	How Textile Fibers Contribute to the Living Standards of the Industrial World *321*
8.10.2.1	Technical Textiles *322*
8.10.2.2	Textiles as Rubber Reinforcements *322*
8.10.3	General Properties of Elastomers (Rubber) *323*
8.10.4	Treatment of Textiles for Rubber Reinforcement *324*
8.10.4.1	Reinforcement Fibers *324*
8.10.4.2	Adhesive Systems Used to Treat Textiles *324*

8.10.4.3	Effect of the Fiber Type on the Adhesive System	326
8.10.4.4	Direct Adhesion Through Addition of the Adhesion Promoter to the Rubber Mixture	328
8.10.5	Adhesion Mechanisms	328
8.10.6	Aging of RFL-Dipped Textiles	330
8.11	Footwear Industry	332
8.12	Road and Highway Engineering	333
8.12.1	Traffic Signs	334
8.12.2	Pavement Markings	335
8.12.3	License Plates	337
8.13	Surface Design	337
8.13.1	Surface Design and Surface Protection	337
8.13.2	Fleet Graphics	338
8.13.3	Public Transport Advertising	338
8.13.4	Building and Floor Graphics	340
8.14	Tamper-Indicating Labels	341
8.15	Medical Sector	343
8.15.1	Medical Devices with Short-Term, Superficial Body Contact	343
8.15.2	Medical Devices with Several (up to 30) Days Body Contact	343
8.15.3	Medical Devices with Long-Term (Over 30 Days) Body Contact	344
8.16	Adhesion in Nature	346
8.16.1	Design Principles of Biological Attachment Devices	346
8.16.2	Gluing Under Water	347
8.16.2.1	Byssus Adhesion in Mollusks	348
8.16.2.2	Tube Feet of Starfish	351
8.16.2.3	Adhesion in Barnacles	353
8.16.2.4	Temporary Adhesion in Cladoceran Crustaceans	354
8.16.2.5	Adhesion in Larval Fish and Ascidians	354
8.16.3	Adhesion in Terrestrial Insects	354
8.16.3.1	Permanent Adhesion in Insects and Spiders	354
8.16.4	Adhesion in Plants	355
8.16.5	Surface-Replicating Mechanism Based on Growth	356
8.17	Little-Known Applications	356
8.17.1	Foils for the Rotors of Wind Turbines	356
8.17.1.1	Erosion-Protection	357
8.17.1.2	Lightning Protection and Performance Enhancement	358
8.17.2	Adhesive Elastomer Composites	358
8.17.3	Light Bulbs	359
8.17.4	Bonding in Art, Jewelry Making and Archeology	361
8.17.4.1	Art	361
8.17.4.2	Jewelry Making	363
8.17.4.3	Archeology	364
8.17.5	Removable Adhesive Joints	365
8.17.5.1	Commercially Available Removable Adhesive Systems	365
8.17.5.2	High-Strength Removable Adhesives	367

9	**Perspectives** *371*	
9.1	Economic Trends *371*	
9.2	Technical Trends *371*	

Appendix *375*

References *393*

Index *403*

Preface

Today, there is a large body of international literature on the basics and manifold applications of adhesive bonding. The interdisciplinary nature of the subject, however, contributes towards publications being entered into this body from a wide variety of sources, making information extremely difficult and time-consuming to find. Moreover, the interactions between adhesives and adherents, as well as the long-term durability of adhesive bonded joints when exposed to hostile environments, have yet to be fully explored, leaving considerable gaps in our knowledge. Both, a solid knowledge of the basics and an empirical-based, expert knowledge gathered from many years of practical experience, are needed to create solid and reliable, longlasting adhesive bonds. The current literature, however, is rather uninformative in this regard.

The second, revised edition of this Handbook provides such comprehensive knowledge in a compact and well-structured form for the English reader. Specialists in a variety of disciplines, from both academia and industry, have compiled a store of knowledge about the performance and potential of adhesive bonding over a wide range of applications.

The reader will first become acquainted with the basics of adhesive bonding, followed by some practice-relevant issues such as the correct selection of adhesive, and the design and manufacture – as well as quality assurance of – adhesive bonds. A variety of applications of adhesive bonding is extensively presented, from structural high-performance bonding in aircraft and automobile manufacture and in the construction and building industry, to nonstructural bonding applications, for example in the packaging and lamination industry. The details of some amazing bonding methods found in Nature are also described, as are adhesive bonded joints between steel and rubber, which form an integral part of our everyday life, without being noticed as such. Subsequently, the reader will learn about 'bonding on demand', bonding in the medical and electronics sectors, and many other areas of application. This Handbook is the very first to provide such comprehensive knowledge about a joining technique that stands out for its versatility. It is both instructive and fascinating to learn about this wealth of experience gained with adhesive bonding since the very beginning of mankind.

Adhesive Bonding: Materials, Applications and Technology
Walter Brockmann, Paul Ludwig Geiß, Jürgen Klingen, and Bernhard Schröder
Copyright © 2009 WILEY-VCH Verlag GmbH & Co. KGaA, Weinheim
ISBN: 978-3-527-31898-8

Unfortunately, despite bonding having an enormous achievement potential when the bond is properly designed and implemented, people are often skeptical towards adhesive bonding. This Handbook provides valuable guidance for the successful implementation of adhesive bonded joints.

Our thanks go to Bettina Mikhail for translating, and to the publisher for releasing this Handbook for the English reader. May it contribute greatly to the advancement of adhesive bonding technology.

Kaiserslautern, September 2008 *The Authors*

List of Contributors

Prof. Dr.-Ing. Walter Brockmann
Brunnenstrasse 71
67661 Kaiserslautern
Germany
Chapters 1–4; Sections 5.2, 5.4,
5.9–5.11, 7.7, 8.2.1–8.2.4, 8.14
8.17.4, 8.17.5.1, 9.2

Prof. Dr.-Ing. Paul L. Geiß
Technical University (TU)
Kaiserslautern
AWOK
Postfach 3049
67653 Kaiserslautern
Germany
Chapter 6; Sections 7.1–7.5, 8.1,
8.3, 8.17.2, 9.1

Dr. Jürgen Klinger
3M Deutschland GmbH
Corporate Research Laboratory
Carl-Schurz-Strasse 1
41453 Neuss
Germany
Sections 5.1, 5.6, 5.8, 8.6, 8.7, 8.12,
8.13, 8.17.1

Dr. Bernhard Schröder
NMI Natural and Medical Sciences
Institute at the University of Tübingen
Markwiesenstrasse 55
72770 Reutlingen
Germany
Sections 5.3, 5.5, 8.9, 8.11, 8.15,
8.17.3, 8.17.5.2; Chapter 10

With the cooperation of:

Prof. Dr. Jürgen von Czarnecki
Bundeswehr Research Institute for
Materials, Explosives, Fuels and
Lubricants
Institutsweg 1
85435 Erding
Germany
Section 7.8

Dipl.-Ing. Siegfried Göb
Corporate Materials & Process
Laboratory Adhesives
3M Germany Laboratories
Carl-Schurz-Strasse 1
41453 Neuss
Germany
Sections 5.5.4, 5.6.4, 5.7.3

Adhesive Bonding: Materials, Applications and Technology
Walter Brockmann, Paul Ludwig Geiß, Jürgen Klingen, and Bernhard Schröder
Copyright © 2009 WILEY-VCH Verlag GmbH & Co. KGaA, Weinheim
ISBN: 978-3-527-31898-8

Dr. Stanislav Gorb
Max-Planck-Institut für
Metallforschung
Evolutionary Biomaterials Group
Heisenbergstrasse 3
70569 Stuttgart
Germany
Section 8.16

Dr. Dirk Hasenberg
Corporate Materials & Process
Laboratory Corporate Research
3M Deutschland GmbH
Tech. Center 035M
Carl-Schurz-Strasse 1
41453 Neuss
Germany
Section 7.6

Dr. Jens Kiesewetter
MEP-OLBO GmbH
Edelzeller Strasse 44
36043 Fulda
Germany
Section 8.10

Dr. Stefan Mattle
APM Technica AG
Max-Schmidheiny-Strasse 202
9435 Heerbrugg
Switzerland
Section 8.8

Dr. Jörg Naß
AACURE ADHESIVES GmbH
Universitätsklinikum
Gebäude 7
66421 Homburg/Saar
Germany
Section 5.6

Dr. Herrmann Onusseit
Henkel KgaA
Henkelstrasse 67
40191 Düsseldorf
Germany
Section 8.5

Dr. Willi Schwotzer
Collano AG
6203 Sempach-Station
Switzerland
Section 8.4

Dr. Detlef Symietz
Dow Automotive
Wollerauer Strasse 15-17
8807 Freienbach
Switzerland
Section 8.2.2.1

Dr. Stefanie Wellmann
Wellmann Technologies GmbH
Hauptstrasse 96
67159 Friedelsheim
Germany
Section 5.7

1
Adhesive Bonding as a Joining Technique

Bonding is the surface-to-surface joining of similar or dissimilar materials using a substance which usually is of a different type, and which adheres to the surfaces of the two adherents to be joined, transferring the forces from one adherent to the other. According to DIN EN 923, an adhesive is a nonmetallic substance capable of joining materials by surface bonding (adhesion), and the bond possessing adequate internal strength (cohesion). Bonding is a material joining technique that, in the traditional sense, cannot be broken without destruction of the bond. Recently, specific bonding-on-demand techniques have been developed (see Section 8.16.5), for example as an assembly tool without further function, or for recycling based on a separation of materials, a method that today is becoming increasingly important.

Bonding is by far the most universal joining technique. Virtually all technically useful materials can be joined with each other, and one with another, by means of this surface-to-surface and material-joining technique.

Adhesive bonding technology offers great design flexibility as it can be easily integrated into almost all available industrial sequences of single-piece work or mass production. Historically, bonding has long been recognized as a high-performance joining technique. The large majority of original natural binding materials have now been replaced by synthetically prepared adhesives. For example, phenolic resins were first introduced in the late 1920s, while during the 1940s epoxide resins and polyurethane were developed which have since made possible the production of synthetic adhesives (see Chapter 2).

As polymer chemistry has advanced in terms of knowledge, specific adhesives have been developed that bind very strongly to organic or inorganic materials. With regard to adhesive strength and deformation, these adhesives meet very specific requirements that result from the configuration of the adhesive joint. Meanwhile, high-strength adhesive assemblies have been created with quite short curing periods. In fact, the longstanding problem of extensive curing times necessary to obtain high-strength joints has been almost completely resolved with the introduction of new chemical developments in the creation of adhesives. Moreover, skepticism is no longer justified as to the long-term durability of adhesive joints exposed to adverse environments, provided that the bonding is properly conceived.

Adhesive Bonding: Materials, Applications and Technology
Walter Brockmann, Paul Ludwig Geiß, Jürgen Klingen, and Bernhard Schröder
Copyright © 2009 WILEY-VCH Verlag GmbH & Co. KGaA, Weinheim
ISBN: 978-3-527-31898-8

Table 1.1 Characteristic features of adhesive joints.

Advantages	Disadvantages
• The adherents are not affected by heat	• Limited stability to heat
• Uniform stress distribution	• Long-term use may alter the properties of the bond-line
• Possibility to join large surfaces	
• Possibility to join different materials	• Cleaning and surface preparation of the adherents is necessary in many cases
• Possibility to join very thin adherents	• Specific production requirements to be met
• Gas-proof and liquid-tight	• Specific clamping devices are often required to fix the joint
• No crevice corrosion	• Nondestructive quality testing is only possible to a certain extent
• No contact corrosion	
• No precise fits of the adherent surfaces are necessary	
• Good damping properties	
• High dynamic strength	

Bonding rarely competes with other joining techniques used in industry. For example, one would not consider bonding a steel bridge or a gantry, but for the lightweight construction of car bodies using steel, aluminum, glass and plastics, adhesive joining offers extremely interesting applications. Adhesive joining is particularly well suited to the joining of large-sized surfaces of different materials, such as in the construction of sandwich assemblies.

The possibilities, advantages and disadvantages of adhesive bonding compared to other joining techniques are summarized in Table 1.1

One of the many advantages of bonding is that little or no heat is needed to create the joint. As a result, the material structure of the adherents to be joined is not macroscopically affected, and deformations or internal stress – which generally are related to the application of heat – rarely occur. Even those materials with finished surfaces or coated materials can easily be bonded without any heat supply. From this point of view, there are no limits with regard to the combinations of materials that can be joined.

One important disadvantage of adhesive bonding, however, is the relatively poor heat resistance of the bond-line as compared to inorganic materials such as metal or glass. Hence, in order to obtain high-performance assemblies the production parameters must meet the specific requirements of the material used. This applies not only to the manufacturing sequences but also to the ambient conditions in which the joints are produced, because adhesion generally develops only during the production process, and the production parameters can have a decisive effect on the quality of the bond. The same more often than not applies to the cohesion of the adhesive layer. The technical properties of cohesion only develop during the course of the production process (with the exception of pressure-sensitive adhesives) after different setting processes. In this case, too, the production parameters mostly have a considerable effect on the quality of the final joint. By way of contrast, the joining process itself has only a minimally significant effect on the quality of traditional joining techniques, such as screw joints.

As the mechanisms of adhesion and the long-term behavior of adhesives are not yet completely known, it has not been possible to develop strict mathematical models for adhesive joints. Although this may be considered to be a disadvantage of bonding, the empirical values obtained with adhesive joints have meanwhile made it possible to conceive safe and sufficiently reliable bonded structures.

2
The Historical Development of Adhesive Bonding

To our best knowledge, bonding has been used as joining technique since the most ancient of times. As long ago as 4000 BC, the Mesopotamians are known to have used asphalt for construction purposes, while in 3000 BC the Sumerians produced glue from animal skins and called the product 'Se-gin'. The discovery of a wooden, gold-plated Ram statue that contained adhesive made from asphalt indicates that, in Ur, about 1000 years later, glue was used for the manufacture of statuettes. Gluing was also known to have been used in Egypt in 1475 BC, since in the tomb of Rekhmara in Thebes the process is depicted on a mural painting dating from that time. The ancient Egyptians presumably used animal glues. In the tomb of Nebanon and Ipuki, which dates back to the same period, another painting shows the gluing of a shrine.

A glue tablet was found in a cave at the upper level of the mortuary temple of Queen Hatshepsut at Deir-el-bahari, i.e. in the tomb of Tutankhamun. The glue's properties were investigated in the 1920s and shown to be identical to those of the skin glue still in use at the time of the archeological investigations, despite the tablet being 3500 years old. These findings give rise to the assumption that, over millenniums – and with few exceptions – adhesives have been subject to only insignificant further developments.

In the Talmud – a collection of Jewish after-Bible laws and religious traditions – reference is made to casein that had been used many years previously by the Israelites as a binder for pigments, although the large-scale use of casein glue was not introduced until much later.

Glue was also clearly well known in Greece, as the famous legend of Daedalus and Ikarus – which falls into the period between 2000 and 1600 BC – is based on the failure of adhesive bonds produced using wax (Figure 2.1). In *A History of Plants*, Theophrastus (371–287 BC) wrote that, for the carpenter's purposes, fir could best be glued together and that it was said not even to crack when glued.

In 79 AD, Pliny the Elder (Gaius Plinius Secundus), in his book *Naturalis Historia* (*Natural History*), wrote that the entrance-doors of the Temple of Artemis (or Diana) at Ephesus, 400 years after having been reconstructed in 324 BC after the burning of the temple which had taken place in 356 BC, still appeared to be new. He noted that it was remarkable that the wings of the door were left in the glue clamp for four years. In

Adhesive Bonding: Materials, Applications and Technology
Walter Brockmann, Paul Ludwig Geiß, Jürgen Klingen, and Bernhard Schröder
Copyright © 2009 WILEY-VCH Verlag GmbH & Co. KGaA, Weinheim
ISBN: 978-3-527-31898-8

Figure 2.1 The fall of Ikarus. A copper plate in Ovid's *Metamorphosis* from the 18th century (Bildarchiv Preußischer Kulturbesitz, Berlin).

other chapters, too, he made remarkable notes on glues [1]. In Volume 3, Book XVI, for example, when describing different kinds of wood, he wrote: "Magna autem et glutinatio propter ea, quae sectilibus laminis aut alio genere operiuntur." ("Gluing is also important because of those things covered with cut boards or by other means.") Shortly after that, he made a remarkable note (still worthy of consideration nowadays): "Quaedam et inter se cum aliis insociabilia glutino, sicut robur, nec fere cohaerent nisi similia natura, ut si quis lapidem lignumque coniungat." ("Some (woods), e.g. robur, cannot be glued with other woods of the same kind or any other, and they almost never adhere to each other, unless they have a similar nature. The same applies to stone which would never adhere to wood.")

In Ancient Greece, woodworking glue was called Xylokolla (ξυλοκολλα), ox glue Taurokolla (ταυροκολλα) and fish glue Ichthykolla (ιχθυκολλα), while the glue maker was called the Kollepsos (κολλεπσος). As early as 530 BC, Theodorus of Samos mentioned the 'gluing of metals', which probably was figurative for a firm joint. It can be concluded, however, that the experience gained with adhesives was quite positive.

In 1122–1123 AD, Theophilus, in his books *De diversis artibus*, described different adhesives and mentioned casein glue which, apparently, was known of by the ancient Israelites. Somewhat amazingly, this glue was still in use when the first bonded technical structures of modern times were created, such as the construction of rigid airships by Schütte-Lanz between 1908 and 1919. The girders of these airships were made from wood glued with casein (see Figures 2.2 and 2.3).

Figure 2.2 The girder of a rigid airship made from glued asp wood (Schütte-Lanz, ca. 1915).

The rough and damp environment, in which naval airships were operated, often resulted in a failure of the glued joints. Their resistance was only slightly improved even when the glued assemblies were exposed to formalin vapors as postcure treatment (see Section 5.9.3.2), or when they were coated with lacquer.

Theophilus also described ground hartshorn that was added to animal glue as a filler, and Isinglass glue, the beluga quality of which was particularly praised in later time. Last, but not least, it is interesting to quote from the Merseburg charms originating from the early Middle Ages: "Bein ze beine, bluot ze bluote sollen sie gelimida sin." (Bone to bone, blood to blood should be glued."), which referred to general joints created by means of bonding for reparative purposes in medicine.

The bonding industry began to develop rapidly during the seventeenth and eighteenth centuries (Figure 2.4). In particular, an interesting treatise, *L'Art de Faire Différentes Sortes de Colle*, was written by Henry Duhamel du Monceau in Paris in 1771. This was published one year later in Germany, with the title *Die Kunst verschiedene Arten von Leim zu machen* (*The art of making different kinds of glue*) by the Prussian Academy of Sciences. In his treatise, Duhamel du Monceau provides a variety of recipes for the manufacture of different types of glue, and notes that garlic can be used as adhesion promoter when rubbed on wood before applying the glue.

Figure 2.3 A junction between three-dimensional airship girders made from glued wood (Schütte-Lanz, ca. 1915).

Figure 2.4 A glue factory from the mid-18th century (according to Duhamel du Monceau).

These details indicate that the existence of adhesion promoters was already recognized in very early times (the chemical structure of garlic extracts gives rise to the assumption that it is very efficient, in that they contain molecular configurations capable of building up chelate complex bonds; see Section 5.10.2).

It must be said that the development of adhesives took a long time, despite their widespread use. However, more important progress was made in the nineteenth century, when self-adhesive substances and tapes for use in medicine were developed and employed from the middle of the century onwards [2]. Horace H. Day was the inventor of pressure-sensitive adhesives based on natural rubber, while in 1845 William H. Shecut and Horace H. Day were granted a US patent for the improvement of adhesive tapes [3]. In Germany, a patent for a tape coated with a pressure-sensitive adhesive based on natural rubber was granted to the druggist Paul C. Beiersdorf in 1882.

For a long time, early pressure-sensitive adhesives were unable to meet the requirements of industrial applications, although their properties could be improved by these inventions and by various other developmental activities. The breakthrough was achieved by Richard G. Drew of the American Minnesota Mining and Manufacturing Company (3M) in the early 1920s, when he developed the first adhesive masking tapes based on crepe paper for lacquering processes in the automobile industry [4]. In 1926, Drew opened the first 3M adhesive tape laboratory, but later developed a cellophane adhesive tape to be used for packaging. This was the first transparent tape on the market, and it became an all round-solution for the office and the home [5]. Whereas, until the

1950s, pressure-sensitive adhesives did not attract much attention, today they have one of the most important growth rates of all groups of adhesives (see Section 5.1).

Meanwhile, in 1889 at Hannover, Germany, Ferdinand Sichel invented the first ready-made plant glue. However, the age of synthetically produced polymers truly began when Leo Hendrik Baekeland was awarded a patent for phenolic resins in 1909. In 1914, Victor Rollett and Fritz Klatte were granted a patent for the production of polyvinyl acetate (PVA), a synthetic raw material that is widely used to the present today, despite not gaining commercial importance until the 1920s. Although, urea resin had been known since 1919, it was not until 1929 – when a curing technique was developed – that it was used in the production of adhesives. Carboxymethyl cellulose and methyl cellulose have been used in painter's size and wallpaper pastes since the 1930s, while in 1931 Wallace Hume Carothers reported details of the production of polychlorobutadiene, through this material did not become important until the 1950s and 1960s.

Alexander Matting described various old recipes [6], notably the so-called 'marine glue' which was particularly famous and was produced from rubber solutions to which shellac or asphalt had been added. Another interesting recipe was used for the production of putties made from 60% lead oxide (PbO), 30% liquid phenol–formaldehyde resin and 10% magnesium carbonate. According to present knowledge, it was possible to achieve bond lines of high durability.

In the history of adhesives, another famous example was the so-called 'Atlas-Ago bonding technique', which used adhesives based on celluloid, was patented in 1912, and used in the shoe industry. From the late 1920s onwards, the former IG Farben produced a hot-curing and cold-curing urea condensation product which is still known by the name of 'Kaurit glue' (BASF). In 1928, Goldschmidt introduced an adhesive based on phenolic resin called 'Tegofilm'; with this fully synthetic, hot-curing adhesive it was possible to create absolutely water-resistant plywood joints. The same effect was achieved later with similar cold-curing systems, many of which are still in use today, having largely displaced the former casein and blood albumin glues (see Section 5.9).

In the early 1940s, the invention of phenolic resins modified with polyvinylformal by Norman Adrian de Bruyne marked a breakthrough in structural bonding not only of wood, but also of metals [7]. These resins contributed greatly to the improvement of aircraft structures, and are still appreciated today for their extraordinary durability, especially in case of aluminum bonding (see Section 8.2.1). They are referred to as 'REDUX', the term being derived from their place of origin (Research at Duxford). Remarkable among de Bruyne's concepts was that he used a combination of different polymer systems that previously had been used systematically only in pressure-sensitive adhesives.

Another name of historical value in the field of epoxy resins is that of 'Araldit' (CIBA). This was the merit of Eduard Preiswerk who, in 1944, discovered that epoxy resins cured with phthalic anhydride, produced synthetically by Pierre Castan in 1937, offered a wide variety of possibilities to create bonded joints, even between metals [8]. Castan intended to use these epoxy resins for dental purposes but, according to reports, he abandoned his experiments because the adhesion between those resins and other materials did not have sufficient water resistance. Both, hot-curing and cold-curing epoxy resins have been considered as standard products of structural adhesive bonding technology (see Section 5.5.1).

The patent for polyurethanes, awarded to Otto Bayer as early as in 1937, marked a milestone in the history of adhesives. Although, for adhesives production polyurethanes were taken into consideration only during the 1950s, they are at present one of the most important raw materials used in the adhesives industries, due mainly to the wide variety of ways in which they can be modified (see Section 5.6).

An interesting note with regard to the early bonding technologies can be found in a book published in 1933 [9]: "The owners of really good recipes for the production of specialty adhesives will be beware of disclosing them urbi et orbi (to the city and to the world)." Indeed, the adhesives industry has been adhering to this concept to the present day.

It is self-evident that this short historical review cannot be complete. Hence, further detailed information with regard to the development and properties of the different groups of adhesives is provided in the following chapters.

3
Adhesion

3.1
Introduction

Adhesion, when considered macroscopically, is the two-dimensional (surface-to-surface) adherence of two similar or different types of material to each other. It is one of the most important material phenomena in nature and technology. In the realms of technology, a building erected by the Romans will still safely be held together by the adhesion between the mortar and bricks. However, a car tire will only function if adhesion holds together – in an absolutely safe manner – the rubber and fabric made from organic substances or steel cords. Corrosion-preventing lacquers can only be applied to cars because there is adhesion, and even paper – which still today is the medium on which books are printed – is made from an adhesive bonded fiber composite. If adhesion is considered as the *sine qua non* of all forms of adhesive joining techniques, it is reasonable to limit the general view of adhesive phenomena to the study of organic substances which, in most cases, are higher-molecular materials (the majority of adhesives), technically useful *inorganic materials* such as metals, glass, stone and ceramics, and *organic materials* such as plastics, wood and textiles.

Many years of experience gained from adhesion research have shown that it is wise to investigate the different aspects involved in the creation of adhesive joints on the one hand, and their behavior within the assembly on the other hand. Strictly speaking, adhesive systems virtually can never be destroyed nor fail where adhesion has built up in advance. There will not be any of the so-called 'interfacial adhesion failure' (as so often described in the literature) if the adhesion partners approach each other in sufficient order so as to build interactions between atoms or molecules, thus creating the possibility to produce a strong assembly.

Today it is well known that mechanical 'interlocking' has no significance in the bonding of the majority of nonporous, technically used materials. Consequently, at this point the subject of adhesion will not be analyzed by focusing on mechanical 'interlocking' as an origin of adhesion, although such an approach may be very useful when, for example, creating bonds between wood or paper and porous, swellable substances.

Table 3.1 Physical and chemical interactive forces in interfaces.

Forces	Physical bonds			Hydrogen bridge bonds	Chemical bonds	
	Permanent dipoles	Induced dipoles	Dispersion forces		Covalent	Ionic
Range (nm)		0.3–0.5		0.3–0.5	0.1–0.2	
Bonding energy (kJ mol^{-1})	<20 (Keesom energy)	<2 (Debye energy)	0.1–40 (London energy)	<50	60–700	600–1000
Mathematically calculated adhesion forces (MPa)	200–1750	35–300	60–360	500	17500	5000; 30
Strength of the assemblies as measured experimentally (MPa)			15–25			

Technically useful interactions between atoms and molecules only occur if and when the adherents involved are brought together in very close approximation, generally less than 1 nm (see Table 3.1).

Other, more important, distances can only be covered by adhesive interactions in the form of *charge transfers* that can contribute to the formation of an electrical bilayer. The orders of magnitude of bonding energies that occur in physical bonds, hydrogen bridge bonds and chemical bonds are listed in Table 3.1. When the adhesion forces are calculated from these bonding energies, the theoretical values (given in MPa in relation to a given surface unit) are, in most cases, clearly higher than the strength of adhesives made from organic substances. Their theoretical strength, which is calculated from the bonding energy of carbon–carbon bonds, is expected to have an order of magnitude of 500 MPa for a polymer packing with the highest density. Of course, the values of strength actually measured not only in adhesive assemblies but also for polymer materials, are lower than those calculated from theory because no ideal material combinations or structures are to be expected. If we suppose that the ratio between theoretical and practical values is approximately the same for both the adhesion and the adhesive, it is clearly possible to create a reliable bonded joint with the majority of potential adhesive interactions. Over many years of experience with bonding technology, this very simple postulate has been confirmed by the fact that, in a loaded adhesive assembly, there is virtually never any failure of adhesion (proper), as noted above. Strictly speaking, this winds up the study of 'adhesion' to be something of a phenomenon, although it is possible today to both understand and systematically optimize the (often very complex) behavior of adhesive systems. For this, the following additional basic knowledge is required.

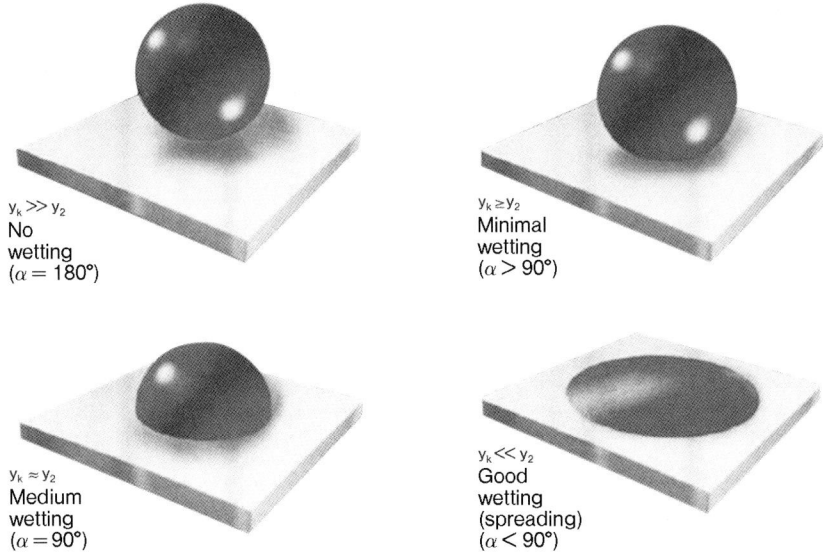

$y_k \gg y_2$
No wetting
($\alpha = 180°$)

$y_k \geq y_2$
Minimal wetting
($\alpha > 90°$)

$y_k \approx y_2$
Medium wetting
($\alpha = 90°$)

$y_k \ll y_2$
Good wetting (spreading)
($\alpha < 90°$)

Figure 3.1 The principle of wetting.

When creating an *adhesive bond*, no or only very little energy is added to the system. Only if the adhesive wets the solid objects to be bonded do the two come into sufficiently intimate contact. *Wetting* is a phenomenon that can be observed when a drop of liquid spreads over the surface of a solid-state material. According to the surface condition, the type of liquid used and the matter to be found in the environment (which often is not taken into consideration), the drop of liquid forms a contact angle (α) between its surface and the surface of the solid-state material. This angle ranges, in theory, from 0° (complete spreading) to 180° (no wetting at all) (see Figure 3.1).

When the drop of liquid comes to rest, interfacial tensions are acting between the solid-state material and the environment, the liquid and the environment, and the liquid and the solid-state material, respectively. These interfacial tensions (or energies) thermodynamically determine the angle of contact. When the drop of liquid is at rest, there is equilibrium between the interfacial tensions. Only if α is 90° or less can a sufficiently intimate molecular or atomic contact (see above) be created, and this results in adhesion between the wetting liquid and the solid-state material, which can be utilized in technical terms. This condition is only fulfilled if the interfacial energy between the solid-state material and the environment is equal to or higher than that between the liquid drop and the environment. It is worthy of mention at this point that the surface energies of common organic adhesives range between 30 and 60 mN m^{-1}; water has a surface energy of 72 mN m^{-1}, whereas inorganic materials such as glass and metals have surface energies in excess of 500 mN m^{-1}.

Nonpolar organic materials, such as polyethylene and polypropylene, have surface energies of less than 30 mN m^{-1}. The surface energy of polytetrafluoroethylene (PTFE) is the lowest known among solid-state materials (17 mN m^{-1}).

Table 3.2 Surface energies of different materials.

Material	σd (mN m^{-1})	σp (mN m^{-1})	σ (mN m^{-1})
Metals	–	–	1000–5000
Iron	–	–	1,400
Ceramics	–	–	500–1500
Mercury	–	–	484
Glass	–	–	300
Mica	27.3	39.8	67.1
PA6	36.8	10.7	47.5
PVC	37.7	7.5	45.2
Polyoxymethylene	36.0	6.1	42.1
Polystyrene	41.4	0.6	42.0
PETP	37.8	3.5	41.3
HDPE	35.0	0.1	35.1
Epoxy resin	19.5	13.2	32.9
PP	30.5	0.7	31.2
Paraffin wax	25.5	0	25.5

HDPE = high-density polyethylene; PA6 = polyamide 6; PETP = polyethylene terephthalate; PP = polypropylene; PVC = polyvinyl chloride.

Consequently, no wetting takes place between adhesives and some plastics if there are no auxiliary modifications taken; hence, these plastics are classified as being 'difficult to bond'. With a view to wetting, several others, such as epoxy resins, phenolic resins and polyester, which have surface tensions of about 60 mN m^{-1}, are better suited to bonding. With regards to the *wetting criterion*, inorganic substances do not represent any problems, provided that they are not covered with contaminants characterized by low surface energies. Table 3.2 provides values of the surface energies of different materials; here, for the sake of completeness, the nonpolar and polar force contributions to the surface energies have been included and will be discussed later.

With regard to bonding technology it can be summarized that, in the first instance, adhesion will only take place if the adhesive wets the solid-state material. This explains why it is possible to deduce the potential adhesion between a solid-state material with unknown surface energy and a liquid with a defined surface tension, from the simple measurement of the wetting characteristics involved. It must be stated, however, that wetting is not the only prerequisite for adhesion. The reasons for this will be provided in the following section.

3.2
Classical Adhesion Theories

Numerous theories have been proposed for a long time, and referred to in even the most recent literature, to explain observed phenomena of adhesion. In the following

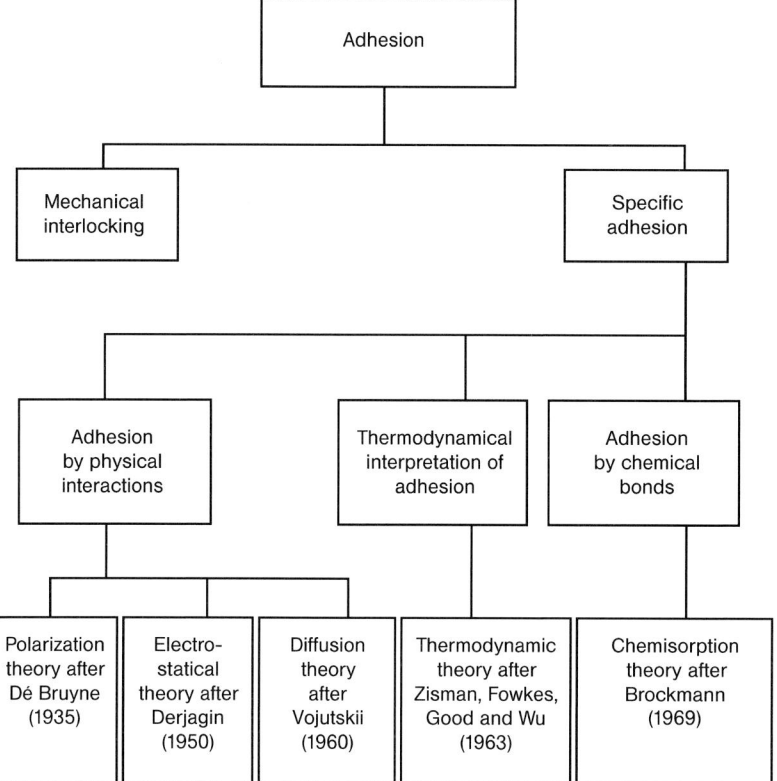

Figure 3.2 Classical adhesion theories.

sections these theories will be briefly assessed, taking into account that each of them only partly explains the true behavior of adhesive bonds. The most important adhesion models are summarized in Figure 3.2. Although, for the sake of completeness, the theory of 'mechanical interlocking' has also been included here, it will not be discussed further because it is quite insignificant for bonding technology today. In contrast, the concept of 'specific adhesion' will be presented in detail, summarizing all other adhesive phenomena based on physical and chemical interactions.

At the time when natural science first began to analyze the phenomenon of adherence, no generally valid knowledge was available regarding the physical and chemical interactions which take place between matter particles. The only possible approach was to calculate the energy balance of the process taking place when a solid-state material was wetted by a liquid, assuming that the wetting characteristics – which could be measured macroscopically – were ideal. This thermodynamic approach is based on the fact that, in terms of physics, the surface tension of liquids and solid objects is a type of energy; that is, work is done in order to create the surface of a material. Consequently, techniques to measure this energy were developed first for liquids, and then adapted for use with solid-state materials.

The calculation of adhesion was based on the contact angle (α; see above) which is formed when a drop of a low-viscosity liquid is applied to a solid-state material. Young's equations of wetting [1] can be calculated by using the surface tensions of the materials when this system, surrounded by its ambient gas saturated with the vapor of the liquid, is in a state of equilibrium (i.e. at rest). As surface tension is a quantity of energy, the Dupré equation [2] makes it possible to calculate the work of adhesion liberated when a liquid wets a solid-state material. The concept of solid surface free energy was later introduced into the Dupré equation to make it more accurate [3]. Subsequently, Wenzel [4] took into consideration that the shape of the solid surface also influences the contact angle, by introducing a factor of roughness into the equation. *Wenzel's roughness factor* is defined as the ratio of the true area of a surface (actually present on the microstructure of a surface) to the geometrically projected area.

The aim of these calculations was to deduce the strength of adhesive assemblies from the work of adhesion. This is possible as long as there is no mixing of the substances during the wetting process – that is, there are no penetration or chemical interactions that would resulting in a reversibility of the wetting process. In this case, the work necessary to separate the two materials after wetting is equal to that liberated during the wetting process. If such work is simply assumed to be constant over a defined range, it can be converted into strengths that are quite similar to those that can be determined with modern-day measurement techniques. In reality, however, the energy balances involved in the creation and restitution of surfaces or interfaces cannot be applied, even if the interfacial tensions are subdivided into their dispersion force components and their polar force components. In terms of thermodynamics, the wetting process virtually cannot be assumed to be reversible: it can be shown by measurement techniques that the original surfaces characterized by defined surface tensions will not be restored when the surfaces are separated from each other. If, for example, a drop of epoxy resin is cured on polyethylene, its surface tension is approximately 35 mN m^{-1} on the side that faces the air. However, when the drop is lifted from the polyethylene surface, its surface tension will be similar to that of the polyethylene surface because, despite poor wetting, some slack polyethylene molecules from the solid surface are most likely sticking to the drop material at the moment it is stripped from the surface.

Furthermore, when explaining adhesion by means of the concept of interfacial energies, surface tensions are rather assumed to be material constants. It is assumed that, after separation of the surfaces, the surface tensions of the liquid and the solid material, respectively, are still the same as before wetting. However, present knowledge indicates that, for various groups of materials, this assumption does not correspond to reality, even if there are no components of one material sticking to the other. Polymers in particular have a distinct ability of restructurization, as has been demonstrated in various recent studies [5]. Restructurization means that polymers are capable of adapting their surface condition – and hence their surface energy – more or less rapidly to the ambient conditions that surround them, even in the solid state. Moreover, the surface of each material – no matter how it is defined – depends on how it came into being. Therefore, it is not logical to proceed on the assumption that materials have constant surface properties, including the assumption of a defined surface energy.

This may be easily demonstrated with the following model. Let us suppose that a conical container is filled with molten polyethylene. When, after cooling, the polyethylene cone is removed from the container the solid material will present two different surfaces: (i) the surface that was formed on the side facing the container; and (ii) the surface that freely solidified in the air. A third, imaginary surface can be produced by cutting the cone into pieces with a knife. On those surfaces which formerly faced the container the polyethylene chains will be arranged in the form of loops, following the contours of the container. But, on the surface exposed to the air, they virtually will not present any roughness at all; they may even have slightly begun to oxidize due to the presence of oxygen in the air – a process that takes place more rapidly in the molten state owing to an improved input of energy. On the surface cut with the knife, however, the molecules could not arrange; they were separated in the solidified state, and those molecules hit by the knife were shortly transformed into radicals before being saturated by reactions taking place between them and residues of the knife or the air, resulting in the creation of stable systems.

Consequently, the polyethylene pieces produced have at least three different surface tensions. If this experiment is performed using containers of the same form, but made from different materials (e.g. Teflon® or silver), the polyethylene side that faced the wall of the container will again have different surface tensions depending on the container material used. In the case of a silver container, the polar polyethylene groups (created in the melting process) will probably have arranged towards the container, whereas in the case of Teflon®, they will have arranged towards the material interior.

To sum up, the simple thermodynamic approach is based on the postulation that adhesion is a two-body problem where the involved partners do not influence each other. But this has been shown not to correspond to reality.

By no means can the measurement of interfacial energies be used to explain adhesion. However, some important surface tension quantities that can easily be measured, may indeed rapidly provide some first indications of how adhesion will take place, or which problems may prevent adhesion from being built up.

3.2.1
Polarization Theory

During the late 1950s it was known to be possible to create high-strength adhesive bonded joints between smooth metal surfaces using organic adhesives under well-defined conditions. However, this represented an important problem for adhesive research because the formation of chemical interactions between those different types of material had long been considered impossible. It was known during the late nineteenth century that in matter, physical, intermolecular forces existed as well as chemical interactions. Indeed, if those forces did not exist it would not have been possible otherwise to explain various well-known phenomena, such as the high boiling temperature of water, or the fact that up to 100 °C, water is a liquid at all. To sum up, these intermolecular interactions have been attributed to the existence of permanent or oscillating dipoles, which interact one with another in chemically saturated systems, but which are also able to induce dipoles in other materials – a process that is particularly obvious in the case of metallic partners.

Those interactions are generally characterized by lower binding energies than chemical interactions; neither do they change the nature of matter to the same extent. Independently of the chemical compatibility, they can, however, take effect between materials of different types. De Bruyne's polarization theory [6], for instance, was an important approach for understanding the phenomenon of adhesion. Today, those intermolecular interactions are called 'physical bonds' and can be divided into three important groups:

- *Permanent dipoles*: These are found in molecules in which one atom with a higher atomic number (e.g. oxygen) is bound to another atom with a lower atomic number (e.g. hydrogen). This homopolar linking is due to the fact that the statistical probability distribution of the electrons in the bonding orbital is shifted towards the larger nucleus, inducing electronegativity and hence a dipole. The permanent dipole is able to build electrostatic attraction forces in the form of a dipole interaction with another permanent dipole; such forces can be calculated. Hydrogen bridge bonds are dipole–dipole bonds, and have been included in the schematic representation of Table 3.1 between physical and chemical bonds. Owing to the relatively high bonding energy and impact, the efficiency of hydrogen bridge bonds is actually assumed to approximate that of chemical interactions. Hydrogen bridge bonds are created by a hydrogen atom located between two very electronegative atoms that, in organic chemistry, essentially can be nitrogen, oxygen or fluorine having added a hydrogen atom to one of their valencies [7].

- *Induced dipoles*: In adjacent molecules without any own permanent polarity, permanent dipoles are able to induce counterdipoles with which they build up static attraction forces characterized by lower bonding energy compared to dipole–dipole bonds. As already mentioned, this approach especially plays a role when trying to explain the adhesion of organic substances with polar groups on bright metal surfaces.

- *Dispersion forces*: These may exist between molecules with nonpermanent dipoles, and are attributed to the fact that weak oscillating dipoles can be found between the involved atoms because, at least depending on time, the statistical probability distribution of the binding electrons is not completely uniform. This in turn may induce weak interactions that take effect between all types of material, including also similar gas molecules. However, their bonding energy is generally lower than that of permanent dipole bonds.

The validity of the polarization theory for interpreting adhesion phenomena is indisputable as it can easily be observed, for example, that organic polymers with permanent dipoles (so-called 'polar groups', such as polyvinyl chloride or epoxy resin) better adhere to smooth metal surfaces than do nonpolar molecules (e.g. the components of polyethylene or polypropylene). In contrast, an adhesive which generally contains polar groups can be shown to adhere better to a polar substrate or solid-state material (e.g. a cured epoxy resin) than to a nonpolar material such as polytetrafluoroethylene.

It may be assumed, therefore, that dipole effects contribute to all of the adhesive processes of specific adhesion. However, in some cases, their bonding energy can only barely explain high-strength adhesion, and it must be appreciated that those bonds which are responsible for physical adsorption can also be destroyed as soon as materials of a higher polarity penetrate the system and break the adhesive bonds by competitive adsorption, taking the place of the polar groups of the adhesive. Water, in particular, is one such substance with high polarity that is able to penetrate all polymers in defined quantities, owing to its short molecular dimensions and extremely high polarity. Water also migrates to the adhesive zone and considerably impairs or even destroys adhesion based on physical bonds, a process which can often be observed macroscopically.

Adhesion with a high resistance to water may occur, however, and is actually often observed between substances of very different types. Therefore, it must be assumed that there are further types of bond that are exposed to a much lesser extent (or not at all) to the attack of water. Primarily, these are chemical interactions that will be discussed later.

In terms of physics, it is indisputable that between two materials presenting different electronic configurations (e.g. a solid object made from an organic material and a metal), an electric bilayer is created as soon as both materials come into intimate contact with each other. Occasionally, for example, if a polyethylene layer is applied onto a metal, the bilayer can even be observed in the form of an electron enrichment in the polyethylene layer [8]. If the two adhesive partners are separated from each other, refined mathematical calculations indicate that, at the moment of breakdown of the bi-layer, the order of magnitude of the energy needed for separation is similar to that of the adhesive energies measured. It can easily be shown that electrostatic or luminescent phenomena are occurring when adhesive systems are separated.

If a simple, single-sided, pressure-sensitive adhesive film, as is commonly used in every household, is peeled from a solid substrate which is close to a pocket radio and its antenna, and the tuning dial is kept in the medium waves range (AM), then some interesting 'information' will emerge from the loudspeaker. At peeling, there is a crackling noise, but if the peeling is performed more or less rapidly the crackling changes. If the assembly is joined together again, followed by a peeling-off, there is no longer any crackle. Alternatively, in a photographic dark room, if a modern envelope where the overlaps are pasted together with a pressure-sensitive adhesive (see Section 8.5) is opened, a distinct light emission may often be observed with the naked eye. Both, autographic and photographic methods have been used scientifically to demonstrate this effect [9].

However, until now it has not been possible to provide unquestionable proof of the effectiveness of this bilayer in adhesion. If the bilayer was the dominant factor, the adhesive strength of the bonded joints created or tested in an intense electric field would differ from those created in neutral surroundings, but as far as we know this is not the case [10]. Although there is clearly no doubt that the bilayer exists, it probably does not predominate the adhesive strength in the majority of cases.

3.2.2
Diffusion Theory

A special case in adhesion is to some extent represented by the strong adherence that can often be observed between two surfaces of polymer materials of similar or dissimilar types that are in intimate contact with each other. Owing to their structure, physical low-energy interactions are thought to be effective here, while chemical interactions are hardly assumed to be involved. During the 1960s, Voyutskii explained this phenomenon [11], his diffusion theory being based on a fact that has often been forgotten, or not been taken into account. In terms of the molecular structure of their components, polymer materials are 'living' entities, especially in a state of low-grade crosslinking or no crosslinking at all, rather than being static systems. Between the points of entanglement, an intense motion of the molecular chains takes place in the form of atomic rotation, at least above the glass-transition temperature, because the free volume allows them to do so, although the polymer as a material itself is in a state of stiffness and rigidity. When this movement takes place, it cannot be excluded that molecular parts leave the original surface and diffuse into an adjacent polymer.

This situation is very easy to demonstrate. Suppose you remove the protective cover of a modern piece of eraser that is made from a highly flexible plastic material to which a filler was added in order to provide for friction. You then place the rubber onto the body of a cheap ballpoint pen or (if you have one) an older fountain-pen made from celluloid. If you then leave this arrangement for about a year you will (at least in the case of the celluloid object) discover, surprisingly, that the parts have become completely welded together 'by themselves'.

Joining well flashed-off adherents coated with contact adhesive or dispersion adhesive on both sides, respectively is most effective if, even for only a short period of time, a high contact pressure is applied. Here, the thermoplastic polymer layers with very smooth surfaces created from the liquid come into intimate contact with each other, and a seamless diffusion process between the molecules from both sides will take place (see Section 4.2). This results in a 'green strength' – that is, the strength of the bond is determined immediately after assembly but will increase during time (e.g. hours or days) as diffusion is a relatively slow-speed process. The same applies to pressure-sensitive adhesives characterized by high molecular mobility, as both by means of yield effects taking place on rough surfaces and by diffusion processes taking place between polymers, they build up adhesion over the course of hours, days or weeks. Ultimately, they will have created a green strength that is quite considerable from the outset, leading to an increase in peel strength of 50–70%.

It appears, therefore, that diffusion theory may be used to explain some interesting adhesion phenomena. Later in this chapter, we note that the molecular mobility which occurs in solid-state polymers, as postulated by Voyutskii, might also be very important for other types of adhesive systems (see Section 3.4).

3.2.3
Chemical Reactions

In 1927, long before the development of present-day industrial high-performance bonding techniques, W. McBain and W. B. Lee [12] published the results of the following experiments. Polished steel and aluminum parts were joined together using an organic adhesive made from (all components by weight): 50 units of shellac, 5 units of creosote (oil obtained from wood tar), 0.5 unit of ammonia and 2 units of turpentine. Although, based on today's knowledge, this was a curious binder, McBain and Lee obtained tensile strengths of $31.71\ N\ mm^{-2}$ and shear strengths of $24.15\ N\ mm^{-2}$ between these steel adherents, and values of $11.73\ N\ mm^{-2}$ and $16.05\ N\ mm^{-2}$, respectively, between aluminum surfaces. In virtually all cases, the authors referred to 'cohesive failure' in the adhesive, with the adhesion forces having proved to be stronger. At the time, it was assumed that the binder had a chain structure that was influenced by the surface and reached deep into the structure of the adhesive layer. This resulted in remarkable strength values, which were clearly higher when thin adhesive layers (<0.1 mm) rather than thicker adhesive layers were used. No clear mention was made by McBain and Lee of any chemical interactions with the metal as being responsible for this high degree of adherence.

It is known today, however, that the most important ingredient of McBain and Lee's adhesive was *shellac*, the two main components of which are *aleuritic acid* and *shelloic acid*. Aleuritic acid is an aliphatic organic acid characterized by long chains and OH-groups. In contrast, shelloic acid is extremely compact, has a cyclic inner structure, while the outer surface is occupied by CH_2-, OH- and CH_3-groups. *Creosote* consists mainly of guaiacol and other phenols and phenolic ethers. Currently, it is assumed that these substances form chelates with metal oxides or hydroxides or that, owing to their electronic configuration, the ring structures of shelloic acid are able to create a strong chemical bond with the metal surfaces.

These early results were largely ignored because, until the early 1960s, the opinion was that the origin of adhesion – at least between organic, higher molecular substances and metals – could not possibly be attributed to chemical interactions. In retrospect, however, this is difficult to understand since aluminum chelates (which today are known form when organic molecules bind to aluminum hydroxide) had long been used as adhesion promoters in the dyeing of cotton, as alizarin (so-called 'vat dye'). Iron–gallate ink is also known to be a chelate of iron and gallic acid, with such types of chelate being formed on well-cleaned metallic surfaces. Experimentally, if a piece of well-degreased steel is dipped into a solution of 10 g gallic acid dissolved 250 ml ethanol, and then removed and left in the air for some minutes, its surface becomes bluish-gray. This effect, which had been described many years ago by Pliny the Elder in his *Natural History*, suggested not only that a chelate metal complex had been formed but also that the surface would show excellent adhesive properties. Finally, when *chromatography* – especially on aluminum oxide – was used to separate chemical compounds, a high retention time was attributed to a chemical interaction. Likewise, despite there being only a rudimentary knowledge available of the surface condition of metals, it was also realized that metals would become covered with oxide layers whilst, under humid ambient conditions, hydroxides could also be created that would be much more chemically reactive in terms of their chromatographic capacity.

The role of chemical interactions in the interpretation of adhesion phenomena changed greatly during the early 1960s, however. Among the emerging high-performance bonding technology, the adhesion behavior of metal surfaces was of particular interest and again subjected to a variety of investigations. Sandstede et al. [13] noted that it was possible to desorb only a small proportion of acetic acid when it had been adsorbed onto aluminum from the vapor phase. This indicated that, in terms of the Langmuir adsorption isotherm, acetic acid must bind irreversibly to the metal surface, a process which could only take place provided that a chemical interactions had occurred. Later, Kautsky and Barusch [14] stated that aminoamide, when adsorbed onto an oxidized iron surface, also presented low desorbability. By using an infrared spectroscopy technique that had been specially developed for the investigation of interfaces [Fourier transform infrared spectroscopy (FTIR) had not yet been invented!], Dunken [15] suggested that copper stearate was most likely formed when stearic acid was adsorbed onto copper. In a similar study, Dimter and Thinius [16] described the result of applying phenolformaldehyde resol to aluminum; the subsequent rise in temperature to approximately 130 °C indicated that an exothermic reaction must have taken place, which could be explained by the formation of aluminum phenolate.

As reported by Lewis and Forrestal as early as in 1964 [17], the adhesion of plastic materials (e.g. polypropylene) which adhered only loosely to metals was considerably improved when the plastic was grafted with chemically reactive groups; this was the case for polypropylene grafted with dihydroxy-boranyl or epoxy groups, respectively. This provided further evidence for effects caused by the formation of chemical bonds at the interface between metals or their oxides and organic substances.

As the irreversible sorption process can, in theory, be attributed to a chemical primary valency bonding to surface atoms [14], the above-described examples of chemisorptive bonds between organic materials and metal surfaces clearly give rise to the conclusion that chemical interfacial reactions do exist. This is also valid when the heat of adsorption measured is lower than the energies of chemical bonds, since chemically active adsorbate molecules must dissociate before adsorption can take place, thus consuming a portion of the energy liberated in bond formation.

Presumably, the heat of adsorption supplies the activation energy necessary to launch the formation of chemical bonds. Since virtually no other energy sources are available when an adhesive is applied, chemisorption or chemical adhesion first requires an exothermic adsorption. In fact, while adsorption takes place rather rapidly (requiring only a few minutes), chemisorption takes longer (between 20 min and 1 h) [3]. As the energy required to separate chemical bonds is considerably higher than the physical bonding energy, the heat produced during further adsorption is not adequate to separate the chemical bonds, and this explains why chemisorbed quantities may not be desorbed according to Langmuir's adsorption theory [18].

During the late 1960s, Brockmann [19] extended the results obtained from adsorption and chemisorption experiments (most of which were based on the measurement of sorption isotherms) to adhesion taking place in bonded joints.

Brockmann showed that the sorption properties of low-molecular-weight phenolic resins on metal surfaces were clearly in analogy with the strength behavior of adhesion. He has already shown that there was no true adhesion failure when metal adherents were separated from a cured phenolic resin adhesive, despite the macroscopic appearance that the separation occurred within the adhesive layer. Rather, by using available measurement techniques, the residues of the adhesive were seen to be spread over almost the entire metal surface. These results were explained by Brockmann on the basis of his investigations into chemical interactions. In experiments with epoxy resins applied to aluminum surfaces (as described in a later publication), Brockmann also detected chemical interactions that occurred to a lesser extent [20] than in the case of phenolic resins. Based on these results, it could also be shown that the water resistance of adhesion could be improved considerably when specific chelate complexing agents (morin, hydroxychinoline) were used which would adsorb to the metal surfaces from alcoholic solutions, before the adhesive had been applied [21].

By using specific measurement procedures, such as FTIR spectroscopy, it was shown later in a variety of experiments that chemical interactions took place at least between chemically reactive adhesives – that is, adhesives curing within the bondline – and solid material surfaces. Although, today this has been established as a foregone conclusion, it has not been shown clearly whether chemical interactions predominate the effect of the adhesion between solid material surfaces and nonreactive adhesives. Whereas, Langmuir's investigation methods made it possible only generally to detect the existence of chemical or chemisorptive bonds, current analytical techniques allow the nature of those bonds to be determined, at least in some cases. A sufficiently high bonding energy is not a prerequisite to determine adhesion strength in a simple model. However, in order to determine the resistance of an adhesive joint against moisture, knowledge of the nature of the chemical bonds can be of decisive importance. If acid–base bonds, alcoholic bonds (e.g. between an opening epoxy ring and aluminum oxide) or salt bonds (phenolates) are formed between the polymer adhesive and the solid material surface, the chemical reactivity is of no benefit to the water durability because those bonds are water-soluble. Chemical bonds only play an important role at the interface if they are resistant to hydrolysis (e.g. chelates). Indeed, this is the case if phenolic resins are used without any auxiliary agent, because hydroxymethyl phenol (an important low-molecular-weight component of this adhesive) is able to form chelates with aluminum oxides or hydroxides by itself. This is considered the main reason for the high durability of aluminum joints bonded with phenolic resins (see Sections 5.5.3 and 8.2.1).

In order systematically to optimize the adhesion process, it is essential to utilize currently available analytical methods for determining the chemical nature of substances present on the surfaces to be joined. Today, surface-sensitive X-ray photoelectron spectroscopy (XPS) analysis is particularly well suited to this purpose, because it allows not only the presence of specific atom groups to be detected but also the identification of the type of bond involved. It is important that such data are known if the type of chemical reaction that may occur with the binder is to be predicted.

One example where chemistry is used specifically in the optimization of adhesion and durability is that of *glass*, a material that has been grafted with reactive silane for adhesive purposes for almost 50 years. Silanes contain organoreactive groups that are thought to react chemically with polymers as soon as their silanol groups become firmly adsorbed to the glass surface [22]. Similar procedures have been successful with other materials, for example on zinc. Both, reactive silanes and chelate complexing agents (stearates) lead to considerable improvements in adhesion and durability, and consequently reactive silanes have been used as adhesion promoters and mixed in various adhesives destined for the bonding of inorganic adherents (see Section 5.10) [23].

However, it is not absolutely safe to assume that the adhesion-promoting effect of this type of adhesion promoter is based simply on bifunctionality and the specific formation of water-resistant bonds with the solid material and the adhesive. Even if these promoters do not present any groups that may react with the adhesive, they do improve durability. *Organosilanes*, for example, promote adhesion on a glass surface without presenting any organofunctional groups – that is, without being able to react chemically with the polymer. The same must be assumed for chelate complexing agents (e.g. hydroxychinoline or alizarin) when using epoxy resins as adhesives. Their positive effect with regards to a higher durability may be due to reaction with the thermodynamically instable solid material surface of glass or even metals, thereby reducing instability. This can easily be demonstrated on oxidized aluminum: if an alizarin chelate is created on pickled or anodized aluminum, scanning electron microscopy (SEM) can be used to show that, owing to the formation of chelates, the tendency of the oxide to hydrate in humid ambient conditions (one of the most important factors which may lead to adhesion failure) is drastically reduced.

Whilst adhesion and durability must be considered as a problem of pure adhesive bonding, it is also important – especially in the case of inorganic materials – to realize that the formation of chelates as well as the acidity or basicity of the adhesive system used may reduce (or increase) the instability of the layers covering the basic material. With weak acids, aluminum oxides resist hydration relatively well. Thus, even in the case of cured phenolic resin adhesives, weak acidity promotes not only chelate formation but also stability within the assembly by improving the resistance of the oxides. The same applies to iron oxides that are very resistant to hydration under alkaline conditions (pH 10–11). This is the reason that, as long as no acid rain penetrates the system, there is a good resistance of adhesion between concrete and Monier steel. Otherwise, the pH will fall to 10 or even 9.5, the oxide will be allowed to hydrate, and the result will be destruction of the adhesion. In fact, this is the most important reason why long-term damage has frequently been observed in concrete structures.

3.3
Adhesion in Real Systems

The standard adhesion theories described above often do not allow predictions to be made of the actual behavior and failure of adhesive bonded joints. This can be illustrated by the following examples that derive from the author's experience (no bibliographic references given).

3.3.1
Gold

Although gold is commonly believed to be a noble (i.e. low-reactivity) metal, it has fascinating adhesion properties. For example, if a sheet of gold is rolled up in normal writing paper and left for a certain time, the glue residues originating from the paper will adhere strongly to the gold surface such that it cannot be removed without causing mechanical impairment to the metal surface. This phenomenon has been recognized for many centuries, one example being the process of *goldbeating*, where the metal leaves are stored between layers of Japanese paper that do not contain binding glue. Alternatively, when applied in leaf form, gold adheres very well to many natural materials which have been sized with 'gold oil' to create very long-lasting and weather-resistant surfaces. The adhesion of gold leaf is very resistant, even in humid ambient conditions, despite the gold leaf itself being soft and having a low resistance. Consequently, the regilding of historical monuments is only necessary when erosion has led to damage of the gold leaf (see Section 8.16.4). This effect cannot be explained by standard adhesion theories; rather, the only possible explanation is that during the early stages of its manufacture, organic residues from the goldbeater's skin (the outer membrane of ox intestine between which the gold leaf is beaten) are bound tribochemically to the gold leaf in quantities sufficient to provide adhesion by polymer diffusion when the gold leaf is applied to the gold size.

3.3.2
Polyethylene and Polypropylene

The adhesion of polyethylene or polypropylene can be demonstrated in a simple experiment. Both materials can be carefully melted without considerably exceeding the melting point, such that the surface is prevented from oxidizing. A well-cleaned aluminum surface is then coated with the molten polyethylene. As might be expected, if the surface is well wetted the polyethylene can easily be stripped from the aluminum sheet, such that no polyethylene residues can seen macroscopically to remain on the sheet. Measurement of the surface tension of the sheet surface will be approximately $30\,\text{mN}\,\text{m}^{-1}$ inside the wetted area, but $72\,\text{mN}\,\text{m}^{-1}$ outside the wetted area (i.e. there is more absorbed water). Hence, it is concluded that although the polyethylene could be stripped off with ease, its residues remained on the aluminum surface, and that cohesive failure had taken place within the polyethylene near the interface between the two materials. This situation cannot be explained by either of the adhesion theories described in Section 3.2, thus confirming an observation made earlier [20, 21], namely that adhesion failure is rare and the properties of the bond are controlled by the surface influencing the condition of the polymer near the interface.

3.3.3
Glass

In the case of glass, without using any adhesion promoter (at least with reactive structural adhesives), no long-term durability can be obtained when bonding this adherent. At present, this is assumed to be due to the fact that in a humid environment at the interface, the glass surface is irreversibly modified by corrosion that may be accelerated by the adhesive, resulting in macroscopically observable adhesion failure [22]. However, if the glass surfaces are bonded with acrylic pressure-sensitive adhesives, even without using any adhesion promoter, a relatively high water resistance (compared to structural adhesives) is obtained. This situation does not apply to adhesion, which common exists as a static, material condition. At present, pressure-sensitive adhesives are considered to build up adhesion by a variety of physical interactions rather than by any firm chemical binding to solid-state materials (this would not be expected due to their macromolecular, saturated condition). It has been noted that, when other difficult-to-bond materials (e.g. untreated aluminum, stainless steel or relatively low-energy polymer surfaces) are bonded with this type of adhesive, the assembly has a surprisingly high durability. The polarities of the adherents have a certain influence on adhesive strength (as can be determined by measurement), although as noted above, physical interactions can be destroyed by water penetration. Hence, the positive adhesive properties of pressure-sensitive adhesives can be explained only by making the assumption that, owing to high molecular mobility (even in the solid state) they progressively replace the adhesive bonds that have been destroyed by water penetration, thus 'fighting' against destruction. Today, we refer to this capacity as 'dynamic adhesion', the existence of which can be demonstrated with macroscopically conducted strength and durability testing [24]. There is a clear relationship between the degree of molecular mobility, which in turn can be controlled by the degree of crosslinking, and water resistance. It can also be shown that the build-up and regeneration of adhesion – both of which factors are dependent on time and are very pronounced in the case of pressure-sensitive adhesives – support the validity of the hypothesis of dynamic adhesion.

3.4
New Concepts in the Field of Adhesion

Recently, it has become clear that new concepts are needed to explain the behavior of adhesive bonds, especially as the 'standard theories' initially proposed to help in our understanding of adhesion phenomena have been shown inadequate.

Polymer dynamics, which appear to be a dominant factor in the adhesion capacity of pressure-sensitive adhesives, are most likely also essential for the adhesion of not only contact adhesives but also hot-melt adhesives, which generally do not form crosslinks (or only very loose crosslinks) and are not considered to present any chemical reactivity.

3.4 New Concepts in the Field of Adhesion

The results of recent experiments [25, 26] have given rise to the assumption that, in addition to chemical bonds, polymer dynamic effects are also responsible for the overall adhesive behavior of reaction adhesives. For example, in epoxide resin adhesives, at the adherent interface there is a distinct influence of the surface condition on the kinetics and density of the crosslinking, respectively. It is remarkable that on surfaces which attribute a good age resistance to an adhesive bonded joint (at least as far as we know), such as steel or aluminum surfaces blasted with corundum, the degree of crosslinking near the interface is less than on surfaces which attribute a poorer age resistance to the assembly. This effect is not explained by surface morphology, but rather by the chemical properties that are influenced by contamination of the surface by the blast medium.

The fact that better, more durable adhesion results from a lower degree of crosslinking is contradictory to standard concepts, as the latter is probably expected to involve a higher rate of water transport.

It has long been assumed, and clearly demonstrated in some cases, that the material surface does have an effect on polymers, although the details of this phenomenon are not yet understood [26]. However, in the case of a lower degree of crosslinking a better polymer mobility can be postulated, which in turn has a stabilizing effect in terms of the dynamic adhesion mentioned above. The results of recent studies [27, 28] have indicated that the corrosive mechanisms in adhesive-bonded metal joints are also controlled by modifications of the polymer condition due to the effect of the surface. Indeed, it is becoming increasingly evident that not only in the non-aged and aged conditions but also in the dried condition, the deformation behavior of adhesive bond-lines is considerably influenced by the surfaces of the adherents.

Frequently, 'softer' (i.e. highly elastic) adhesives with plastic deformation behavior show a better durability than rigid systems, a property which speaks well for the importance of dynamic effects. Previously, adhesion was thought to be irreversibly weakened by humidity, but it is increasingly realized that this effect is in fact not absolutely irreversible. If, after an aging process, the recovery time is sufficiently long then there is a reimprovement of the adhesive bonded joints in many cases [28, 29].

At present, the modification of a polymer near the interface is attributed to chemical effects exerted by the surface [25, 26, 28]. The microscopically measurable roughness of the surface clearly has no influence either on these mechanisms or on the long-term durability of the adhesive bonds [25].

It is known, however, that nanostructured surfaces which can easily be produced on aluminum by pickling, or alternatively by anodizing on zinc or stainless steel, may have a considerable effect on adhesive strength, and particularly on durability when compared to non-nanostructured surfaces of an identical chemical composition. This phenomenon cannot be explained by simple concepts such as the postulate of micromechanical adhesion because, once again, the failure of adhesion occurs not within the nanostructures but rather within the polymer near the interface. Thus, it is safe to assume that there is an effect of the surface morphology that can be explained not only by the steric hindrance of adsorption or segregation within the nanostructures, but also by orientation effects. The results of these investigations are presented in Section 7.5.

It becomes clear that, when attempting to understand the build-up and behavior of adhesion, one reaches a dimension which lies between the molecular one and the dimensions of matter, the characteristic technical properties of which only take effect in case of a total quantity of approximately two orders of magnitudes higher than the molecular dimension. Although it has become possible to analyze this dimension only recently, the prevailing – albeit sometimes unsatisfactory – uncertainty with regard to the behavior of adhesive systems is expected to be overcome within the next few years.

3.5
Conclusions

Although, to date, the phenomenon of adhesion in bonded joints has not yet been completely explained by means of scientific concepts, our existing knowledge allows some simple conclusions to be drawn that are of considerable technical importance:

- The wetting of adherent surfaces by the adhesive is an important – but insufficient – prerequisite for the build-up of adhesion. Wetting experiments may provide the first indications with regards to the bonding characteristics of a surface and its polarity, the latter referring to the quality of adhesion. However, it is not possible to determine the exact characteristics of adhesion by means of wetting measurements.

- In addition to physical interactions, it is particularly desirable to have adhesives (especially reactive adhesives) build up chemical interactions with the adherent surfaces in order to obtain good adhesion. The formation of chemical interactions may be enhanced by surface pretreatment, the use of adhesion promoters, and with specific formulations of adhesives.

- Chemical interactions modify the nature of the adhesive and the adherent surfaces, respectively, near the interface. This may have positive effects on the water resistance of adhesive assemblies, for example by the stabilization of metal oxides as a result of chemical reactions, or by modification of the polymer dynamics within the adhesive near the interface.

- Adhesion hardly ever fails where it has developed. Invariably, so-called 'adhesion failure' will actually occur within the adherent or the adhesive near the interface – an area which will have been influenced by the build-up of adhesion.

- In practical terms, the microscopic roughness of the adherent surface does not have any influence on the quality of adhesion.

- Nanostructures or nanomorphologies on the adherents may improve the quality of adhesion, although the reasons for this are not yet known.

4
Survey and Classification of Adhesives and Primers

Although a systematic survey of available adhesives according to their properties, such as their technical performance and processability, would be greatly appreciated by potential users, this does not exist to date. While various attempts have been made to systemize adhesives, it is virtually impossible due to their nature. In contrast to other joining aids, such as welding electrodes, adhesives have not been standardized since, according to current knowledge, standardization would represent an inhibitory factor for further development.

At present, there is available a most varied array of adhesives that often differ considerably one from another in terms of their processing properties, their strength and their durability characteristics. In order to be effective as an adhesive, a substance must fulfill two basic requirements:

- At a specific point of the joining process the adhesive must have a high molecular mobility – that is, it needs to have the properties of a more or less viscous liquid in order to approach the uneven, rigid adherent surfaces up to molecular dimensions (nanometer range), and to allow material joining.
- Within the adhesive layer, the material must have as low a molecular mobility as possible so as to allow tensile, shear and peel forces to be transmitted, respectively; that is, in technical terms, it needs to behave like a solid object.

In the following sections the most important adhesive systems are classified according to their mode of setting (also called 'curing mechanisms'). A general survey of the types of adhesive available, based on their curing properties and mode of setting, is provided in Figure 4.1 and Table 4.1.

4.1
Noncuring (Pressure-Sensitive) Adhesives

Extensive development has led to the successful use of permanently tacky, nonsetting pressure-sensitive adhesives in aircraft and bodywork construction, as well as in other fields [1]. Pressure-sensitive adhesives are high-viscosity polymer systems that at least partly retain the properties of a liquid in their final state. This allows them to

Table 4.1 Classification of the different types of adhesive according to their mode of setting.

Mode of curing	Type of adhesive	Typical base resin (example)	Characteristic properties
No curing	Pressure-sensitive adhesives	EVA, acryle resins	• rapid processing • not harmful to the environment
Physical curing	Contact adhesives	Neoprene rubber	• easy processing • low price • solvent-containing
	Adhesive dispersions		• easy processing • long flash-off time • very low solvent content
	Heat-sealing adhesives	Similar to contact adhesives	• rapid processing • activation energy needed • solvent-containing • dispersion in water possible
	Hot-melt adhesives	PVC copolymers	
		Polyamide	• very rapid processing • solvent-free
Physical and chemical curing	Pressure-sensitive adhesives with postcure	Silane crosslinked polyester Acryle resins	• under development
	Contact adhesives with addition of a curing agent	Neoprene rubber with isocyanate curing agent	• easy processing • low price • solvent-containing

Setting mechanism	Adhesive type	Example materials	Properties
	Hot-melt adhesives with postcure	PUR, epoxy resin	• very easy processing with postcure • solvent-free
Chemical reactions (polyreaction) after the mixing of different components	Two-part and multi-part adhesives	Epoxy resin, polyurethane rubber, UP	• costly processing • high price • solvent-free • reactive source material • no application of heat needed for curing
Via application of heat	One-part reactive adhesives	Epoxy resin, phenolic resin	• can be exposed to the most extreme conditions among all adhesives • high price • solvent-free • less reactive substances in the noncured condition as compared to two-part adhesive systems
Setting by the conditions prevailing in the bond-line	Fast-curing, one-part adhesives	Cyanoacrylates, acrylic diester	• rapid processing • high price • reactive source material • no application of heat needed

EVA = ethylene vinyl acetate; PUR = polyurethane resin; PVC = poly(vinyl chloride); UP = unsaturated polyesters.

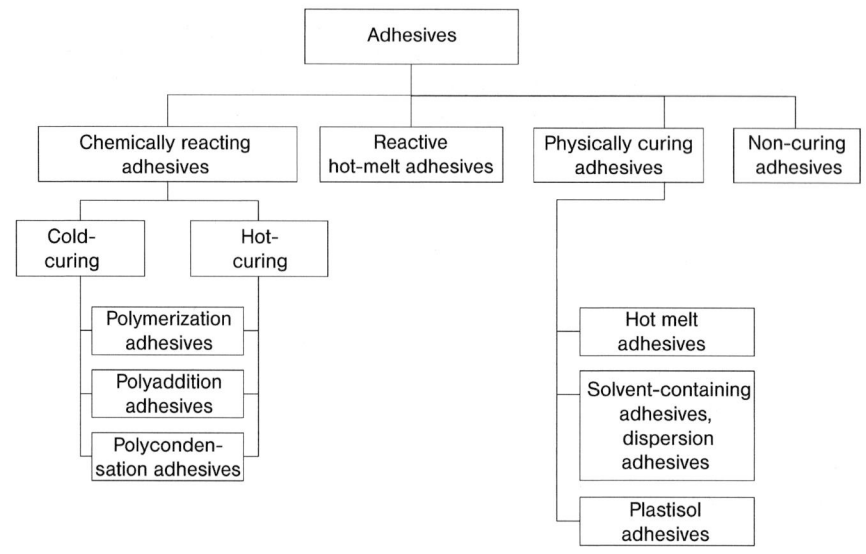

Figure 4.1 Classification of adhesives according to their mode of setting.

completely adapt to the roughness-induced surface contour of the adherents, thereby forming adhesive bonds.

Pressure-sensitive adhesives are not only used in the form of adhesive tapes coated with binders on one or both sides, but also as transfer systems which are applied to the solid material, together with a base material which is later stripped off. Furthermore, pressure-sensitive adhesives can be applied in the molten state at temperatures of approximately 80 °C, allowing them to be solvent-free from the start of the application. Another possibility is to apply them to the adherent as aqueous dispersion, the aqueous component evaporating later. Finally, at present, other pressure-sensitive adhesives are available which are applied to the surfaces as solvent-free or solvent-containing, low-viscosity systems and later are shortly irradiated with ultraviolet (UV) light or heated to develop adhesive capacity (see Section 5.7.2.5).

Pressure-sensitive adhesives are generally applied to one of the surfaces to be bonded. A good bond is only obtained by uniform short pressing of the adherent surfaces, which must be exactly positioned because the bond has a relatively high strength immediately after the pressing. In generally, an increase in strength by 50–100% of the initial value is observed during the course of several hours or days.

Pressure-sensitive adhesives have very good peel strength and, in comparison to chemically setting systems, a relatively low shear strength. *Tack* does not necessarily have a direct connection with peel strength occurring later; indeed, systems with a relatively low tack may have very high peel strength in the final state, or *vice versa*. Compared with various other adhesive systems (chemically setting high-performance adhesives included), pressure-sensitive adhesives develop very good aging resistance, often even on surfaces which are difficult to bond. The reasons for this behavior have been discovered only recently [2].

In general, pressure-sensitive adhesives are solvent-free and almost monomer free, they are easy to process, and have relatively low susceptibility to processing flaws. Processing aids are rarely needed, and any processing speed is possible. After joining, the bond is immediately ready for mechanical loading.

4.2 Physically Setting Adhesives

If the transition from the low-viscosity state (in which the adhesive is capable of wetting the adherent surface) to the solid state takes place by physical processes only (e.g. evaporation, solidification of the melt or diffusion processes) without chemically changing the polymer components of the adhesive, this is called 'physical setting'.

4.2.1 Contact Adhesives

Contact adhesives are made from polymer components which are already in a high-molecular state, but which are not yet chemically crosslinked. By means of the addition of solvents, a low-viscosity state is obtained which allows the adhesive to wet the surface of the solid material. Contact adhesives must be applied to both adherents, and solidification takes place by a drying process before the materials are joined. The adherents are pressed together as soon as the adhesive layers are dry to the touch.

During the application of pressure, and even afterwards, a diffusion process takes place between the adhesive molecules of both coats, firmly joining the adherents together. The strength of the bond can increase, often within hours, after the application of pressure. If at least one adherent is porous, the adherents can be joined before the drying process has taken place at all or only partly, because the remaining solvent is capable of evaporating through the porous adherent. In this case, it is also possible to reposition the adherents after having joined them because, in the semi-dry condition, the adhesive still largely presents the properties of a liquid.

In the dried or cured condition, the adhesive coat is generally in the *thermoplastic state* – that is, the molecules are not three-dimensionally crosslinked. As a consequence, the adhesive coats applied do not present the same resistance to heat and creep as chemically cured adhesives. It is possible, however, to add slowly acting hardeners to these polymer systems; this induces a three-dimensionally wide-meshed crosslinking of the systems over the course of several days. While the addition of hardeners does not alter the processing properties of the contact adhesives, the durability and strength of the cured system is considerably improved. One disadvantage of contact adhesives is an occasional very high solvent content (up to 80% by weight), so that appropriate protective measures must be taken during processing.

As a consequence, solvent-containing contact adhesives are now being replaced, when possible, by so-called 'dispersion adhesives'. These consist of a dispersion of the adhesives in water, during the evaporation of which the dispersed polymer particles combine by diffusion (see Section 3.2.2). Unfortunately, dispersion adhesives have the disadvantage that the drying process of water takes considerably longer than that of solvents, and an acceleration can only be obtained by the application of heat, which requires additional energy input. Furthermore, during the drying process, dispersion adhesives develop inadequate tack, a property which is desirable in many cases. A drawback of this approach is that the strength and durability of dried adhesive coats are generally poorer than those of solvent-containing contact adhesives; this occurs because the polymer layer created is not absolutely homogeneous and has a tendency to redisperse under the influence of moisture.

4.2.2
Plastisol Adhesives

Plastisol adhesives, which have been used widely in the bodywork construction for automobiles, represent another group of physically setting adhesives. A plastisol adhesive is formed by the dispersion of a polymer, mostly polyvinyl chloride (PVC), in a plasticizer. During the curing process, when the dispersion is heated, the polymer dissolves irreversibly in the plasticizer that acts as a solvent, transforming into soft PVC. When the cured adhesive coat is in the thermoplastic state its resistance to heat and its creep strength are poorer than those of chemically setting systems. An improved resistance to heat can be obtained by adding chemical setting components to the plastisol adhesives, such as epoxy systems. The advantages of plastisol adhesives are that they are relatively cheap, and high adhesion and durability are obtained even on nonpretreated steel surfaces or on body sheets that have been oiled for the purpose of corrosion protection.

4.2.3
Hot-Melt Adhesives

Hot-melt adhesives, which are also physically setting adhesives, exist in a solid, high-molecular state and do not contain any solvents. They are applied in the molten state or as a powder or foil that is applied between the adherents and then melted under contact pressure and heat in order to obtain an appropriate wetting of the adherents. Immediately after cooling, hot-melt adhesives are capable of transmitting forces. By using high heating and cooling rates, manufacture processes using hot melts can be carried out with split-second timing.

When bonding metals care must be taken that, due to the good thermal conductivity of the adherents, the hot-melt adhesive must not solidify too rapidly near the interface; otherwise, complete wetting of the surfaces would be prevented. It is therefore advisable to preheat the adherents when this type of adhesive is used. An increase in the strength and long-term durability of hot-melt adhesives can be obtained if the adhesives continue to crosslink after the melting point has been reached; this can be achieved by the addition of specific substances or by the specific structure of the polymers. Either approach is possible with epoxy systems and polyurethane hot melts with *postcure*; this is the most widely used method at present, in which the subsequent chemical reaction is initiated by moisture from the outside (see Section 5.6.1). These systems are used on a large scale to bond windscreens in automotive bodyworks (see Section 8.2.2.1). A slight disadvantage of hot-melt adhesives is that, by their nature, when used without postcure they are only resistant to heat up to the processing temperature. However, an advantage of such a bond is that it can be debonded and reconstituted by heat. Hot-melt adhesives with postcure can be applied at relatively low temperatures (about 60–80 °C), while being able to resist temperatures of 120–150 °C in the cured state.

4.3 Chemically Setting Adhesives

A wide array of chemically setting adhesives is available. Initially, there are low-molecular (and consequently flowable or low-viscosity) substances in which the reactive groups may react with each other under specific conditions (see Sections 5.4 to 5.8); this transforms the low-molecular systems into polymer substances with a high relative molecular mass and high mechanical resistance [3].

In the chemical setting process, the build-up of strength by means of chemical reactions must take place only after application of the adhesive and after joining of the adherents; otherwise, the adhesive would lose its wetting capacity. The chemical reaction can be initiated by mixing two or more reactive components shortly before applying the adhesive. After application, the materials transform into a macromolecular substance within the adhesive layer at room temperature. The adherents must be fixed until this substance has at least been partly formed because, in the low-molecular state, the adhesive is not able to transmit forces. In most cases, the time needed for the setting process can be shortened by the application of heat.

It is possible to force the reaction to start under specific conditions in the bond-line, for example when heat is applied, by using reactive components with specific chemical structures. These adhesives are known as 'one-part adhesives', although from a chemical standpoint they mostly consist of two or more reactive components.

4.3.1 Two-Part Adhesives

The most important adhesives of this group are polyester, cold-setting epoxy resins, polyurethanes and acrylic adhesives. Common two-part adhesives consist of several components that are mixed in a specific ratio before application. For acrylic adhesives of the 'second generation', it is possible to apply each of the components separately to one of the adherent surfaces, and then to press the adherents together, thereby initiating the curing process.

In addition to the specific mixing ratio, the so-called 'pot life' must also be respected; this is the period of time during which the adhesive can be used (i.e. be applied) after mixing the components, and is dependent on the type of adhesive used and the volume prepared. As the crosslinking process gradually starts during the pot life, and the viscosity of the adhesive gradually increases during this time, exceeding the pot life will cause a lack of wetting of the adherent surfaces by the adhesive and result in poor adhesion properties within the bond-line. The crosslinking of two-part or multi-part adhesives, respectively, almost always takes place as an *exothermic reaction*, so that the adhesive will be heated up during this process. When a small volume is prepared it will warm up more slowly than a larger volume, for which the ratio 'volume to heat-emitting surface' is less favorable. This effect represents a particular risk. It is difficult to recognize the moment when crosslinking begins by the degree of viscosity because the heat emitted will lead to a decrease in viscosity. This in turn will neutralize the effect of the crosslinking process, which increases the viscosity of the mixture. If such a warmed system is applied to cold adherents, the partly crosslinked system will solidify immediately, without adequately wetting the surfaces.

4.3.2
Hot-Setting, One-Part Adhesives

Although, one-part systems are considerably easier to process than either two-part or multi-part adhesives, on most occasions it will be necessary to apply heat. One-part adhesives consist of low-molecular and plastified substances. In general, the base resins are phenolic resins, epoxy resins and, for the creation of high-temperature adhesives, polybenzimidazoles and polyimides.

With regard to phenolic resins and polyimides, the crosslinking process takes place in the form of *polycondensation*, and consequently additional pressure must be exerted on the bond-line during the setting process. Heat must also be applied in order to remove water from the bond-line, while the pressure must be higher than the water-vapor pressure at the setting temperature (i.e. generally >500 kPa).

In contrast, contact pressure is dispensable with hot-setting epoxy resins, because the setting reaction takes place as *polyaddition*. Only in the case of extended-surface bonds need contact pressure be applied to the adherents because, in the noncured state, these adhesives are not capable of transferring any mechanical forces, and any deformation of the adherents may result in destruction of the bond.

4.3.3
Cold-Setting, One-Part Adhesives

Besides setting via heat, several other physical effects like absence of oxygen in the glue line or presence of moisture at the adherent surfaces can be utilized in order to initiate a curing reaction. The most common one-part adhesives are based on an initiation by irradiation (e.g. UV light), by an absence of oxygen (anaerobic), or by exposure to moisture (see Section 5.7). Owing to the precise control of the curing process, systems setting under the influence of UV light or light (most of which are based on acrylates or epoxy resins) have been increasingly used to date. In low-molecular base resins, photoinitiators are dissolved or chemically incorporated which initiate the crosslinking process under irradiation with UV light or visible light. Two such groups of adhesive may be distinguished: (i) those that set exclusively under continuous irradiation (radical reaction mechanism); and (ii) those in which crosslinking is initiated by short irradiation of the open bond-line and continues to take place after joining, without further irradiation (ionic reaction mechanism; see Section 5.7.2).

Cold-setting, one-part adhesives which crosslink under the effect of moisture (e.g. one-part polyurethane and silicon resin systems) are often used as sealing compounds. In the noncured state these adhesives are low-molecular and noncrosslinked. However, at the bond-line they are set by the effect of moisture diffusing from the outside, which results in a three-dimensionally, chemically crosslinked system with high plasticity. Silicon resins, for example, have a high plasticity and are characterized by the fact that their strength and deformation properties are virtually unaffected over a temperature range from −55 to +250 °C (see Section 5.6.3.1).

4.3.4
Microencapsulated Adhesives

One specialized group of chemically setting adhesives incorporates microencapsulated, chemically setting systems that are used mainly as locking devices for screws. Both, anaerobically setting adhesives (diacrylic acid esters) and two-part or multi-part reaction adhesives may be microencapsulated. These components can be applied to the adherents, for example, in a latex solution. The capsules are only destroyed by relatively high mechanical pressure, and the reactive systems contained will subsequently set at room temperature, resulting in the formation of polymers. In the case of screws or threads the manufacturer may apply the microencapsulated system directly to the adherents.

4.4
Primers

A *primer* is a coating system that is additionally applied to a surface to be bonded prior to the application of the adhesive in order to protect the surface against uncontrolled, environmentally induced changes, or to improve the adhesion of an adhesive applied subsequently. In general, primers are highly diluted polymer solutions that wet the surfaces very well and sometimes may even remove minor impurities. Currently, due to problems posed by solvents, systems in the form of aqueous solutions or even dry powders are also used, and may be applied by electrostatic means. Primers are generally applied at a dry-film thickness of a few micrometers.

From a chemical standpoint, primers may be similar to adhesives – that is, they may be cured separately when reactive components are used. In this case, their only purpose is to improve wetting and to protect the surfaces until the adhesive is applied. In particular, adherents coated with cured primers may be stored for up to three weeks (and even more) before application of the adhesive, and without any impairment of adhesion quality. From experience, we know that it is less critical if contamination occurs on the primer rather than on freshly pretreated metal surfaces. Moreover, any contamination may be removed, if necessary, by using mild cleansers and without impairing the adhesion properties. As a consequence, primers may serve as useful auxiliary agents that can be used in order to provide for production steps, without any time constraints.

However, the composition of a primer may differ from that of an adhesive in order to obtain a specific effect on adhesion and corrosion resistance. As the primer coats are thin, the deformation properties of the primer do not play a predominant role within the overall joint. Therefore, it is preferable to incorporate any components which would adversely affect the adhesive into the primer; examples are inorganic pigments or brittle resin components, as well as adhesion promoters (see Section 5.10). The use of primers may also have an economical bearing since, owing to the small amounts of primer needed, adding these components to the primer may be cheaper than incorporating them into the adhesive. For example, if curing primers based on epoxy resins are used, then relatively cheap polyurethane adhesives with good plasticity may be applied later. But, without the use of the primer, their adhesion would have a weaker resistance to water, particularly on aluminum and steel alloys.

Although, the various possibilities of optimization that can be achieved with primers are often not well recognized, they may outweigh the disadvantage(s) of introducing an additional production step.

4.5
General Handling Instructions

In general, polymers are not very harmful – or not even harmful at all – at least for short-term human contact, as long as they are present in the form of solid substances within the adhesive layers. However, if the adhesive contains low-molecular substances before the setting process takes place, then the hazard potential may be deemed greater. The same applies to the handling of solvent-containing contact adhesives, and to chemically setting systems in particular. In the noncured state these are often chemically reactive, while some components may even be toxic, such as the amine hardeners of epoxy resins, isocyanate curing agents of polyurethane, and acrylates of the second generation. Besides respiratory toxicity, in some cases an allergic sensitization or hazard of strong caustic reactions may be expected on direct skin contact. Likewise, monomeric or low-molecular substances may be easily dispersed, which may result in an increased hazard potential. Due to the reactivity and potential dispersion risk (most likely in the form of vapors), a regulation specifies that, in the noncured state, chemically reactive adhesives should be treated as special hazardous refuse, while cured adhesives are considered as household refuse.

Low-molecular organic systems also present a greater fire hazard than do polymer materials. At present, classification rules have been introduced for toxicity, ecological hazard potential and fire hazard, the details of which must be specified by the adhesive manufacturer in standardized data sheets.

5
Chemistry and Properties of Adhesives and Primers

The designations of adhesives are often derived from their base polymers, such as 'neoprene contact adhesives', 'phenolic resin adhesives' or 'epoxy resin adhesives'. In this chapter, we will apply this classification principle to all chemically setting systems. Adhesives are rarely made from only one base polymer; more often, they are formulations of several organic or inorganic components because they must combine seemingly mutually exclusive properties to yield a material capable of delivering required functions. Adhesives must not be too viscous, in order to allow pumping, mixing and wetting of the adherent. In the noncured state they must not leak from the bond-line and so must be 'thixotropied'. In the cured state, they must not be too brittle; they must present a plastic deformability with minimum creep under repeated load. Only a combination of components in specific formulations allows specific properties to be conferred to an adhesive; this is the art of adhesive manufacturers who skillfully combine the components to meet the customers' specifications. The base resins confer the essential basic properties to an adhesive, while other characteristics result from a fine-tuned formulation. Criticism is sometimes leveled that adhesives contain too many components, and this is not completely unfounded. Adhesive manufacturers often receive short-term specifications to deliver test products with insufficient time or money to create new formulations. Hence, they select a ready formulated product and fine-tune it to new requirements by adding further components; moreover, the consumer need not know the constitution of the adhesive as long as it functions well. However, this principle cannot be applied to high-performance bonding, where the recommendation is to make a confidentiality agreement with the manufacturer, who must make available the formulation details so that users can perform analytical screenings for quality assurance purposes. Although, occasionally, adhesive manufacturers will not accept this procedure, it has long been widely adopted and proven for various other materials, and has contributed to building confidence in new technologies.

Adhesive Bonding: Materials, Applications and Technology
Walter Brockmann, Paul Ludwig Geiß, Jürgen Klingen, and Bernhard Schröder
Copyright © 2009 WILEY-VCH Verlag GmbH & Co. KGaA, Weinheim
ISBN: 978-3-527-31898-8

5.1
Pressure-Sensitive Adhesives (PSAs)

5.1.1
Introduction

Pressure-sensitive adhesives (PSAs) are permanently tacky at room temperature, and firmly adhere to a variety of surfaces upon contact with finger or hand pressure. They do not require any activation by either water, solvent or heat to build up a strong bond with such materials as paper, plastics, glass, wood or metals. Additionally, some of these products have a sufficiently cohesive strength so that they can be removed cleanly from smooth surfaces, without leaving a residue. In contrast to systems activated by heat, radiation or moisture, adhesion is obtained without changing the chemical or physical condition of the PSAs. They must present an adequate deformability to adapt to rough surfaces, a sufficient wetting capacity towards the adherent surfaces to provide for proper adhesion, and a sufficient inner strength to withstand repeated loads over a prolonged period of time [1].

At present, PSAs are used in a variety of applications, most often in the form of coated products available as adhesive tapes (in roll form), leaves or sheets, or as die cuts. They help to solve various problems in many branches of industry and in a most varied array of everyday applications, be it at home, at the office, or for hobby activities. The following commercialized products are available:

- single-sided adhesive tapes (e.g. for fastening, covering, protecting or packaging)
- double-sided adhesive tapes (for joining, fastening, mounting, splicing)
- self-adhesive labels
- repositionable sheets of notepaper (Post-it®)
- self-adhesive special products (e.g. removable fastening systems, self-adhesive damping products, electrically/thermally conductive adhesive tapes)

Development activities are focused on the creation of a third functionality besides peel and shear strengths, for example to confer thermal, electrical or optical conductivity to PSAs. Another priority is the development of high-performance adhesive systems that not only adhere to glass, metal, ceramics or high-grade plastic materials, but also to low-energy, mostly structured modern polymer materials.

5.1.2
Chemistry of Pressure-Sensitive Adhesives

Although, for a long time, natural rubber [poly (1,4-*cis*-isoprene); see Figure 5.1] was used as base material for PSAs, today they are made from a great variety of elastomers.

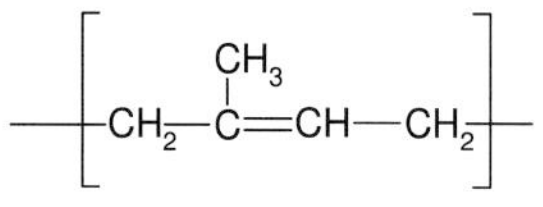

$$\left[-CH_2-\underset{\underset{CH_3}{|}}{C}=CH-CH_2-\right]$$

Figure 5.1 Isoprene unit.

In order to generate tack and adhesion, 'tackifiers' and other materials are mixed with the base material. These formulations stand out for their excellent adhesion features and, particularly when used as masking tapes in painting processes, their excellent removability after the paint curing process. PSAs based on natural rubber, however, have a tendency to yellow and crosslink, thus becoming brittle, due to the double bonds present in the base polymer. Although this problem was partly relieved by the addition of antioxidants, natural rubber-based PSAs remain unstable to long-term exposure to the environment. Several other base resins were therefore developed which did not suffer from these deficiencies [2].

During World War II, there was a shortage of natural rubber and therefore a number of new base resins had to be introduced. It had been known since 1929 [3] that alkyl acrylate ester polymers had the characteristic tack of PSAs, and a short time later it was discovered that polyvinyl alkyl ethers presented the same tacky feature. Due to the shortage of natural rubber, the development of these new materials experienced a rapid boom. Unlike the former formulations, they were transparent and colorless, and they had an excellent resistance to aging processes. However, the first formulations based on polybutyl acrylate and polyiso-butyl ether did not have the required tack and cohesive strength, respectively. However, after many years of successful development acrylic copolymers, for example, were widely accepted for use in high-performance PSA tapes (Figure 5.2).

The two primary acrylates used are 2-ethylhexyl acrylate and iso-octyl acrylate, but polymers based exclusively on these monomers do not perform as PSAs in their own right. During the early 1950s, various research groups found that the performance of these materials could be substantially improved by adding certain amounts of acrylic acid [4].

Generally, acrylate adhesives do not require any tackifiers to perform as well as PSAs; rather, their pressure-sensitive characteristics result from the inherent physical properties of the acrylate polymer. Today, acrylate-based PSAs are used in a wide variety of applications, from transparent tape to medical tape, office tape or foamed adhesive tape for industrial high-performance applications.

Both, natural rubber and acrylate-based PSAs must be chemically crosslinked in order to obtain the cohesive strength necessary to meet all of the requirements of a PSA. Block copolymer-based elastomers meet these requirements through phase separation. A–B–A block copolymers of isoprene with styrene (S–I–S) or butadiene with styrene (S–B–S) were introduced during the 1960s. The structure of a S–I–S block copolymer is illustrated in Figure 5.3.

Following coating, the glassy styrene blocks phase out and yield a phase-separated structure in the elastomer matrix due to incompatibility with the elastomer (i.e. polyisoprene or polybutadiene) (Figure 5.4), and this leads to physical crosslinking. Block-copolymer-based PSAs are used in a wide array of applications, but

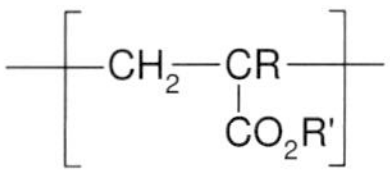

$$\left[-CH_2-CR-\atop |\atop CO_2R'\right]$$

Figure 5.2 Acrylate unit.

Figure 5.3 Structure of a styrene–isoprene–styrene (S–I–S) block copolymer.

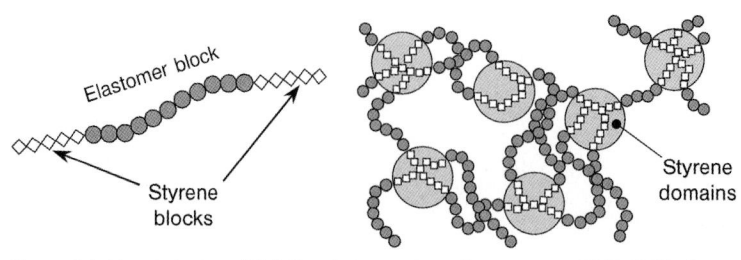

Figure 5.4 Morphologies of S–I–S and styrene–butadiene–styrene (S–B–S) block copolymers.

most commonly in packaging tapes. However, due to the double bonds in the B-block, these adhesives suffer from oxidative instability in the same way as natural rubber-based adhesives. In comparison to acrylate systems that achieve only about 50–80% of their final strength after pressure application but then build up final strength during the course of hours or days, block copolymer systems reach their final strength shortly after compression.

Since elastomers have a low tack and a low adhesion force towards commonly used surfaces, tackifiers must be added to natural rubber-based and A–B–A block copolymer-based adhesives. For this purpose, the viscous component is introduced into the elastic polymer. However, even if the base polymer has adequate viscoelastic properties to perform as well as a PSA, acrylates are tackified in order to improve adhesion on low-energy surfaces such as polyethylene or polypropylene. With regard to the base elastomer, a tackifier must fulfill the following requirements [5]:

- adequate compatibility
- very low molecular weight
- higher glass transition temperature (T_g)

Tackifiers increase the T_g-values of the PSAs and decrease their storage plateau modulus. As illustrated in Figure 5.5, tackifiers decrease the storage plateau modulus of elastomers to such an extent that the polymers fulfill the Dahlquist criterion for PSAs (see Section 5.1.5).

At present, the majority of tackifiers used have a low molecular weight and a T_g ranging from $-20\,°C$ to $+70\,°C$. At room temperature, they generally behave like brittle glass, and are classified into hydrocarbon- and natural resins-based products.

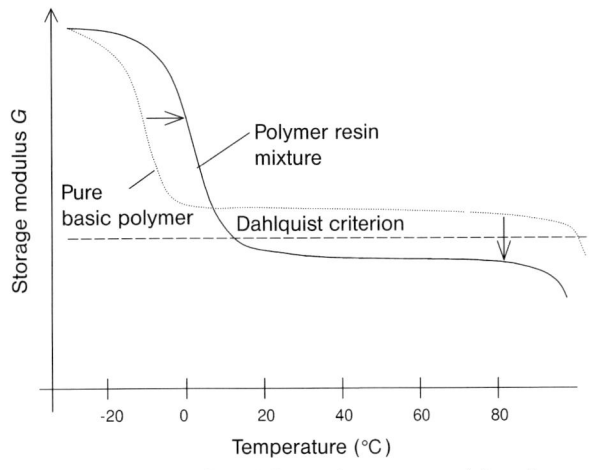

Figure 5.5 Influence of a tackifier on the storage modulus of a polymer. The Dahlquist criterion is discussed in Section 5.1.5.

5.1.3
Physical Properties of Pressure-Sensitive Adhesives

PSAs require a characteristic combination of rheological properties; first, they must wet the adherent surfaces rapidly and completely to build up a good holding force. The quality of the wetting is determined not only by the viscoelastic behavior of the PSA but also by the surface tensions of the adhesive and the adherents.

During bonding, a PSA is required to have adequate flowability to adapt to the surface structure of the adherents. The building up of adhesion may take place slowly, usually on the order of a second or more, so that the adhesive has enough time to deform. Mechanical strains created by compression must be relieved by relaxation within the adhesive in order to prevent detachment of the adhesive from the adherent surface. The finished bond must also resist the peeling and shearing forces that occur during service.

Build-up of the bond takes place at a low deformation rate, while destruction of the bond is a relatively rapid procedure, for example when the adhesive is removed. The time given to the adhesive to yield is on the order of one-tenth of a second. Considerably higher deformation speeds may occur in peel tests. In shear tests, the properties of the adhesive are determined by its creeping behavior at a relatively low load during a long period of time. Only PSAs make it possible to combine these mutually exclusive properties owing to their complex viscoelastic behavior.

The correct synthesis or formulation allows the physical properties to be fine-tuned in order to obtain a balanced combination of tack, peel resistance and inherent strength (cohesion).

5.1.4
Properties of Pressure-Sensitive Adhesives as a Function of Temperature

Dynamic mechanical analysis (DMA) makes it possible to examine the viscoelastic behavior of PSAs over a wide range of temperatures, and to determine the reversible (elastic) and irreversible (viscous) reaction of a polymer to deformation as a function of the temperature or rate of deformation [6]. Sinusoidal mechanical oscillation is applied to an adhesive sample, after which the elastic and viscous component of the modulus may be determined from the amplitude and phase shift of the stress–strain curve (see Section 7.2.2).

The viscoelastic properties of a PSA are characterized by the storage modulus G' (elastic modulus) and the loss modulus G'' (viscous modulus). The phase angle $\tan \delta$ is the ratio of G'' and G', and describes the energy loss in relation to the energy storage per cycle. Over a specified temperature range, the peak $\tan \delta$ is an indicator of the dynamic T_g of the adhesive tested.

The DMA spectra of PSAs, plotted as a function of the excitation frequencies tested, allow the viscoelastic behavior of the adhesives to be determined in the relevant ranges of temperature and deformation rate, respectively. Figure 5.6 demonstrates the storage modulus (G') of a PSA at the relevant frequencies with view to application, peeling, creeping and hot-melt processing. DMA is most useful for both the developers and customers of PSAs.

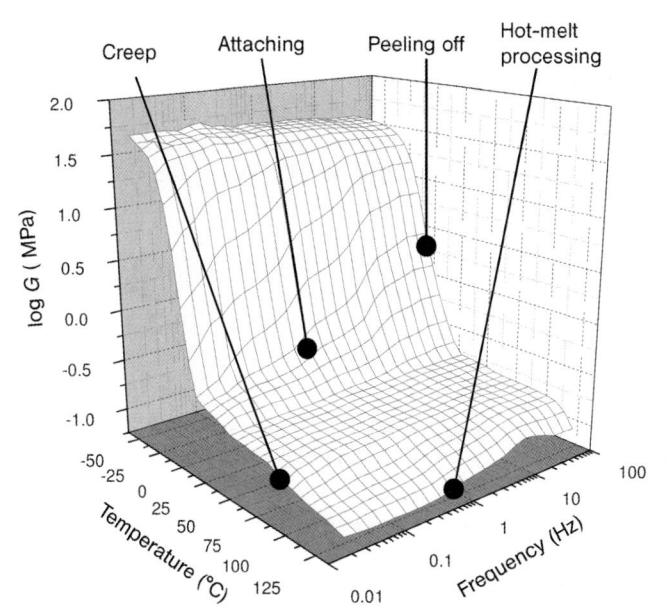

Figure 5.6 The dynamic mechanical analysis (DMA) spectrum of a pressure-sensitive adhesive (PSA).

5.1.5
Tack

Tack is the property of a PSA that enables it to form a bond immediately after the adherent and adhesive are brought into contact; a strong force is then needed to separate the joint. A PSA must be removed cleanly from the surface without leaving a residue. When building up the bond, a PSA is expected to have a very good deformability and to act as a liquid. When the bond is separated, it is expected to have a high modulus and to act as a solid. A modulus of 0.1 to 1 MPa is required for a PSA to have proper tack during the time period of 1 to 100 s which is the decisive relaxation time period during the build-up of the bond. This requirement for tack is termed the 'Dahlquist criterion for tack', after the scientist who first studied this phenomenon [7].

5.1.6
Peel Resistance

The peel resistance of a PSA is an important property which is determined in terms of the force needed to peel back an adhesive tape from a solid surface at a defined rate in a defined peel angle. The peel resistance depends on the adhesion of the adhesive to the surface, the viscoelastic behavior of the adhesive and the support material, the peel rate, and the temperature.

The peel force–peel rate curve can be divided into three segments. At very low peel rates, the failure mode is cohesive within the adhesive. However, when the peel rate is increased the peel failure mode changes from cohesive failure within the adhesive to adhesion failure. This abrupt change occurs with an initial decrease in peel force, but the latter increases with increasing peel rate. The third segment of the curve shows decreasing peel force with increasing peel rate. This region of the peel force–peel rate curve is associated with a glassy response of the adhesive, and a 'shocky' peel occurs where the adhesive is not uniformly peeled from the surface but rather starts to tear off periodically (the 'slip–stick effect').

5.1.7
Creep

Another important property of a PSA is its resistance to creep. This property is specified by the customer, and depends on the field of application. PSAs are expected to resist shearing forces during a very short period at high temperatures (as when masking tapes hold masking paper during a curing process), or over a very long period at room temperature (as when objects are fixed to a wall using double-sided adhesive tape).

The higher the molecular weight of the base polymer or the crosslinking density, the lower the creep. It must be noted, however, that there are practical limits as molecular mobility is an important requirement for wetting ability during build-up of the bond, thus for the adhesive properties of a PSA and last, but not least, for its long-term durability.

5.1.8
Formulations of PSAs

The property first noted about PSAs is their tack. We know what tack feels like, but how do we describe it and, for that matter, how do we generate materials with tack? The addition of certain low-molecular-weight materials to elastomers in a process called 'tackification' generates tack, although due to their low molecular weight these tackifiers decrease the cohesive strength of the elastomer. It is important, however, that a PSA has both tack and sufficient cohesive strength to hold two surfaces together. Indeed, this represents the major challenge when formulating PSAs, namely to generate a balance of properties so that PSAs combine these seemingly mutually exclusive properties to produce a material with sufficient ability to not only hold onto the adherent but also be cleanly removed from the adherent and have adequate shear strength. A combination of high-molecular-weight polymers with low T_gs and low-molecular-weight resins or plasticizer oils can generate the required balance of properties.

Tackifiers form a unique class of materials in that they normally have a low molecular weight and are resinous, while their glass transition and softening temperatures are often much above room temperature. It is this combination that makes the materials useful in the formulation of PSAs. Tackifying resins are usually based on natural products or components obtained from crude petroleum.

5.1.8.1 Typical Formulation of Natural Rubber-Based PSAs
In order to increase the ease of dissolution, natural rubber is masticated by means of rollers or internal mixers prior to dissolution (in gasoline, toluene, hexane or heptane). Typical tackifiers used are resins based on rosin, terpene or hydrocarbon, which comprise approximately 40–50% of the dried product. Occasionally, oils are used as plasticizers (2–5%) to generate a softer formulation. Antioxidants based on amine, phenol and dithiocarbamate are also added (0.5–2.0%) to prevent degradation by oxidation at the polymer double bonds.

5.1.8.2 Typical Formulation of Block Copolymer PSAs
Block copolymer-based PSAs are generally extrusion-processed or melt-processed without the use of any solvents. The molecular weight of these elastomers is considerably lower than that of natural rubber. The two primary block copolymers used for PSAs are A–B–A block copolymers made from polystyrene end blocks (molecular weight 10 000–50 000) and polyisoprene or polybutadiene middle blocks (molecular weight 30 000–150 000). Typical formulations contain about 40% block copolymers, 50% hydrocarbon resin, 9% plasticizer oil, and 1% antioxidants.

Care must be taken in the selection of the resins to ensure that they are compatible with the blocks. Aromatic resins such as coumarone-indene or poly(alpha-methylstyrene) resins are compatible with the aromatic polystyrene phase, and they may have a positive or a negative effect on the T_g of this phase. Aliphatic resins such as terpene and rosin have a better compatibility with the middle block and thus increase tack and shear strength.

5.1.8.3 Typical Formulation of Acrylate-Based PSAs

Acrylic adhesives are usually not tackified; rather, their pressure-sensitive character results from the inherent physical properties of the polymer. The primary monomers used for copolymerization are 2-ethylhexyl acrylate, iso-octyl acrylate and butyl acrylate (about 90% with regard to the dry adhesive). Polymers based on these monomers do not necessarily perform as well as PSAs on their own, but their performance may be improved by adding acrylic acid (ca. 10%) to the copolymer structure.

In order to improve adhesion on low-energy surfaces, acrylic adhesives are tackified with resins (10–30% with regard to the dry adhesive) usually based on rosin derivatives.

In the case of UV-curable acrylic adhesives, the resins must not interfere with the radical crosslinking reaction; neither must they absorb radiation in the UV-range used for crosslinking. For this purpose, hydrogenated and glycerol-esterified rosin is used (e.g. Foral 85, Eastman).

5.2 Contact Adhesives

5.2.1 Composition of Contact Adhesives

Typical base polymers of contact adhesives are polyvinyl acetates, chloroprene rubbers, polyethylene copolymers (with polar components), or poly (vinyl chloride) (PVC) derivatives. In general, the solvents used have different vapor pressures to prevent cobwebbing during the application process or skinning during the drying process. For the bonding of thermoplastics, for example PVC, specific solvents are added to obtain swelling or dissolution of the adherents and, in addition to adhesion, molecular diffusion between the adhesive and the solid object in terms of a welding effect. However, this may lead to microcracking (crazes) in adherents that are sensitive to stress cracking, such as polymethyl methacrylate (PMMA), polycarbonate or polystyrene. Microcracking starts in a localized polymer zone where the molecules are highly stretched, and extraneous molecules penetrate into the free volume, thus inducing the formation of cracks.

Besides solvents, 10–30% of a solvent-containing contact adhesive is composed of solids which may contain added materials such as wax or dissolved rosin to generate tack in the open dried bond-line, or anti-aging agents (e.g. metal oxides) to degrade any hydrochloric acid contamination from damaged base resins containing chlorine, or fungicides and UV stabilizers. Furthermore, slowly and weakly crosslinking prepolymers (e.g. phenolic resins in low concentrations) or, as a second component added prior to application, isocyanates, are added which improve the long-term creeping properties [8].

5.2.2
Properties and Fields of Application of Contact Adhesives

Contact adhesives are easy to process by casting, spraying, or using a doctor blade. When porous materials are joined, the adherents may be repositioned by briefly lifting them if the solvent has not yet fully evaporated. This property is utilized especially in the textile industry and saddlery, as well as in the manufacture of cars.

Some 70 years of experience have taught us that, depending on the formulation, the age resistance of contact adhesives can be very good. When joining surfaces that are difficult to bond, such as stainless steel, the adhesion properties too are generally good. It is not compatible with conventional adhesion theories (see Chapter 3) that the bond-line generally has a high resistance to water; rather, this is most likely due to the dynamic adhesion properties of contact adhesives (see Section 3.4).

Nowadays, the high solvent content (up to 80%) of contact adhesives is hardly tolerable unless proper ventilation systems are available which provide retrograde condensation of the solvents. These systems, in turn, have a high demand for energy, and consequently alternative options such as dispersion contact adhesives are increasingly being used. Nanometric adhesive particles are maintained in a semi-fluid suspension in water by dispersing agents. A water content of 30–40 % necessary to obtain a dispersion in the liquid state which usually contains small amounts (5%) of organic solvents to improve film formation at drying and to enhance diffusion between the particles. Surface-active substances (tensides) are added to maintain the stability of the liquid dispersion, but these may accumulate between the particles and induce inhomogeneities that may adversely affect the resistance to water at a later stage.

As with solvent-based contact adhesives, milky dispersions are applied to both adherent surfaces, and the water is allowed to evaporate from the adhesive layers until they are dry to the touch and tack-free, using the 'fingerprint method'. The adhesive layers are ready to be joined when the milkiness vanishes, and they become translucent. After the evaporation of water, joining is carried out using high pressure for a short period of time, as in the case of solvent-containing contact adhesives. If at least one adherent is made from a porous material, the substrates may be joined before the bond is allowed to dry completely. Compared to organic solvents, water has a high boiling temperature and evaporation energy, and therefore the drying process must take place slowly and may only be enhanced by an input of energy.

As mentioned above, when solvent-based contact adhesives change from the liquid into the solid state, a high tack makes it possible briefly to undo the bond for repositioning purposes, the adhesive being already loadable. This is not possible with dispersion adhesives and is considered to be a disadvantage in the textile industry and saddlery. It also causes problems in the car manufacturing industry when bonding insulating pads.

Unless hardeners (crosslinking agents) are added (as in the case of solvent-based systems) in the cured state, all contact adhesives used so far have been thermoplastic materials with limited resistance to heat and solvents. The primary fields of application of contact adhesives are the bonding of facing materials in car manufacture, the creation of composite panels for furniture and for use as insulating materials, and last – but not least – the textile and leather industries.

At this point it must be noted that it is possible today to create solvent-free and water-free contact adhesives based on modified silane (MS) polymers (e.g. MS polyurethanes). These materials have a high viscosity in the initial state, while tack confers an inherent strength to the bond shortly after compression. The bond-line is cured by the polycondensation reactions of the silane end groups that occur at a later stage, when ambient moisture diffuses into the bond-line.

5.3 Hot Melts

Hot melts meet the criteria of adhesion and inner strength in a unique way. They are applied from the melt that has intense contact to the surface to be bonded and thus allows adhesion to build up. After cooling, hot melts solidify and build up inner strength. Hot-melt adhesives can be categorized as: (i) physically setting adhesives, which are still thermoplastics after cooling; and (ii) chemically setting adhesives, which become thermosets (i.e. three-dimensional) networks. Both types of hot melt are liquids when applied, and solids at room temperature.

The advantages and disadvantages of hot melts compared to chemically setting and solvent-containing adhesives are summarized in Table 5.1.

Postcuring or enlargement of the adhesion area can relieve the problem of limited resistance to heat.

The processing temperatures of different hot melts as a function of time are a good indication of the assessment of their fields of application (see Figure 5.7) [9].

The *softening temperature* induces a deformation of the hot melt; with a further increase in temperature, the adhesive is transformed into a melt (*temperature interval*). As soon as the melt has a sufficient viscosity, an optimum wetting of the adherent surface can take place without adversely affecting the melt. During processing, the typical viscosities of hot melts range between 20 Pa·s (polyamide) and 10 000 Pa·s [ethylene vinyl acetate (EVA) copolymers]. At processing temperature, the bond-line strength is raised with increasing melt viscosity, which in turn increases as the molar mass of the polymers rises [10]. The processing temperatures

Table 5.1 Advantages and disadvantages of hot melt adhesives.

Advantages	Disadvantages
• No solvent required (being melts)	• High viscosity and heat stress of the adherents
• No dosing or mixing required	• Hot-melt equipment necessary
• Short setting periods	• Low resistance to heat (tendency to creep at higher
• Thermally detachable bonds	temperatures or sustained loading)

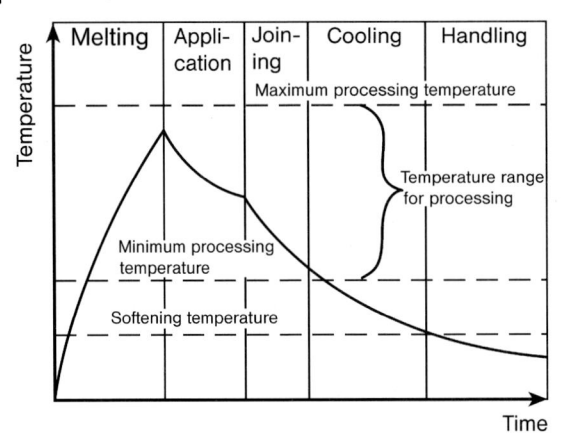

Figure 5.7 Temperatures of hot melts during the application process.

depend on the type of base material used, and range between 120 and 240 °C; for example, high-melt systems based on polyimides have a processing temperature in excess of 260 °C. The maximum processing temperature depends on the resistance of the melt towards thermal or oxidative degradation. Thermal stability is obtained by the addition of stabilizers and anti-oxidants. When, after joining, the adhesive solidifies, the solidification rate is seen as a criterion for the mechanical ability of the joint to take stress.

5.3.1
Thermoplastic Hot Melts

The primary polymers used to generate physically setting hot melts are polyamide resins, saturated polyesters, EVA copolymers, polyolefins, block copolymers (S–B–S or S–I–S) and polyimides [11].

For high-performance adhesives, polyamides, polyesters and polyimides are employed, whereas EVA copolymers and polyolefins are used for commodity and general-purpose hot-melt adhesives.

The most important fields of application for high-performance hot melts are process equipment construction (painted and coated surfaces), in car manufacture (heated seats, hat racks, roof liners), in the building industry (window construction, acoustic ceilings) and the textile industry (textile inserts). General-purpose hot melts are used primarily in the paper and packaging industry (on packaging machines which erect and seal packages, on folding machine gluers, for cigarette paper), for sticky labels (address labels), in bookbinding (adhesive binding) and for the manufacture of hygienic products (baby diapers).

The polymers determine the properties of the bond-line with regard to adhesion, inner strength and temperature behavior, while various additives are required to ensure their functionality and to confer specific properties to the adhesive (Table 5.2) [12].

Table 5.2 Hot-melt additives.

Additive	Function
• Resins like hydrated rosin, tall resin or shorter chain hydrocarbon resins	• Increase of the tack of the melt at processing temperature ('hot tack')
• Long-chain (dibutyl or nonyl)phthalate	• Plasticizer for not sufficiently flexible polymers
• Aromatic amines or phenols	• Stabilizers or anti-oxidants as radical scavengers to prevent oxidative alterations during the processing of the melt under the influence of oxygen
• Chalk, heavy spar or titanium dioxide	• Increase in strength and as extender (cost reduction)
• Waxes	• Viscosity regulators

5.3.2
Hot-Seal Adhesives

Hot-seal adhesives are dispersions or solutions based on polyvinylidene chloride, polyvinyl acetate and polyacrylates. When the water or solvent is allowed to evaporate, a coating is generated on the substrates (e.g. a printed board) that will seal under heat. Hot melts (generally based on EVA copolymers) can also be used as hot-seal adhesives, by depositing them on the substrate with roller coating application or extrusion. Hot-seal coatings are used primarily in the packaging industry (blister packs) or in foil laminating. The bonding process is similar to the processing of hot melts; here, the adhesive layer is melted together with the blister by means of a heated sealing tool applied until the softening point of the coating is reached. After wetting of the adherents, the layer immediately solidifies upon cooling under pressure [13].

5.3.3
Plastisols

Plastisols are pastes containing polymer (PVC) particles dispersed in a very high-boiling liquid (plasticizer), and low-molecular-weight heat-reactive substances. As with hot-melt adhesives, plastisols are heated in order that they are capable of wetting the adherents. The polymer takes up the plasticizer to generate a toughened polymer material that may range from tough and hard to rubber-like and flexible. Plastisols are used for metal bonding purposes in car manufacture industry [14]; they are particularly well suited for the bonding of oil-contaminated surfaces (see Section 8.2.2).

5.3.4
Self-Bonding Varnishes

Self-bonding varnishes, which are used primarily in the manufacture of coreless, self-supporting coils in the electronics industry, contain a dissolved thermoplastic adhesive based on copolyamides. The coils are manufactured by coating the wires with self-bonding varnishes, allowing the solvent to evaporate, winding the wires into coils and thermally bonding them. Such coils are used to create components that must meet high requirements with regard to dimensional stability and strength [15].

5.3.5
Polyurethane-Based Reactive Hot Melts

For the production of reactive hot melts (RHMs) which set under the effect of moisture, a polyurethane prepolymer is generated from solid rather than liquid starting polymers, primarily using polyester polyols (see Section 5.6) [16].

After the application of a reactive hot melt, the solidified bond has sufficient initial strength to allow transportation and further processing of the joined components. The adhesive then reacts with moisture, inducing an increase in molar mass, and a crosslinking takes place. RHMs have a much higher strength and resistance to water than thermoplastic hot melts.

The wide array of different starting polymers available to adhesive manufacturers can be combined with each other at different ratios to create an extensive range of properties [14]. RHMs are used primarily in bookbinding (adhesive binding), in furniture-making (as edge bandings and veneers), in the packaging industry (transport boxes), and in car manufacture (direct glazing).

5.3.6
Epoxy Resin-Based Reactive Hot Melts

Epoxy resin-based RHMs are also applied as a melt, but they solidify to build a thermosetting polymer network that cannot be re-melted (see Section 5.5). Compared to thermoplastic hot melts, they have a higher inner strength and are used primarily in car manufacture.

These adhesives are delivered as one-part reactive pre-polymer systems with a long pot life. Curing at room temperature is prevented by a catalytic system that is only effective at higher temperatures, or by a second reactive amine component that is also only reactive at higher temperatures. Immediately before processing, epoxy-based RHMs are melted in the quantity needed and then applied at temperatures ranging between 60 and 80 °C; assembly may be handled during the solidification. The final strength is obtained (e.g. in car manufacture) in the stoves where the body undercoat and paint system are dried [17].

Depending on the nucleophilic characteristics of the functional groups ($-NH_2$ > $-NH$ > $-OH$), the polyaddition reactions (which are used to generate epoxy resin-based systems) proceed at different rates. For example, at a molar ratio of 1 : 2, the cold-setting reaction of secondary diamines with bifunctional epoxides makes it possible to generate thermoplastic, meltable alkanolamine polymer compounds. At room temperature, further reaction is not expected to take place owing to the low reactivity of the hydroxyl groups. With increasing temperatures, however, these hydroxyl groups react further with the remaining epoxide groups in the form of an additional crosslinking effect, resulting in a polymer structure that is no longer meltable [18].

5.3.7
Trends in Hot-Melt Technology

Following developments reported in 2002, hot melts have been formulated for the paper and packaging industries which allow processing at temperatures around 100 °C, thus alleviating the thermal stress to which the hot melts and substrates to be joined are exposed [19] (see Section 8.5).

If the setting period of hot melts is very short, then the adhesive forces will not fully develop as the melt has insufficient time to optimally wet the surface of the substrate. Hence, in the printing industry a system was developed that allows perfect adhesion and rapid processing speed to be combined. First, a hot melt designed to show excellent adhesion properties is applied to the binding edges of the sheets, causing them to be glued together. A second adhesive is then applied that holds in place and fixes the cover to the book's spine [19].

A new type of UV-reactive hot melt was created from UV-setting polyacrylates which have increasingly been used in the production of sticky labels and adhesive tapes. By modifying the intensity of the UV radiation, very different adhesion and cohesion properties may be conferred to the adhesive film applied from the melt [20].

A water-soluble hot melt was developed for the labels of reusable plastic bottles to facilitate their recycling.

Examples of adhesives with new functionalities are water-swellable hot melts used for example in communications cables. When these hot melts come into contact with water, they swell and thus seal the cable in its longitudinal direction [19].

5.4
Phenolic Resin Adhesives

5.4.1
Chemistry of Phenolic Resins

The starting monomers of phenolic resins are phenol and formaldehyde or paraformaldehyde. These are allowed to transform into low-molecular-weight resins called 'novolaks' (Figure 5.8) by a 'careful' polyaddition reaction under acid conditions, followed by transformation into resol resins under alkaline conditions (Figure 5.9).

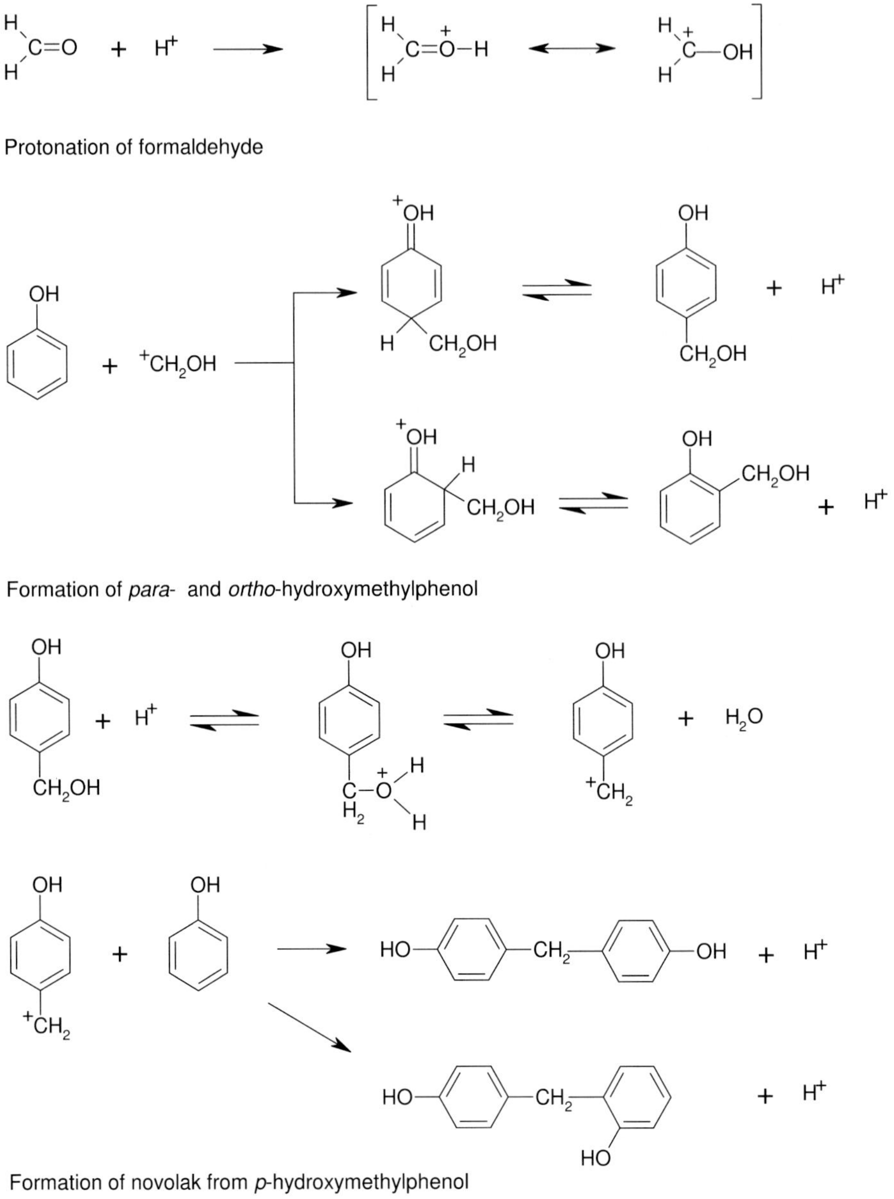

5 Chemistry and Properties of Adhesives and Primers

$$H_2C=O + H^+ \longrightarrow [H_2C=\overset{+}{O}-H \longleftrightarrow H_2\overset{+}{C}-OH]$$

Protonation of formaldehyde

Formation of *para-* and *ortho-*hydroxymethylphenol

Formation of novolak from *p*-hydroxymethylphenol

Figure 5.8 The chemistry of novolak.

Novolak and resol-type resins are still meltable or soluble, and serve as the starting products for phenolic resins and phenolic resin adhesives, respectively, that are generated by polycondensation reactions and the liberation of water. Novolaks condense (cure) under mild conditions to become hard and brittle thermosets; this occurs with the addition of either formaldehyde or substances that split off

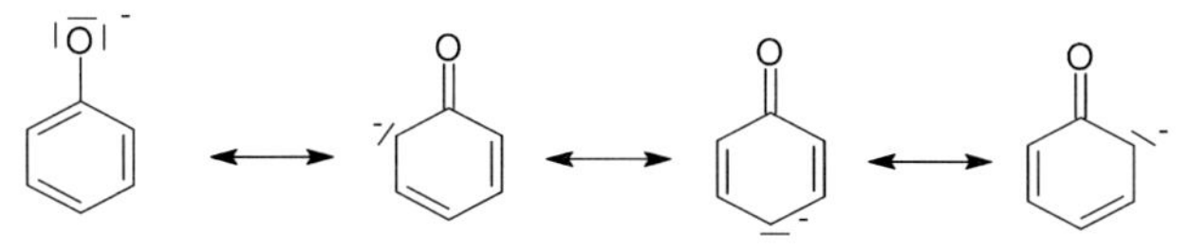

5.4 Phenolic Resin Adhesives

Resonance structures of phenolate ion

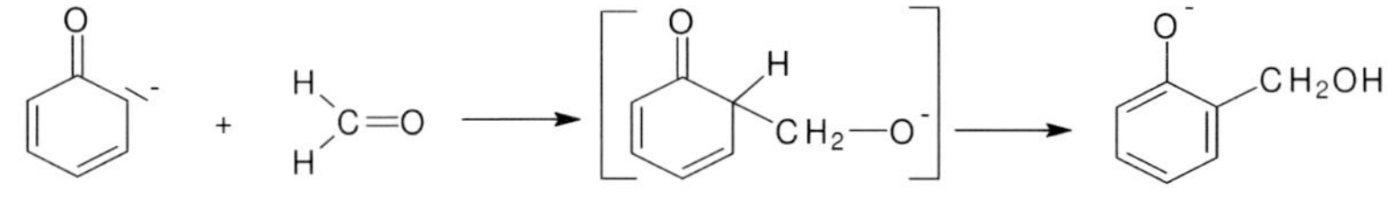

Addition of formaldehyde in ortho position

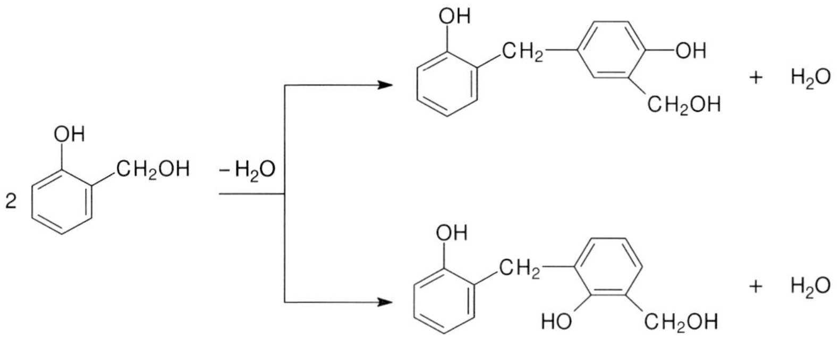

Ortho and para resol

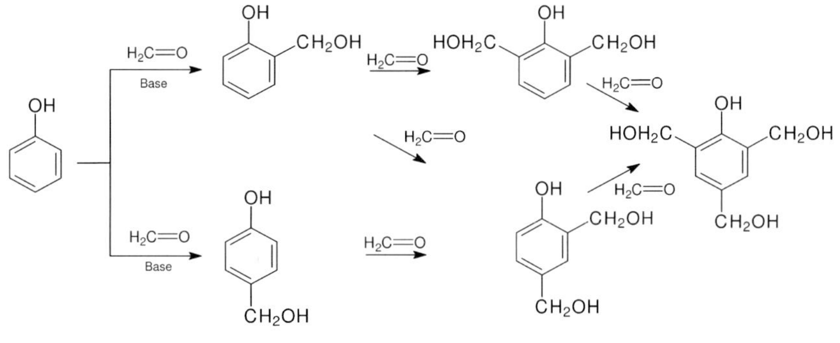

Methylol phenol becomes di or trimethylol phenol, with ortho and para positions

Figure 5.9 The chemistry of resol.

formaldehyde, such as hexamethylenetetramine. This further enhances the condensation reaction owing to the generation of alkalinity (ammonia). Novolaks may be provided with epoxy groups (via the OH-groups of the phenol), thus increasing the heat resistance of epoxy resin adhesives. Novolaks may also be employed as low-temperature base resins for the production of wood adhesives (see Section 8.4) [21].

Initially, resol-type resins were used in the production of adhesives since, after precondensation under alkaline conditions and neutralization, they cure in a self-condensation reaction when heated to about 150–170 °C by reaction of the hydroxymethyl groups with orthohydrogen or parahydrogen atoms of other phenol nuclei and the liberation of water.

In common phenolic resin adhesives there are no dimethylene ether bridges between the two hydroxymethyl groups of two phenol rings which were postulated earlier; rather, there are methylene bridges [22]. Without any modifications, a hard and brittle thermoset is also generated which has a high durability and has been used for joining wood since the 1930s.

5.4.2
Formulation of Phenolic Resin Adhesives

During the early 1940s, with the major breakthrough in the development of versatile phenolic resin adhesives for the bonding of various materials, it was possible for the first time to use these adhesives for the bonding of metals. De Bruyne [23] added relatively large amounts (75%) of high-polymer polyvinyl formal (PVF) to brittle phenolic resin as early as in the resol state. Although at room temperature PVF is a solid material, it melts when the mixture is allowed to cure at elevated temperatures (ca. 170 °C), generating a high-performance adhesive with excellent deformation properties known as 'REDUX'. This adhesive was a marker in the history of adhesive technology, and indeed is still used today, for example in some branches of aircraft manufacture (see Section 8.2.1).

Although the steric hindrance of resol condensation was recognized from the outset as one factor responsible for the plasticizing effect of PVF, another – not fully understood – factor soon became apparent. Important was that when PVF was added to resol as a coarse powder and allowed to cure, it formed domains rather than becoming homogeneously mixed. In the solidified state this led to limitations and control of crack propagation under load, as in the case of ABS plastics. If the resol reacts too much with PVF, or if the resol and PVF are homogeneously mixed in a mixture of ethanol and dichloromethane, the plasticizing effect is lost for the most part.

In the noncured condition, the volatile components of resol-containing adhesives may give cause for attention. Furthermore, due to the relatively large amount of thermoplastic PVF required, the resistance to heat of the cured adhesive was limited to approximately 60 °C, despite the phenolic component having high thermal stability.

In order to identify other possibilities of modification, novolaks (which are free from volatile components) were modified with butadiene nitrile elastomers. If a vulcanization agent (e.g. sulfur), a vulcanization aid (e.g. zinc oxide) and other vulcanization accelerators are added to the adhesive, these elastomers vulcanize when the novolak cures. These so-called 'nitrile-phenolic adhesives' have a T_g of approximately 0 °C due to their high nitrile content (up to 50%) and low content in phenolic components (ca. 20%). Once in service, however, they act as elastomers and barely alter their properties up to a temperature of 200 °C, applied for a short period. Hence, these materials are widely used for the bonding of brake and clutch linings.

5.4.3
Behavior and Applications of Phenolic Resin Adhesives

Phenolic resin PVF adhesives are available as liquid–solid products or in the form of ready-made adhesive films, with or without support material. Both types cure at temperatures above 150 °C and at a pressure of more than $4 \times 10^5 \, \text{N m}^{-2}$, requiring corresponding equipment such as heated presses or autoclaves. The pressure is required to squeeze out any water liberated within the adhesive in the form of vapor during condensation.

The liquid–solid product comprises a resol solution that is first applied, after which PVF powder is then sprinkled on; any excess PVF that does not adhere to the somewhat tacky resol coating is simply blown off. Nitrile-phenolic adhesives are applied in the form of a solution, the solvent is then allowed to evaporate, and the adhesive cures under pressure.

In the cured condition, nitrile-phenolic adhesives and phenolic resin PVF adhesives are extremely durable. Following correct oxidative surface preparation, PVF modified resol-type resins in particular demonstrate the unique property of building up unmatched, water-resistant adhesive forces towards metals (preferably aluminum alloys), without requiring any additives. There are three reasons why this salient feature occurs:

- The molecular weight of the resol component is low (mononuclear, dinuclear or sometimes trinuclear molecules), and the molecular mass distribution narrow; this allows the molecules to easily penetrate into nanostructured aluminum oxides without segregating in relation to stoichiometry.

- In cooperation with the phenolic OH-group, the *ortho* hydroxymethyl group may form absolutely water-resistant chelate compounds with the metal present on the aluminum surface.

- The cured phenolic resin retains a weakly acid character that stabilizes the aluminum oxides against hydration as soon as the moisture penetrates (see Section 3.1) [24].

These are the reasons why, despite a number of disadvantages, phenolic resin adhesives have been widely used since the 1950s and still are used today, virtually without any long-term problems, particularly in aircraft manufacture where aluminum is used predominantly. Although today, epoxy resins with better deformation properties and lower curing temperatures (120 °C) are employed, phenolic resin-containing primers are still indispensable and widely accepted. Nevertheless, due to their difficult processability (a curing temperature >170 °C for only 30 min will alter the long-term resistance of cured aluminum alloys) and rather problematic brittleness and toxicity, phenolic resin adhesives have been replaced by other adhesive systems except in the wood construction industry (see Section 8.4) and in friction lining bonding. Although, during the past 30 years there have been no major advances in the development of this type of adhesive, it might prove very interesting to take advantage of the new approaches with nanoparticles being used for plastication purposes.

5.5
Epoxy Resin Adhesives

5.5.1
Chemistry of Epoxy Resin Adhesives

Epoxy resin adhesives are low-molecular-weight substances with a so-called 'epoxy function' – that is, a three-membered oxirane-ring composed of one oxygen atom linked to two carbon atoms (Figure 5.10).

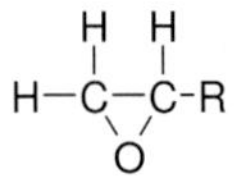

Figure 5.10 The epoxy group.

Among a variety of epoxy compounds available commercially, the most important group for epoxy resins is derived from bis-phenol-A. Owing to the presence of a benzene ring, these resins are particularly stable yet at the same time they have a high polarity due to their electron configuration. Epoxy resins are produced from bis-phenol-A by reaction with epichlorohydrin to generate the diglycidyl ether of bis-phenol-A, a liquid resin used widely in adhesive manufacture (Figure 5.11).

Notably, care should be taken to avoid diglycidyl ethers coming into contact with the skin as they frequently induce allergic reactions (itching, inflammation, rash). Such sensitizing effects may be avoided by producing solid epoxy resins. In this case, the reaction of bis-phenol and epichlorohydrin can be controlled in such a way that some of the bis-phenol molecules react with each other to generate oligomers that carry epoxy groups at their ends. Owing to the bis-phenol-A unit, the resins have a high rigidity and very good adhesive properties. Another group of widely used epoxy resins are the cycloaliphatic diepoxides in which, due to a shorter distance between the reactive groups, intense crosslinking can take place at curing, providing for a high dimensional stability of the bond-line at elevated temperature.

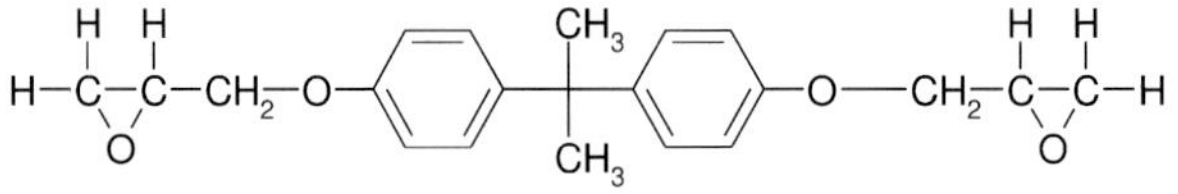

Figure 5.11 The diglycidyl ether of bis-phenol-A.

5.5.2
Reactions of Epoxy Resins

The reaction between epoxides and polar substances may take place in two different ways, both of which are associated with an opening of the oxirane ring.

Polymerization by Polyaddition In the manufacture of adhesives, the most important reaction type is the addition of a highly polar compound (hardener) to an epoxide with simultaneous ring opening (Figure 5.12) [25, 26].

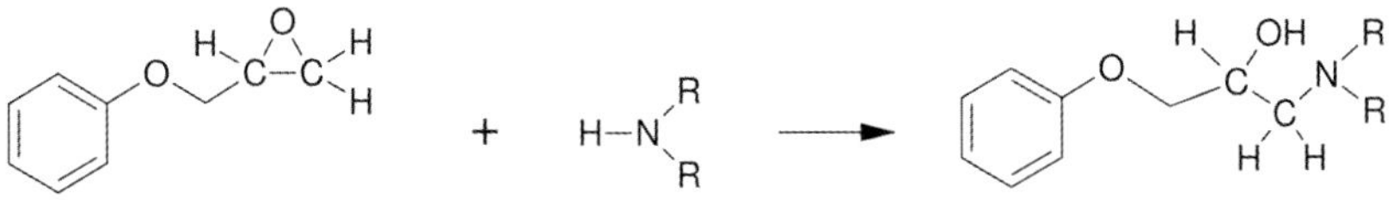

Figure 5.12 Addition reaction of a secondary amine and an epoxy group.

Often, primary or secondary amines are used as hardeners, thus generating a product that contains two groups with high polarities – that is, an amino group and a hydroxyl group which both promote the adhesive properties of the resin. Long-chain polymers are generated starting with diepoxides and diamines; this reaction is known as 'polymerization by addition' or, preferably, 'polyaddition'.

Another group of widely used hardeners is that of liquid, low-molecular-weight polyamides which have amino groups at both ends of their molecules (polyaminoamides). These are relatively nontoxic and have only a slight inherent odor. Their reaction rate is a little slower than that of amines, and the heat development less intense.

Ring-Opening Polymerization The second type of epoxide reaction is the so-called 'ring-opening polymerization' of the oxirane ring, initiated by a substance that activates opening of the oxirane ring by an addition reaction. The ring-opening reaction may be either acid-catalyzed or catalyzed by non-nucleophilic bases (Figure 5.13).

The initial ring opening then starts a chain reaction. In contrast to polyaddition, which requires a defined ratio of the reaction partners, ring-opening polymerization requires only a small amount of the initiator. The polymers thus generated have considerably fewer polar groups and hence less-favorable adhesion properties than those generated from diamines and diepoxides by polyaddition. Another disadvantage of the ring-opening polymerization reaction is that, after addition of the initiator, the reaction takes place very quickly and only a short time is left for processing. Therefore, in terms of the curing of epoxy adhesives polymerization by polyaddition is of minor importance. However, the process is used in cationic photocuring, where epoxy resins are mixed with materials that liberate cations when exposed to high-energy radiation (e.g. UV light). The ring-opening polymerization reaction is initiated by the liberation of cations.

Curing with Dicyandiamide Dicyandiamide is a colorless powder that is poorly soluble in liquid epoxy resins and, therefore, does not react with epoxy groups at

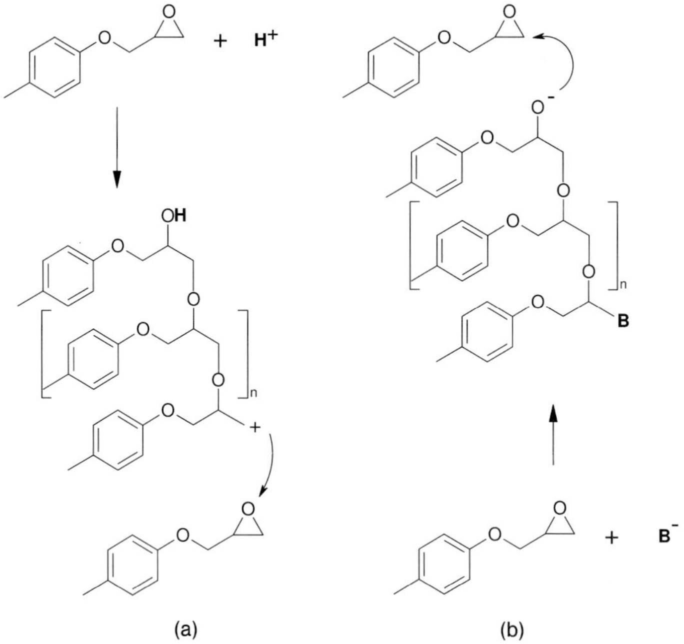

Figure 5.13 (a) Acid-catalyzed and (b) base-catalyzed ring-opening polymerization of epoxides.

room temperature. Only at temperatures above 150 °C will dicyandiamide react, resulting in the curing of the epoxy resin. Hence, dicyandiamide – which is referred to as a 'latent hardener' – is admixed to the resin during adhesive manufacture to generate a one-part adhesive.

The reaction between dicyandiamide and epoxides begins with an addition reaction, the primary products of which initiate a ring-opening polymerization together with further epoxides. Dicyandiamide is a dimer of cyanamide, and in solution there exists an equilibrium between both compounds. Dicyandiamide acts as source of cyanamide in the reaction with epoxides; the cyanamide joins two molecules of epoxide to generate an aminooxazoline as a first intermediate. The amino group of this intermediate then adds to another epoxide to form a tertiary amine that acts as a catalyst for a subsequent base-catalyzed ring-opening polymerization (Figure 5.14) [27, 28].

The curing temperature of epoxides and dicyandiamides may be reduced to 125–130 °C by adding urea derivatives such as monuron or diuron.

5.5.3
Properties of Epoxy Resin Adhesives

Epoxy resin adhesives are characterized by their strength, long-term durability and high resistance to ambient conditions and chemicals. Owing to their excellent

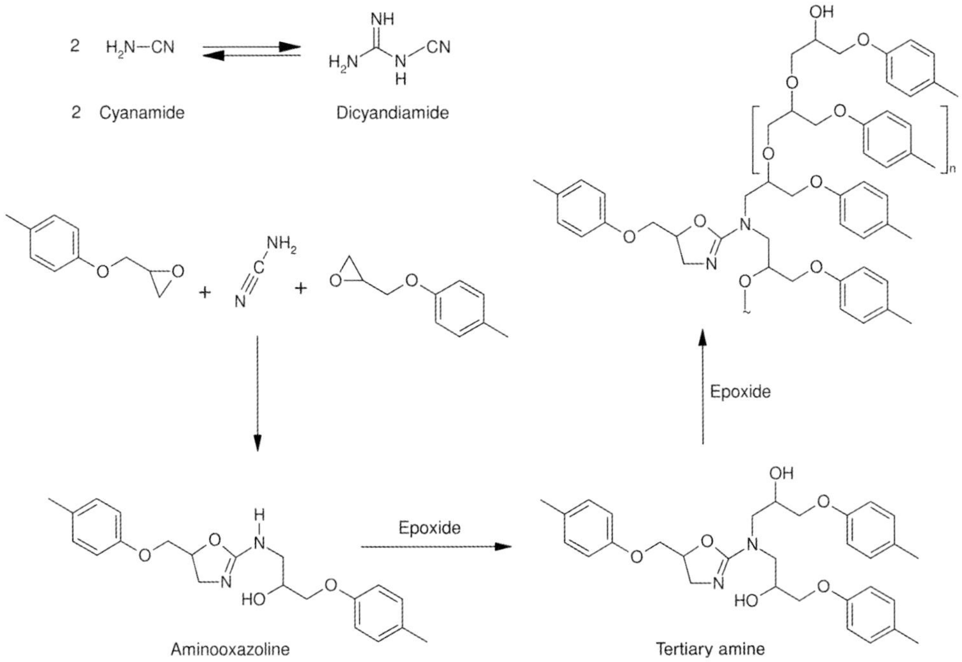

Figure 5.14 The equilibrium between cyanamide and dicyandiamide; the reaction of dicyandiamide (cyanamide) with epoxides.

adhesion properties towards metals, mineral surfaces and wood, they have a wide field of applications in adhesive technology.

In the car manufacturing, aircraft and building industries, the adhesives must meet specific requirements depending on their intended use(s). The mechanical loads imposed may be of static or dynamic nature, and may be acting in different directions, for example compressive stress, tensile stress, shear stress or peel stress. Furthermore, the adhesive joints must be able to withstand different ambient conditions caused by temperature, moisture and various chemical influences.

Epoxy resin adhesives are specially formulated to allow their correct processing, handling and storage stability. The curing conditions depend on the formulation: one-part systems which have a good storage stability at room temperature over a prolonged period of time generally require curing temperatures above 150 °C and curing periods of 15–30 min. In contrast, accelerated one-part epoxy adhesives require temperatures of only perhaps 130 °C for 30 min, but they have to be stored at low temperatures.

Curing at room temperature is only achieved with two-part adhesives. According to the formulation, the curing process may take several minutes to several hours. At curing, epoxy resin adhesives pass through different transformation phases before reaching their full strength. The course of the cure – that is, the course of the conversion of the epoxy resin – depends on several factors, including the reactivity of

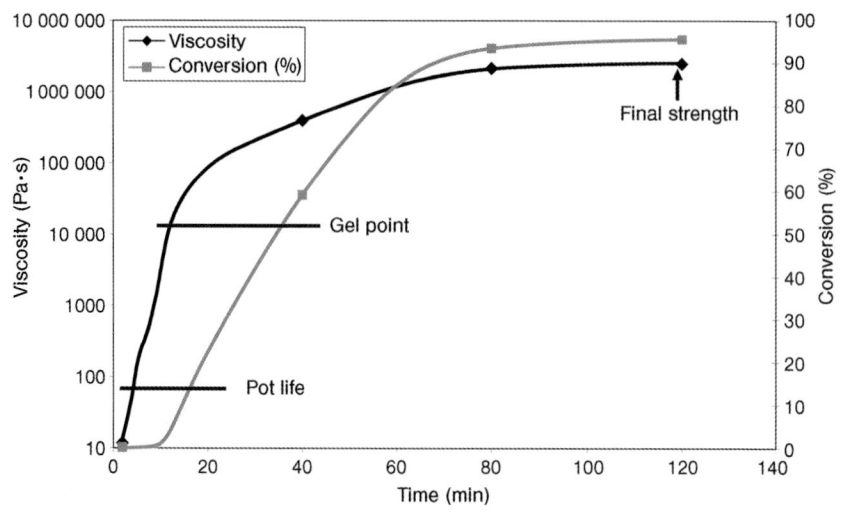

Figure 5.15 Viscosity of a two-part epoxy adhesive, depending on conversion of the epoxy resin.

the components, time and temperature. Here, one important period is the 'pot life', during which the adhesive can be applied and the adherents can be joined. At a curing degree of 50–60%, the adhesive reaches the 'gelation point' where it changes from a liquid state into a solid (Figure 5.15). Although the adhesive is in the solid condition when reaching the gelation point, full strength is only obtained at a conversion rate of 90–100% (Figure 5.15).

With advancing conversion of the epoxy resin, there is an increase in the T_g, inner strength, elasticity modulus, and tensile and flexural strengths, respectively. Epoxy resin adhesives with a curing or crosslinking rate greater than 90% almost reach their full strength (Figure 5.16) and full resistance against ambient influences such as moisture and chemicals.

When curing takes place at room temperature, the mechanical properties are generally inferior, and the temperature resistance is lower as compared to one-part epoxy resin adhesives. However, when two-part systems are directly heated, or heated in addition to curing at room temperature (e.g. to 180 °C for 15–30 °min), both their T_g and strength and resistance to moisture and chemicals correspond to those of one-part epoxy resin adhesive systems.

Present-day, two-part epoxy resin systems have the advantage that they combine very fast curing at relatively low temperatures with a high performance level. For example, their impact resistance, which is relevant for crash resistance (e.g. in car manufacture), is as high as that of one-part epoxy resin adhesives. This provides for innovative and optimized manufacture processes in structural bonding, particularly in the car manufacturing industry. As an example, immediately after application of the adhesive and joining of the adherents, a green strength of 2–4 MPa can be built up within 40–60 s by inductive heating of the adherents to 120–140 °C. This provides for stable and distortion-free assemblies shortly after joining, and also avoids problems with tolerances in bodywork construction. The components may be heated either

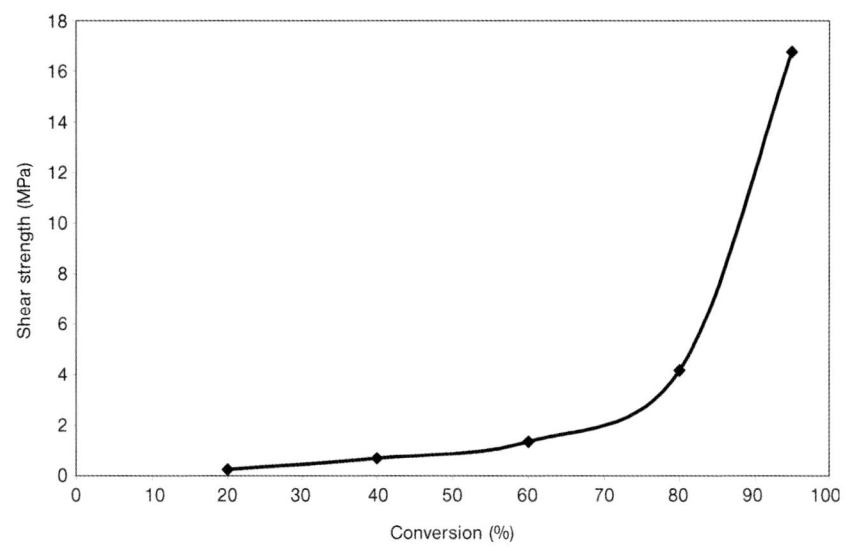

Figure 5.16 Shear strength of a two-part epoxy adhesive, depending on conversion of the epoxy resin.

partially or all over the surface. Two-part epoxy resins reach the gelation point very early and by that time have already built up a high green strength. Moreover, the reaction then advances even without further heating. If during car manufacture, for example in the e-coat stove (170 °C), the component is re-heated, the modulus decreases but no viscous flow phase occurs, as would be the case for one-part epoxy systems (Figure 5.17).

In order to cure one-part epoxy adhesives, heat or energy in the form of radiation is required, with common curing temperatures ranging between 120 and 180 °C, with curing periods of 10–30 min. One-part adhesives undergo a low-viscosity phase, followed by a rapid curing process. Flow at room temperature is prevented by the application of high-viscosity adhesive pastes at elevated temperatures (60–80 °C). Additionally, the adhesives are often pregelated at 120–140 °C to obtain a certain green strength.

5.5.4
Formulations of Epoxy Resin Adhesives

Epoxy resin adhesives are composed of an epoxy resin and a hardener. In order to generate a high-performance adhesive, ten or more additives are included in the formulation; these additives must be carefully adjusted to each other because of their mutual interactions.

5.5.4.1 Epoxy Resins
The primary epoxy resins used are bis-phenol-A diglycidyl ether (BDGE) or bis-phenol-F diglycidyl ether. Reactive low-viscosity diluents (e.g. hexanediol diglycidyl

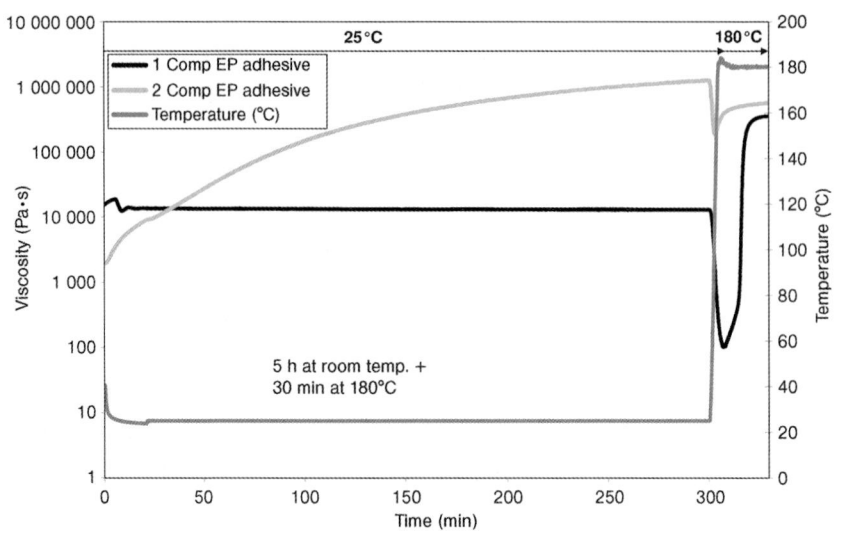

Figure 5.17 Viscosity of one-part and two-part toughened epoxy adhesives depending on time and temperature.

ether) are frequently added to regulate the viscosity, to obtain higher filling rates, and to improve the wetting properties of the substrate.

5.5.4.2 Crosslinking Agents
Higher-functional resins, such as phenol or kresol novolaks, are used to increase the crosslinking density. Higher-functional epoxy resins may also be incorporated, such as tetra-functional N,N′-tetraglycidyl-4,4′-diaminodiphenylmethane. The crosslinking rate may also be increased by the addition of higher-functional hardener-components such as tetraethylene pentamine.

5.5.4.3 Hardeners
Epoxy resin adhesives are cured with primary or secondary diamines. Epoxy-amine adducts are produced by the reaction of bis-phenol-A diglycidyl ether with a six- to eightfold molar excess of diamine. One-part systems are generally cured with dicyandiamide. A flexibilizing effect is obtained using longer-chain diamines such as polypropylene glycol diamine, amino terminated acrylonitrile–butadiene copolymers and amino amides. The latter are produced, for example, by the condensation of fatty acid dimers with diamines. Elastic bonds are obtained using higher-molecular-weight polythiols.

5.5.4.4 Flexibilizers and Additives That Improve Impact Toughness
Flexibility may be improved by using amine compounds as hardeners. Flexibilization and an increase in impact toughness are also obtained by incorporating suitable polymers on the epoxy resin side. Whilst flexibilizers are incorporated into the epoxy matrix and decrease the T_g-values, polymer modifiers that improve impact toughness

are in solution with the epoxy resin from the outset, and then form separate polymer phases during the crosslinking process. Although having only a slight influence on T_g, they confer an increased impact bending strength to the epoxy resins, yielding adhesive systems that are crash-resistant under dynamic loads. Modifiers that increase impact toughness are added at a rate of 15–20% in relation to the total formulation. Polypropylene glycol diglycidyl ether is an example of a flexibilizing epoxy resin. Usually, flexibilizers of epoxy resin adhesives are added to the hardener part, while modifiers of impact toughness are added to the resin part (e.g. acrylonitrile–butadiene copolymers provided with carboxyl groups). In order to ensure their compatibility with the epoxy resin, these copolymers are reacted with an excess of epoxy resins.

The impact toughness can also be improved by adding core-shell particles. The core consists of elastomers (e.g. polybutadienes, polyacrylates or polysiloxanes) with a low T_g-value, and is coated with a thin layer of crosslinked polyacrylates. Core-shell particles have the advantage of having a defined particle size ranging between 0.1 and 100 μm.

5.5.4.5 Fillers and Thixotroping Agents
Chalk, aluminum oxide, silicates and kaolin are used as fillers to reduce costs and shrinkage during the curing process. Pyrogenic or precipitated silicic acids or bentonites allow the flow characteristics of epoxy resins to be regulated.

5.5.4.6 Further Additives
The build-up of adhesive forces towards certain substrates can be improved by using further additives, such as functional silanes. Anticorrosive agents are used to improve the corrosion characteristics, and small amounts of glass beads with defined diameter are used to generate a minimum bond-line thickness. By regulating the density, the glass bubbles ensure that the components of two-part epoxy resin adhesives are in an integer ratio.

5.5.4.7 Typical Epoxy-Resin Adhesive Formulations
A typical impact toughness-modified, one-part epoxy resin adhesive consists of 45% epoxy base resin, 17.5% crosslinking agent (e.g. novolac resin), 27.5% additive to improve the impact toughness, 2.5% pyrogenic silicic acid as rheology additive, 5.5% hardener (e.g. dicyandiamide), and 2% accelerator (e.g. a urea derivative).

The typical parts A and B of a flexibilized, impact toughness-modified, two-part epoxy resin adhesive for use in the car manufacturing industry are composed as follows:

- Part A
  - 48% epoxy base resin
  - 13% reaction diluting agent
  - 9% additive to improve impact toughness (e.g. core-shell polymer)
  - 1.5% adhesion promoter (e.g. a functional silane)
  - 0.5% pyrogenic silicic acid as rheology additive

- ~1.5% glass beads
- ~17.5% filler (e.g. a silicate)
- ~7.8% glass bubbles
- ~1.2% pigment

- Part B
  - ~42% hardener (e.g. an amino amide)
  - ~8.5% catalyst (e.g. a tertiary amine)
  - ~16% flexibilizer
  - ~7% hardener (e.g. a primary amine)
  - ~23% filler
  - ~3% pyrogenic silicic acid

The volume-mixing ratio of part A (epoxy) to part B (hardener) is 4 : 1.

5.6
Polyurethane Adhesives

5.6.1
Chemistry of Polyurethanes

As the chemistry and reactivity of isocyanates are described in detail in several textbooks of organic chemistry, only those reactions relevant to polyurethane adhesives will be detailed in the following sections. The reactivity of isocyanates is based essentially on the electrophilic character of the carbon atom in the cumulated double-bond system of the isocyanate group. The resonance structures are illustrated in Figure 5.18.

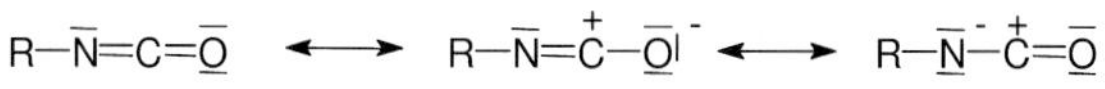

$$R-\bar{N}=C=\bar{O} \longleftrightarrow R-\bar{N}=\overset{+}{C}-\bar{O}|^{-} \longleftrightarrow R-\bar{N}^{-}-\overset{+}{C}=\bar{O}$$

Figure 5.18 Resonance structures of the isocyanate group.

Isocyanates react with other substances either by polyaddition or polycondensation. In the manufacture of polyurethane adhesives, the polyaddition reaction is used; the reactivity of the substances added depends on the nucleophilic character of their reactive group. The reactivity of the attacking group decreases in the following order: aliphatic amines > aromatic amines > alcohols > phenols > thiophenols. The steric hindrance of the molecules also has an influence on the reactivity of the nucleophilic reagents. For example, the reactivity decreases from primary to tertiary alcohols. The most popular method investigated and used commercially to date is the generation of polyurethanes by means of a reaction of diisocyanates or polyisocyanates with primary monoalcohols, dialcohols and polyalcohols (Figure 5.19).

In the presence of moisture, another two-stage reaction takes place. In the first step, an unstable carbamic acid is formed, while in the second step carbamic acid

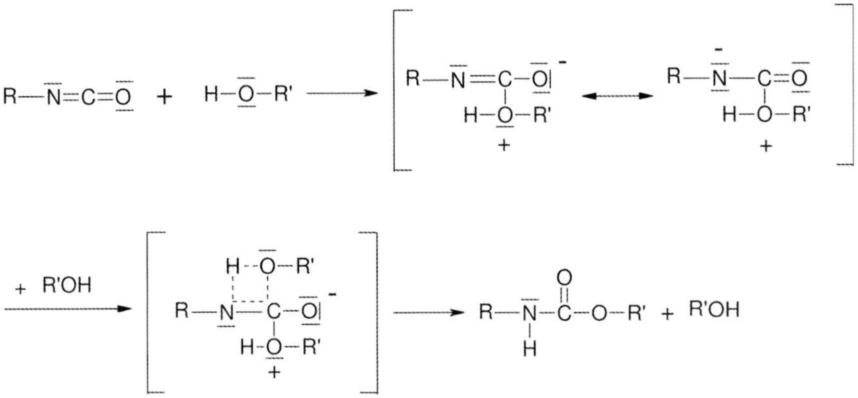

5.6 Polyurethane Adhesives

$$R-N=C=O + H-O-R' \longrightarrow \left[\begin{array}{c} R-N=C-O^- \\ | \\ H-O-R' \\ + \end{array} \longleftrightarrow \begin{array}{c} R-\overset{-}{N}-C=O \\ | \\ H-O-R' \\ + \end{array} \right]$$

$$+ R'OH \longrightarrow \left[\begin{array}{c} H--O-R' \\ | \\ R-N=C-O^- \\ | \\ H-O-R' \\ + \end{array} \right] \longrightarrow R-\underset{H}{N}-\overset{O}{\overset{\|}{C}}-O-R' + R'OH$$

Figure 5.19 The reaction of isocyanate with alcohol.

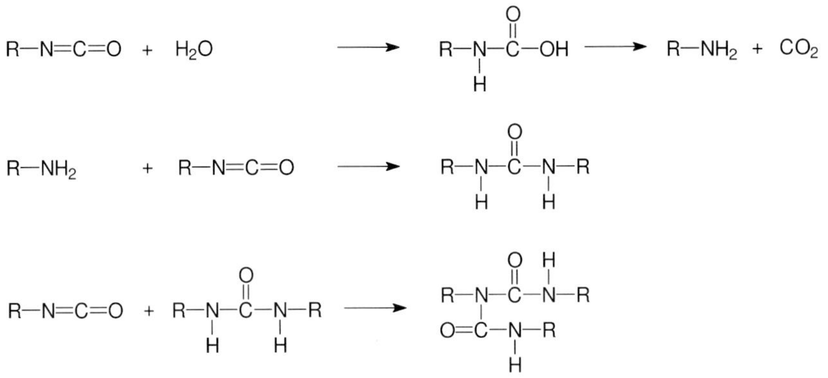

$$R-N=C=O + H_2O \longrightarrow R-\underset{H}{N}-\overset{O}{\overset{\|}{C}}-OH \longrightarrow R-NH_2 + CO_2$$

$$R-NH_2 + R-N=C=O \longrightarrow R-\underset{H}{N}-\overset{O}{\overset{\|}{C}}-\underset{H}{N}-R$$

$$R-N=C=O + R-\underset{H}{N}-\overset{O}{\overset{\|}{C}}-\underset{H}{N}-R \longrightarrow \begin{array}{c} OH \\ \|| \\ R-N-C-N-R \\ | \\ O=C-N-R \\ | \\ H \end{array}$$

Figure 5.20 The reaction of isocyanate with water, and consecutive reactions.

splits off carbon dioxide to give an amine (Figure 5.20). The amine then reacts with a further isocyanate group to give a urea group. At room temperature the urea group in turn reacts with free isocyanates, thereby crosslinking the polymer chains to form biurets, as it has a much higher reactivity than the urethane group [29–32].

5.6.2
Raw Materials

5.6.2.1 Isocyanates
Both, aromatic and aliphatic isocyanates, are used for the synthesis of polyurethanes. The most popular aromatic isocyanates are 2,4-toluene diisocyanate (TDI; an isomeric mixture of 2,4- and 2,6-toluene diisocyanate at an isomer ratio of 80 : 20 or 65 : 35), 4,4-methylene dip(henyl isocyanate) (MDI), and the oligomeric form of MDI, a phosgenized aniline formaldehyde condensate. The most commonly used aliphatic and cycloaliphatic diisocyanates are hexamethylene diisocyanate, isophorone diisocyanate, 2,2,4-trimethyl hexamethylene diisocyanate, and 2,4,4-trimethyl-hexamethylene

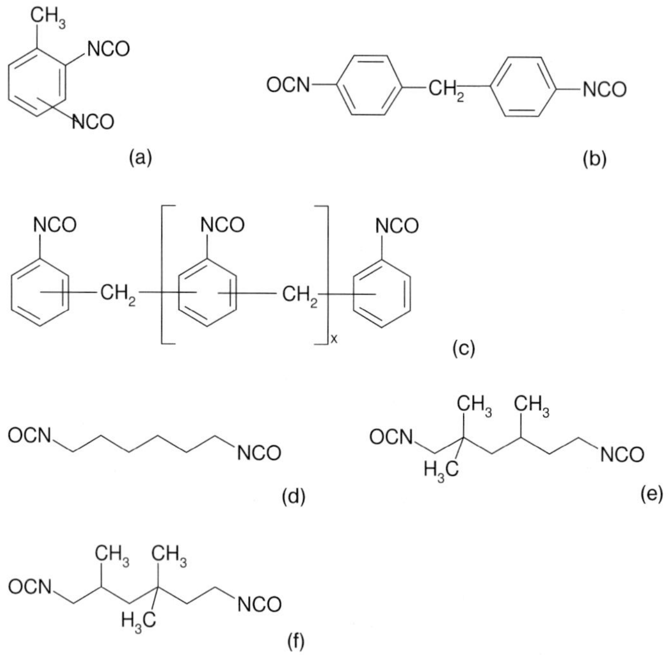

Figure 5.21 Commercialized aromatic and aliphatic diisocyanates.
(a) TDI; (b) MDI: (c) oligomeric MDI; (d) hexamethylene
diisocyanate; (e) 2,2,4-trimethyl hexamethylene diisocyanate;
(f) 2,4,4-trimethyl hexamethylene diisocyanate.

diisocyanate. Some structures of diisocyanates are shown in Figure 5.21. For the manufacture of polyurethanes, aromatic diisocyanates are preferred to their (cyclo) aliphatic counterparts because they are less expensive and more UV-stable.

5.6.2.2 Polyols

The most widely used polyols employed in polyurethane adhesives are mainly polyether polyols and polyester polyols. The synthesis of polyether polyols takes place via a ring-opening polymerization of ethylene, propylene or butylene oxides, respectively, with an initiator in the presence of a strong base.

Polyether polyols can be obtained with different functionalities, molecular weights and hydrophobic characteristics, respectively, by means of specific procedures of synthesis. The most important properties of polyether diols are their low viscosity, good flexibility at low temperatures, and their resistance to alkaline hydrolysis. Polyester polyols are usually generated by a reaction of adipic acid with different glycols. Compared to polyether polyols, polyester polyols have a higher tensile strength and a higher resistance to heat.

5.6.2.3 Catalysts
Tertiary amines, such as triethylamine, N-methyl-morpholine and triethylene diamine, or metal salts such as tributyltin acetate and dibutyltin dilaurate, are used to accelerate the reaction of isocyanates with nucleophilies [29–31]. Tin compounds have a higher catalytic activity than tertiary amines [33].

5.6.3
Structure and Properties of Polyurethane Adhesives

A distinctive feature of the polymer structure of polyurethane adhesives is the presence of hard and soft segments. Soft segments formed by long-chain polyether polyols have a low T_g; at room temperature they are in the entropy-elastic dispersion range. Hard segments are created by the crosslinking of diisocyanates with short-chain diols or diamines. Hard segments associate with each other to generate regions with high T_g-values, sometimes even leading to crystallization, thereby making it possible to generate polyurethane adhesives with two different T_g-values.

While it is possible to generate adhesives with a wide variety of mechanical material properties, it is not easy to separate the properties and application fields of polyurethane adhesives from those of other adhesive systems. Generally, the T_g of polyurethanes is higher than that of silicones, but lower than that of highly crosslinked, structural epoxy resin adhesives; hence, polyurethane adhesives also have other mechanical material properties compared to these adhesive systems. Polyurethanes can be subdivided into either one-part or two-part systems.

5.6.3.1 One-Part Polyurethane Adhesives
Within the prepolymer structure of moisture-setting polyurethane adhesives, there are still isocyanates. These polyurethane adhesives react with the ambient moisture according to the equation shown in Figure 5.20; polyureas or biurets are generated by crosslinking. Moisture-curing polyurethanes are used in the furniture-making industry, in the manufacture of motor homes, and as glazing adhesives in car manufacture.

Hydroxyl polyurethanes are thermoplastics with a hydroxyl group content of 0.5–1.0%, and are produced by a reaction of MDI with polyester diols. Typically they have a mean molar mass ranging from 50 000 to 200 000 g mol^{-1}. For the bonding of shoe soles, hydroxyl polyurethanes are end-modified with a solvent.

Moisture-setting thermoplastic polyurethane hot melts are obtained when hydroxyl polyurethanes are mixed with polyisocyanates. Their resistance to heat is increased by a crosslinking which takes place at room temperature.

Emulsifying or dispensing linear polymers in water produces aqueous polyurethane dispersion adhesives. For this purpose it is necessary either to add emulsifiers, or to incorporate hydrophilic groups or long hydrophilic polyol segments into the polymer matrix. *Dispersion adhesives* are used primarily in the packaging and textile industries.

5.6.3.2 Two-Part Polyurethane Adhesives

Unlike one-part polyurethane adhesives, two-part systems are composed of low-molecular-weight polyisocyanates or prepolymers cured with low-molecular-weight polyols or polyamines. They are used in car manufacture for bonding metals with plastics, for textile foam coatings, and for the bonding of PVC foils to wood in the furniture-making industry.

5.6.4
Formulations of Polyurethane Adhesives

The properties of polyurethane adhesives depend on the isocyanates, polyols and chain extenders used (these may be short-chain amines or alcohols).

Primary TID was used as an *isocyanate source* for many years, but was later found (due to its high volatility) to cause allergic reactions. To overcome this problem, oligomers with TID end-caps were made available. Other diisocyanates used are based upon methylene dianiline, and its dimers and trimers.

The most widely used *polyols* employed in polyurethane adhesives fall into three main classes: polyesters, polyethers, and polybutadiene polyols. *Polyester polyols* are usually based upon caprolactone and are manufactured with a variety of molecular weights. The molecular weights of polyether polyols are generally less than 10 000, and they are usually based on tetrahydrofuran or propylene oxide. *Butadiene polyols* are short-chain butadiene polymers with hydroxyl end groups. Castor oil is a low-priced polyol that is often used together with low-molecular-weight polyethers; however, due to its low molecular mass and high functionality it yields high-strength, but sometimes brittle, products.

Chain extenders are part of the mechanism which provides an improvement in the toughness of polyurethane-based adhesives. For this, materials such as ethylene glycol, 1,4-butanediol, 1,4-bis(hydroxymethyl)cyclohexane or ethylene diamine can be added. The chain extender reacts with the isocyanate, which then phase-separates into domains enriched in isocyanate/chain extender blocks or 'hard segments'; the polyol phase then separates into 'soft segments'. This phase-separated structure is the source of the extraordinary toughness displayed by polyurethane adhesives [34].

Unlike the case with other adhesives, the addition of fillers and rheological additives is a 'must' in the formulation of polyurethane adhesives. For cost-reduction purposes, carbonates and silicates such as talc, silica or clay are added (up to 60%). In addition, the rheological and mechanical properties of the adhesive can be improved by means of active fillers such as carbon black or pyrogenic silica (up to a maximum of 20%). The use of *plasticizers* (mostly based upon phthalates, phosphates or aromatic oils, up to a maximum of 30%) allows the processing viscosity and swelling behavior of the finished adhesive to be controlled. *Molecular sieve desiccants* are often employed to dry the added fillers, plasticizers and polyols.

A typical formulation for a two-part polyurethane adhesive based on polyether and castor oil contains about 33% polyols, 10% castor oil, 2% rheological additive, 50%

filler, and 5% desiccant. If, for the formulation, this represents 100 parts of part A, then about 40 parts of polyisocyanate are added as part B [35].

A typical formulation of a moisture-setting, one-part polyurethane adhesive contains 49.3% isocyanate-terminated prepolymer, 49.3% filler, 1% polyol, 0.27% rheological additive, and 0.13% high-boiling solvent [35].

5.7 Acrylate Adhesives

5.7.1 Physically Setting Acrylates

5.7.1.1 Solvent-Containing Acrylates

The primary solvent-containing acrylate adhesives used are thermoplastic methyl methacrylate copolymers dissolved in common organic solvents. The adhesives reach their final strength as soon as the solvent has evaporated or migrated into the substrate. Solvent-containing acrylates are mainly used for the bonding of paper or paperboard and for diffusion bonding (solvent welding, cold welding) of thermoplastic materials such as PVC.

5.7.1.2 Dispersion Acrylates

In the wake of growing ecological awareness, low-solvent or solvent-free adhesive systems are rapidly gaining importance. Acrylate dispersions can be adjusted to the majority of applications owing to the wide range of raw materials available. They have an excellent age resistance and a reduced tendency towards yellowing, and can also be used as PSAs for the fabrication of labels and adhesive tapes, and as permanent adhesives for the fixing of tiles, floorings and wall coverings.

5.7.1.3 Acrylate-Based Contact Adhesives

Acrylate-based contact adhesives make it possible to obtain bond-lines of relatively high strength within a very short period of time after application, subsequent drying and joining under pressure. These adhesives are available as either solvent-containing systems or dispersions. Traditionally, chloroprene or butadiene–acrylonitrile rubber is used, but acrylate homopolymers and copolymers are also employed, for example in the bonding of floor coverings using acrylate–styrene copolymers. These copolymers are highly flexible but have a low shear stress strength on tensile loading. Acrylate-based contact adhesives are also used in car manufacture and in the fabrication of mattresses and shoes.

5.7.2 Chemically Setting Acrylates

Reactive acrylate and methacrylate adhesives are solvent-free 100% polymer systems that comprise one-part and two-part systems, as well as anaerobic and UV-setting

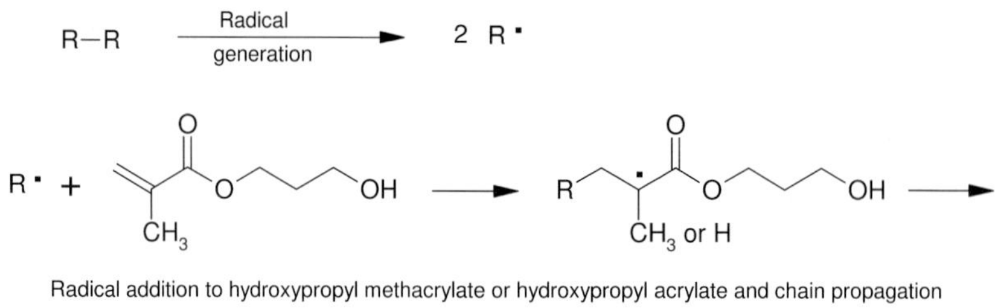

5 Chemistry and Properties of Adhesives and Primers

$$R-R \xrightarrow{\text{Radical generation}} 2\ R^\bullet$$

Radical addition to hydroxypropyl methacrylate or hydroxypropyl acrylate and chain propagation

Figure 5.22 Free-radical polymerization of acrylates/methacrylates.

adhesives. Reactive acrylate/methacrylate systems differ from physically setting acrylates/methacrylates by their curing mechanism, although they have a similar chemical base.

Acrylates and methacrylates are monomers or prepolymers with vinyl group-containing double bonds; these double bonds are subsequently attacked by free radicals to initiate the polymerization reactions in different ways (Figure 5.22).

The monomers are clear, colorless, low-viscosity liquids, while the prepolymers are highly viscous (sometimes solid) resins the homopolymers of which have characteristic properties that are used to generate the final properties of an adhesive system. The adhesive developer is able to identify a wide variety of monomer and prepolymer types which allow the final properties of the adhesive to be tailored for desired applications.

With regard to their chemistry, methacrylates differ from acrylates only by the presence of an extra methyl group (Figure 5.23).

The physical properties of the methacrylate homopolymers differ considerably from those of acrylate polymers (see Table 5.3). The main difference between one-part and two-part acrylate/methacrylate adhesives and UV-setting adhesives on the one hand and anaerobic adhesives on the other hand is the functionality of the monomers. While anaerobic adhesives are mostly based on bifunctional, trifunctional or tetrafunctional systems, the one-part, two-part and UV-setting adhesives are mostly based on monofunctional compounds. Owing to their high functionality, anaerobic systems react to yield highly crosslinked, brittle polymers with high dimensional stability, whereas the properties of the polymers of the one-part, two-part and UV-setting systems range from extremely strong and impact-resistant to highly flexible. These systems therefore belong to the class of structural adhesives.

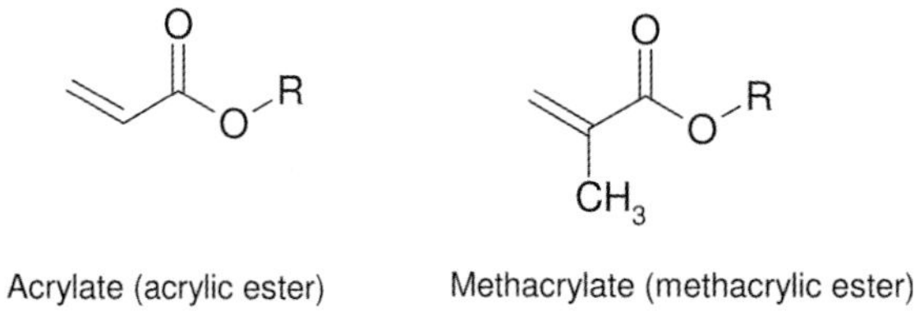

Acrylate (acrylic ester)   Methacrylate (methacrylic ester)

Figure 5.23 Structures of acrylates and methacrylates.

Table 5.3 Comparison of the properties of acrylate polymers and similar methacrylate polymers.

Physical property	Acrylate polymer	Methacrylate polymer
Velocity of reaction	High	Low
Glass transition temperature	Often lower	High
Strength	High	Very high
Chemical resistance	Good	Very good
Flexibility	Good	Moderate
Prize of the monomers	High	Low

5.7.2.1 Monomers

For one-part, two-part and anaerobic adhesives, mostly methacrylates are used, while UV-setting adhesives are mainly based on acrylates. UV-setting adhesives are required to cure very rapidly, whereas one-part and two-part systems are often employed as high-performance systems the mechanical properties of which are considered to be more important than the curing rate. The monomers are mainly chosen according to their T_g-value that represents a parameter for the mechanical performance at different temperatures. Appropriate combinations allow the flexibility, strength and durability characteristics to be customized. Some examples of the wide array of monomers available, together with their T_g-values, are listed in Table 5.4.

Acrylate/methacrylate adhesive groups are described in detail in the following sections.

Table 5.4 Glass transition temperatures (T_g) of different monomers [36].

Product	T_g (°C)
Polybutadiene dimethacrylate	−75
Tridecyl acrylate	−55
Tridecyl methacrylate	−40
Polyethylene glycol (400) dimethacrylate	−21
Polyethylene glycol (400) diacrylate	+3
2-Phenoxy ethylacrylate	+5
Tetraethylene glycol diacrylate	+23
Glycidyl methacrylate	+41
1,6-Hexane diol diacrylate	+43
Trimethylol propane triacrylate	+62
Triethylene glycol diacrylate	+70
Isobornyl acrylate	+88
Isobornyl methacrylate	+110

5.7.2.2 Two-Part Acrylates/Methacrylates

The typical base monomer of this adhesive group is methyl methacrylate (MMA). This has a very good wetting behavior and easily 'etches' plastic materials, thus allowing excellent diffusion bonding. MMA is also capable of wetting metals and glass. MMA homopolymers have outstanding mechanical properties and are highly durable; however, in order to reduce the brittleness of these adhesive systems, acrylonitrile–butadiene copolymers, liquid isoprene rubber systems or chlorosulfonated rubber systems are often added to toughen them [37]. Impact toughness-modified adhesives easily reach a shear stress strength by tensile loading of 30 N mm^{-2}. A peroxide (e.g. benzoyl peroxide or t-butyl perbenzoate) is typically added as a hardener in combination with an amine accelerator (e.g. N,N-dimethyl-p-toluidine) as an accelerator (Figure 5.24) [38].

The adhesives are available at a ratio of 1: 1 or 10: 1 in double-chamber cartridges, ready to use. The adhesive parts are mixed in a static mixing tube and applied as if it were a one-part system. As the reaction takes place as a polymerization, and not as a polyaddition, the systems are rather tolerant to incorrect mixing. For automated applications, the adhesive parts can be applied using dose dispensers for two-part systems. Generally, the pot life of these systems is rather short – that is, the adhesive must be applied rapidly after mixing (Figure 5.25).

Two-part systems are also available as 'no-mix' adhesives where the hardener is applied to one adherent surface, and the resin to the other. The substrates are then joined to each other, thereby mixing the adhesive parts, and the curing takes place. The advantage of these systems is that no pot life must be observed, although perfect bonding is only obtained with thin bond-lines.

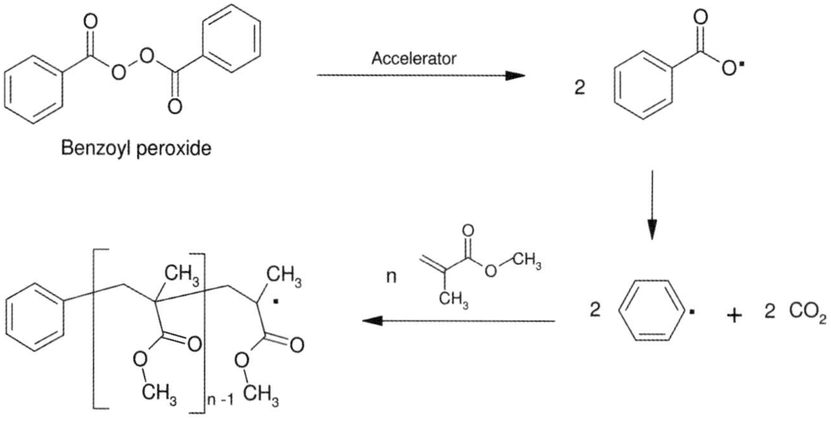

Figure 5.24 Reaction of MMA with dibenzene peroxide.

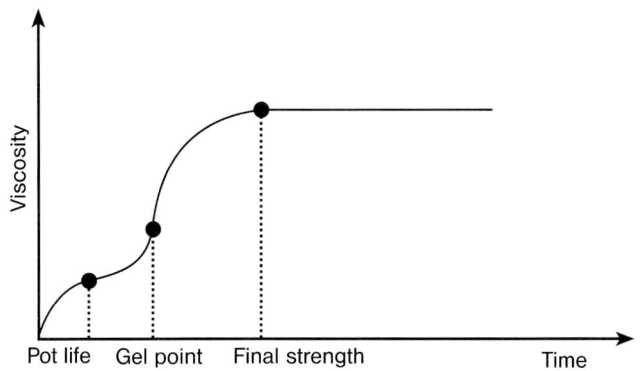

Figure 5.25 Pot life, gel point and final strength and their dependence on time. It is important that the adhesive is applied during its pot life.

5.7.2.3 One-Part Acrylates/Methacrylates

With one-part acrylate/methacrylate adhesives, the reaction is initiated by peroxides, as in the case of two-part systems. Since, with one-part adhesives, the peroxides do not decompose under the influence of the hardener but rather under the effect of temperature, the storage stability and desired reaction time must be aligned with each other. In order to allow storage and transportation under normal conditions, the adhesives have curing temperatures from 120 °C upwards. The reaction temperature is determined by the half-life of the peroxides (Table 5.5).

The lower the desired reaction temperature, the less storage-stable the adhesive.

5.7.2.4 Examples of Application

One-part and two-part adhesives are used mainly for structural bonding as they have a very good adherence to metals, glass and plastic materials, a high tolerance to slightly contaminated and oiled surfaces, and a relatively short curing time compared to epoxy resin-based or polyurethane-based reactive adhesives. Some typical applications are listed in Table 5.6.

MMA-based adhesives can be used in a temperature range from −50 °C to 100 °C and, for a short period, they can even support temperatures of up to 180 °C. The viscosity of mixed two-part methacrylates can be customized so as to allow large gaps to be bridged. Various degrees of elasticity are also easily generated.

Table 5.5 Temperature requirement for a 10-h decomposition half-life for different peroxides [39].

Product	Decomposition temperature (°C)
Dilauroyl peroxide	62
Dibenzoyl peroxide	72
tert-Amyl peroxyneodecanoate	47
tert-Butyl peroxybenzoate	104

Table 5.6 Applications of two-part and one-part acrylate adhesives.

Industrial branch	Application
Electric motors	Fixation and bonding of magnets
Glass	Bonding of solar cells, door hinges, glass furniture, mirrors
Audio	Bonding of loudspeakers and accessories
Automobile	Bonding of subsequently assembled sunroofs, bonding of brake lights
Commercial vehicles	Bonding of panels in busses and rail vehicles, bonding of reinforcement elements in ships

5.7.2.5 Anaerobic Acrylates

Anaerobic adhesives have been available commercially since 1953, the first anaerobic adhesive based on dimethacrylate having been developed that year by Vernon Krieble, in the United States. This was commercialized under the name Loctite [40] and, indeed, the early success of the Loctite Company – which today is the leading manufacturer of acrylate adhesives – is owed to the use of acrylates in adhesive bonding technology.

'Anaerobic' is a designation borrowed from biology which signifies that the adhesive remains liquid while it is in contact with the air, but cures to generate a polymer in the absence of air and in the presence of metal ions, as for example in a metallic bond-line.

Typical applications are the securing of screws and threads, the fixing of adherents, ball bearings and bushes, and the sealing of surfaces. Anaerobic adhesives with improved resistance to temperature and the ability to take up oil are suitable for use in motors. Anaerobic adhesives are one-part systems that are easily and economically applied. The adhesive only cures in the bond-line, and noncured adhesive is easily removed. Owing to the low activation energy of the curing process, the polymerization takes place at room temperature. Since, in general, starting systems with higher reactivity are used, the polymers have a high degree of crosslinking and a very good resistance to different media and temperatures.

One prerequisite for a successful curing process is the presence of metal ions on the surface of the substrate, because they act as catalysts. Their influence on the curing rate depends on their degree of activity (copper > brass/bronze > iron > steel > zinc > aluminum > special steel). Typical starting monomers include triethylene glycol dimethacrylate or tetraethylene glycol dimethacrylate (Figure 5.26).

Tertiary amines, sulfimides or hydroperoxides are used as hardeners. A typical hardener system consists of saccharin, N,N-dimethyl-p-toluidine and cumene hydroperoxide (Figure 5.27).

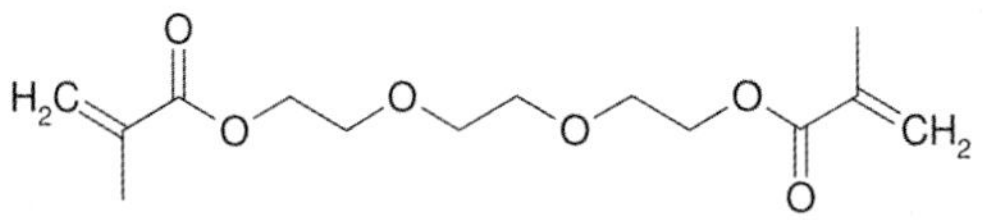

Figure 5.26 Triethylene glycol dimethacrylate as a starting monomer.

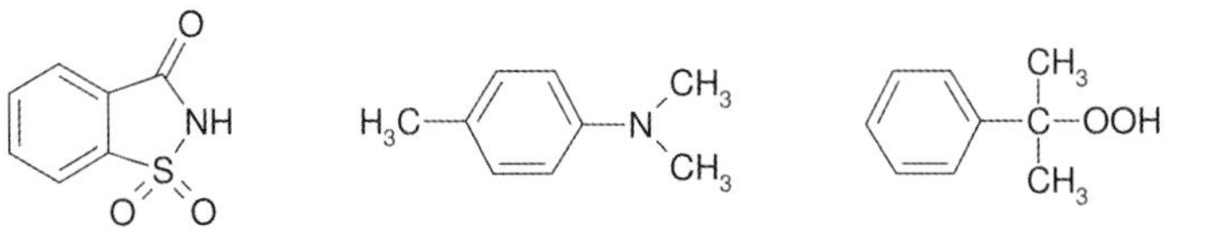

Saccharin N,N-Dimethyl-p-toluidine Cumene hydroperoxide

Figure 5.27 The components of the curing system.

The radical curing of these adhesives is initiated by cumene hydroperoxide. Especially active metal ions are regenerated in a cycle by saccharin and amine, providing for a rapid and full cure (Figure 5.28).

Anaerobic adhesives are mainly used to take advantage of their chemical interlocking effect rather than their bonding effect, owing to the characteristic properties of their polymers. However, anaerobic adhesives may be impact-toughened to become suitable for surface-to-surface bonding in the case of very narrow gaps. When combined with UV-setting adhesives (see below), anaerobic adhesives can also be used for other applications.

5.7.2.6 UV-Setting Acrylates

UV-setting acrylate adhesives are also radically initiated systems. In contrast to other adhesive systems, they have an extremely high curing rate that is generally on the order of a second or a millisecond [40, 41]. UV-setting acrylate adhesives are therefore extremely well suited for bonding purposes in the production of high-precision, high-quality goods manufactured in high quantities. In particular, highly automated processing fits in well with the manufacturing requirements of the electronics industry, where development continues to be extremely rapid.

Curing Through Irradiation The electromagnetic spectrum (Figure 5.29) comprises different regions of radiation which may be either low or high in energy. Radiation with a very long wavelength (e.g. radio waves, microwave radiation) is very low in energy, while radiation with a short wavelength (e.g. X-rays, cosmic radiation) is very high in energy. The lower the energy of the radiation, the less its impact on matter.

Microwaves and infrared (IR) radiation have an accelerating effect on chemical reactions as they induce a rotation or vibration of the molecules. Radiation-based technologies may therefore be used to accelerate or initiate the curing reactions of standard epoxy resin adhesives, for example. To the left of the IR radiation band there is the region of UV radiation which is higher in energy and induces an excitation of the electrons. Initiators which absorb specific wavelengths in the UV/visible band are placed in a state that is high in energy, and this allows them very rapidly to initiate chemical reactions. This technology is used for UV-setting adhesives and will be discussed in detail in the following section.

Absorption and Reaction In order to obtain an effective curing of UV-setting adhesives, the wavelength of the radiation emitted by the UV light source must correspond to the absorption wavelength of the initiator.

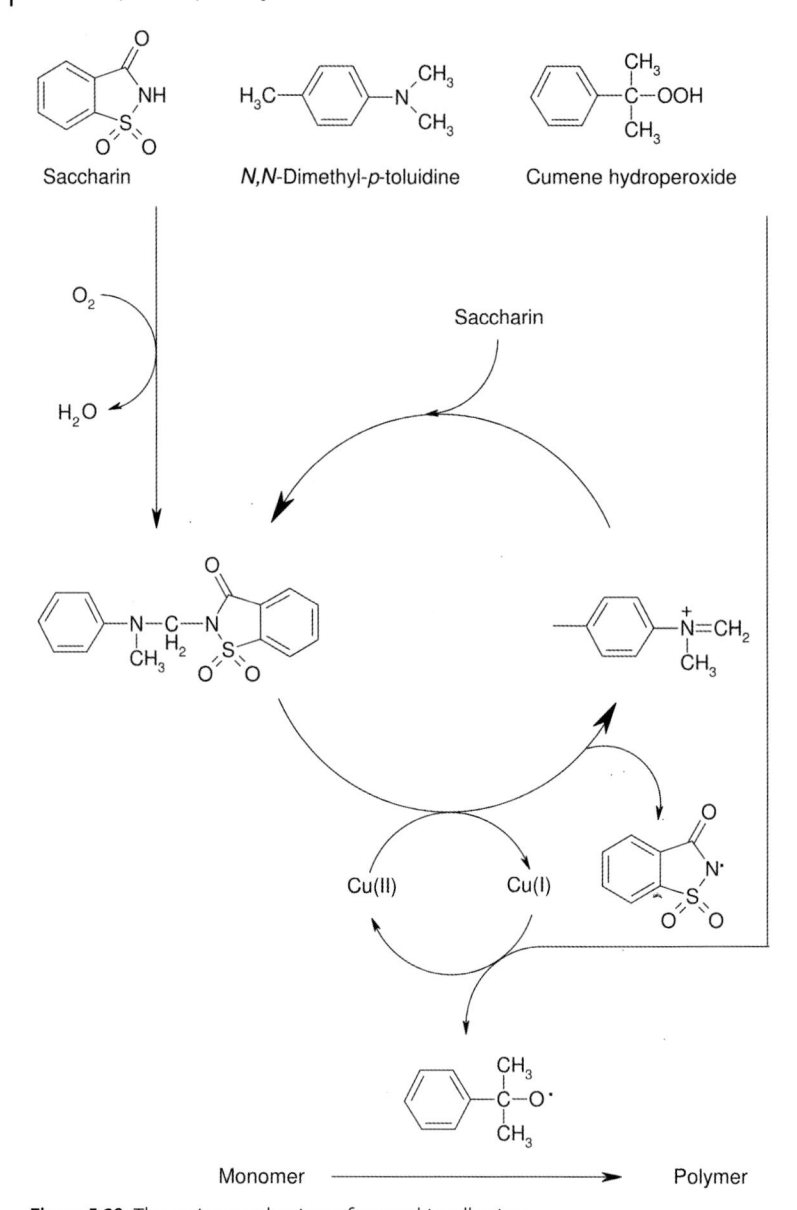

Figure 5.28 The curing mechanism of anaerobic adhesives.

In the example shown in Figure 5.30, only a small portion of the large spectrum of the radiation emitted by a light source is overlapped by the absorption wavelength of the photoinitiator. The light source and adhesive must match as perfectly as possible to obtain the best possible curing effect.

Another factor to be considered is the adherent material. For example, metal substrates reflect radiation, thereby accelerating the reaction and influencing the

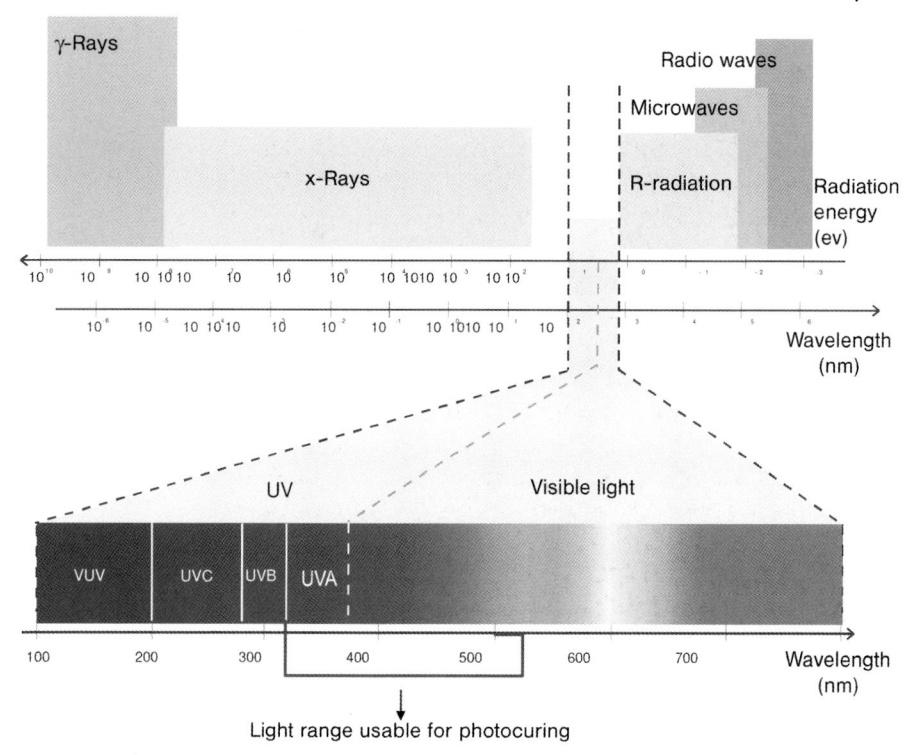

Figure 5.29 Electromagnetic spectrum [41].

degree of crosslinking near the adherent surface. Adherents that are transparent in the visible range (i.e. they can be seen through) are not necessarily translucent with regards to UV radiation. For example, polycarbonate and PMMA (Plexiglas®) contain UV stabilizers that slow down, or even block, the UV curing process.

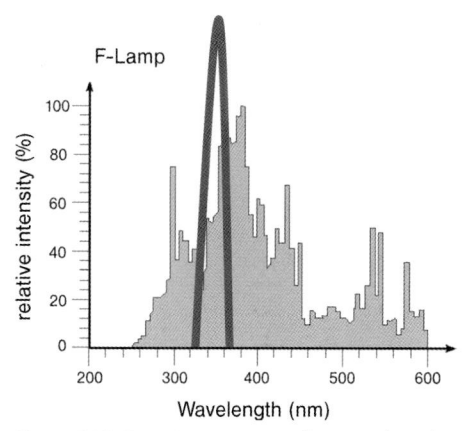

Figure 5.30 Emission spectrum of an iron-doped gas discharge lamp and maximum absorption of the photoinitiator [41].

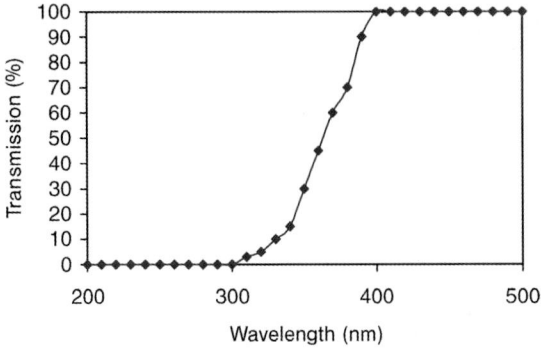

Figure 5.31 Transmission curve of a 5 mm-thick polymethyl methacrylate (PMMA) screen (Plexiglas®, Röhm).

In Figure 5.31, a 5 mm-thick PMMA screen is seen to absorb 60% of UVA radiation, 10% of visible light and 100% of UVB/UVC radiation.

Reaction Mechanism With radical curing acrylates, at least one adherent must be translucent for UVA/UVB radiation because these adhesives only cure upon irradiation with light of the necessary wavelength. The reaction stops immediately when the light is removed (Figure 5.32).

An adequate period of irradiation must be determined in order to obtain full cure. Radical curing systems have several advantages, in that they have not only a very rapid curing rate but also a wide range of almost infinitely variable properties such as viscosity, color, flexibility, hardness and strength, all of which allow customization of the adhesive system.

UV-setting adhesives develop their properties fully provided that a suitable light source is available (Figure 5.33). Today, important advances have been made in light technology, and improved systems with different dimensional characteristics are now available to the user. The wavelengths to be used can also be confined to a specific range. In generally, the UV radiation is generated by gas discharge lamps, concentrated by

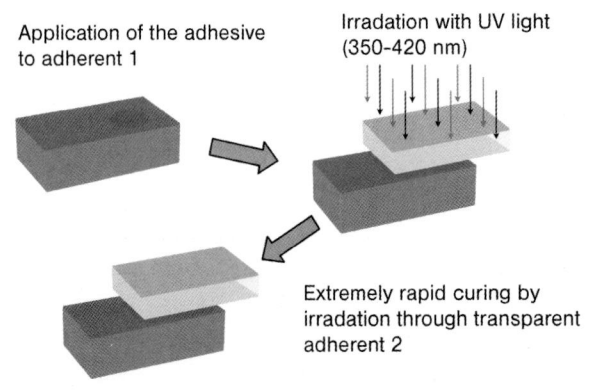

Figure 5.32 Schematic of the radical curing of UV acrylates.

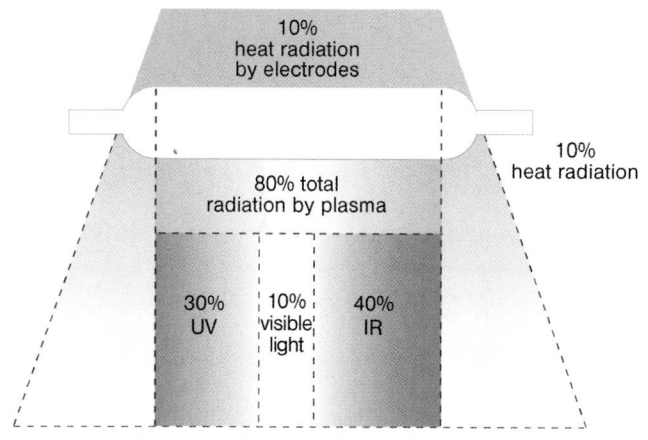

Figure 5.33 Balance of performance of a gas discharge lamp [41].

Table 5.7 Characteristic parameters of UV curing.

Parameter	Characteristic	Unit
UV dose	Calculated from intensity × radiation period	$mJ\,cm^{-2}$
UV intensity	Depending on the distance from the adherent, the adherent material, the lamp reflector, and the age of the lamp	$mW\,cm^{-2}$

special reflectors and then transformed into a homogeneous and focused radiation. The radiation emitted by gas discharge lamps, however, has a high percentage of IR rays that may cause damage or deformations in some adherents, especially in plastic materials. By reducing the IR portion while maintaining the UV intensity, special reflectors make it possible to reduce the adherent temperature by up to 50%.

The UV intensity and dose are important curing parameters (Table 5.7). Typically, the UV dose defines the energy density needed to cure an adhesive system, whereas the UV intensity data provide more detailed information, taking different spectral ranges and radiation periods into consideration.

Examples of Application UV-setting adhesives can be used in a wide range of applications, as short curing periods provide for high production rates and short cycle times. Owing to the radical curing mechanism, UV curing can easily be combined with heat curing by adding peroxides, or with anaerobic curing, making it possible to use UV-setting adhesives for bonding sites with shadow regions that are inaccessible to the radiation source. UV-setting adhesives are particularly well suited to the industry sectors, as summarized in Table 5.8.

When UV-setting adhesives are used in industry (Table 5.8) they must meet various different criteria (Table 5.9), for example a rapid curing rate. It has become clear that modern UV-setting adhesives will, in time, take over many very demanding functions that will lead to the creation of new fields of application.

Table 5.8 Fields of application of UV-setting adhesives.

Industry sector	Application
Electronics	Fixing and protection of electronic components
Optics	Bonding of lenses, optic filters
Glass	Bonding of furniture, glass cabinets, tables
Audio	Bonding of loudspeakers and accessories
Automobile	Sealing of sensors
Design and fashion	Scratch-resistant coating
Medical technology	Bonding of disposables

Table 5.9 Requirements of UV-setting adhesives in different industry sectors.

Industry sector	Requirements
Electronics	• Good adherence to noble metals, ferrites, FR-4, polyamide • Ionic purity • Solder resistance • Low coefficient of thermal expansion • Flow control • Dosability • Defined dynamic mechanical properties
Optics	• Defined refractive index • Defined transmission • Low degree of shrinkage • High thermal stability
Glass	• High transparency • UV resistance • Structural bonding requires high strength and durability on glass and combinations of glass with other materials • Compensation of stress differences
Audio	• Good adherence to unusual surfaces such as impregnated paper • No impairment of acoustic properties
Automobile	• Resistance to different media such as oil, gasoline • Durability over a wide range of in-service temperatures (e.g. $-40\,°C$ to $+150\,°C$) • Weatherproofness • Defined dynamic mechanical properties
Design/Fashion	• High transparency • UV resistance • High abrasion and scratch resistance
Medical devices	• Biocompatibility • High transparency • Good adherence to plastic materials • Compatibility with sterilization methods

5.7.2.7 Cyanoacrylate Adhesives

In the case of cyanoacrylate adhesives, which are known as 'Super glues', curing is initiated by the moisture adsorbed onto the adherents and takes place within the bond-line within a few seconds or minutes. Although cyanoacrylates are easy to process, they cannot overcome gaps which have a width in excess of 0.1 mm. These materials are thermoplastics after curing, and cannot (or can only barely) be plastified; hence, they are mainly suited to the joining of small rigid parts. Some elastomers can also be bonded easily by cyanoacrylate adhesives, for example in the production of packing rings.

The structure of the basic monomer, as well as the anionic polymerization of cyanoacrylate adhesives, is shown in Figure 5.34. The double bond, due to its proximity to two electron-withdrawing groups (the nitrile and the ester group), is extremely sensitive to nucleophilic attack. Such susceptibility provides a monomer the polymerization of which can be initiated by species that are as weakly nucleophilic as water. Various bases can also be used to induce the polymerization of cyanoacrylates, although acids inhibit the process. The extreme reactivity of the monomer results in a fast cure. Cyanoacrylate monomers may contain ester groups from methyl, ethyl to isobutyl, and ethoxy ethyl; acrylates or methacrylates, which are soluble in cyanoacrylate monomers and present a high molecular mass, are frequently added to the basic monomers to increase the viscosity.

It is important to note that cyanoacrylate adhesives are thermoplastics after cure, as this makes them susceptible to creep as well as to attack by moisture. In order to overcome these problems the formulations are sometimes modified by means of crosslinking agents or thermally stable polymers.

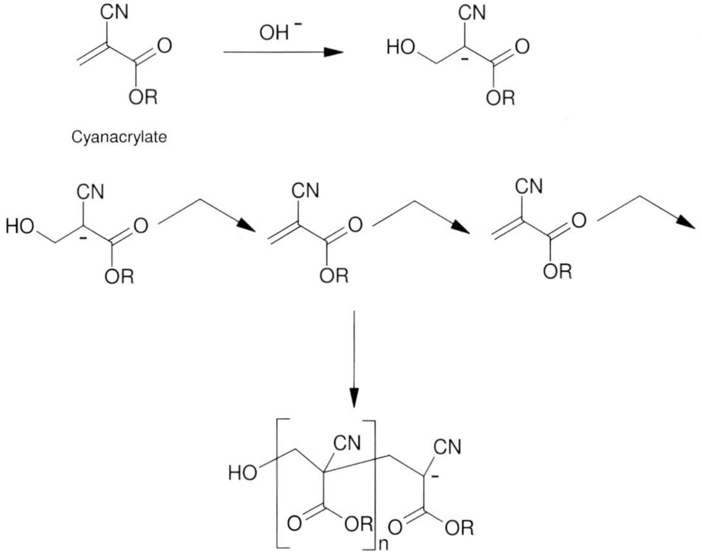

Figure 5.34 Structure and anionic polymerization of cyanoacrylates.

Cyanoacrylates were first developed in the United States in 1957 and marketed under the trademark 'Eastman 90'. Three years later, Sichel introduced them in Germany under the trademark 'Sicomet', which now belongs to the Henkel Group. Cyanoacrylate adhesives are used for a wide variety of applications, particularly for the industrial bonding of plastic and rubber materials. They are also employed as tissue adhesives in medicine, or as 'Super glues' in home applications.

5.7.3
Formulations of Acrylate Adhesives

Formulations of acrylate adhesives are mainly based on MMA, which is generally copolymerized with methacrylic acid or acrylic acid. These polar components make it possible to improve not only the adherence of the adhesives on many surfaces, but also their thermal resistance. Problems arising with objectionable odors can be overcome by the use of acrylate monomers that boil at higher temperatures, such as cyclohexyl methacrylate or tetrahydrofuryl methacrylate.

Acrylonitrile–butadiene copolymers, liquid polyisoprene, chlorosulfonated rubber or chlorosulfonated polyethylene are added to reduce the brittleness of the adhesives. The incorporation of 10% bifunctional monomers such as ethylene glycol dimethacrylate provides for crosslinking, thus significantly improving the thermal resistance. Organic peroxides (e.g. cumene hydroperoxide) are used as initiators or hardeners, allowing adequate storage stability to be combined with a fast cure.

5.7.3.1 Formulation of a Two-Part Acrylate 'No Mix' Adhesive

A typical formulation of a two-part acrylate 'no mix' adhesive (also known as a 'hardener lacquer system') with good thixotropic properties and excellent impact toughness comprises, in the resin component, 45% methyl methacrylate, 8% methacrylic acid, 1% dimethyl acrylate (e.g. ethylene glycol dimethacrylate), 45% toughening polymer (e.g. chlorosulfonated polyethylene), and 0.5% peroxide (e.g. cumene hydroperoxide).

In the no-mix procedure, the activator (hardener lacquer) that is contained in an organic solvent is applied to one of the surfaces to be joined. The activator slowly releases an amine that later acts as an accelerator. Surfaces coated with hardener lacquer can be stored for any period, and remain to be activated for subsequent bonding. The peroxide needed for polymerization is contained in the acrylate monomer component (resin component) and only activated by the amine when the adherents are fitted together.

Although, when working with the no-mix procedure, the pot life does not need to be considered, the adhesive system can only overcome gap widths of up to 0.5 mm. However, this procedure has been widely adopted owing to easy processing and process reliability.

5.7.3.2 Formulations of UV-Setting Acrylate Adhesives

The formulations of UV-setting acrylate adhesives are similar to those discussed above, except that a photoinitiator is responsible for the curing process. Under the influence of UV radiation, the photoinitiator generates radicals that initiate the polymerization. The resin to be cured contains reactive monomers or oligomers that are based, for example, on urethane, as well as additives and the photoinitiator. Depending on their structure, the monomers shape the specific adhesion properties. Strength and elasticity are owed to the oligomers. A typical formulation for a UV-setting acrylate adhesive contains 33% urethane acrylate, 33% isobornyl acrylate, 25% trimethyl pentane triacrylate, 5% pigment, and 4% photoinitiator.

5.7.3.3 Formulations of Cyanoacrylate Adhesives

Methyl esters of cyanacrylic acid are used to generate fast-curing, high-strength cyanoacrylate adhesives for metals, while ethyl esters are used to generate adhesives for metals and plastic materials. In order to overcome problems with the odor of these two monomers, higher esters are employed. but these lead to the generation of adhesives with lower strengths and longer curing periods. Cross-linking agents based on bifunctional cyanoacrylates can be used to improve the thermal resistance.

All of these monomers have a low viscosity that may in time cause problems with absorbent adherents and vertical applications. Consequently, up to 25% thickener may need to be added, mostly as polymers such as PMMA or PVA. The additional incorporation of hydrophobic pyrogenic silicic acid results in a formulation that provides for structural viscosity, easy processing on vertical surfaces, improved bonding of absorbent surfaces, and enhanced gap-filling characteristics. Due to their high T_g-values, cyanoacrylate adhesives are very brittle, but the impact resistance can be improved by adding 10–15% of an elastomer such as acrylonitrile–butadiene–styrene (ABS) copolymer.

The polymerization of cyanoacrylate monomers takes place very easily, both in the form of anionic and radical polymerization. To prevent uncontrolled reactions and too-short a storage life, acid stabilizers such as sulfur dioxide or phosphoric acid, and radical scavengers such as hindered phenols, are used. Owing to its gaseous condition, sulfur dioxide is released immediately after application of the adhesive and joining of the adherents. This stabilizer therefore provides for a quick onset of polymerization, which is a desirable feature in many cases.

A typical formulation for a cyanoacrylate adhesive contains 88% cyanoacrylate monomer, 9% thickener, 3% rheological additive, and 0.02–0.03% acid or radical stabilizer.

5.8
Silicones

5.8.1
One-Part Silicone Adhesives

One-part silicone adhesives are known as RTV-1 adhesives (RTV = room temperature vulcanizing). They contain a polyorgano-siloxane with terminal –OH groups, and a crosslinking agent (e.g. methyl triacetoxy silane) which is sensitive to hydrolysis – that is, it reacts with water. When the silicone is applied, the crosslinking agent reacts with

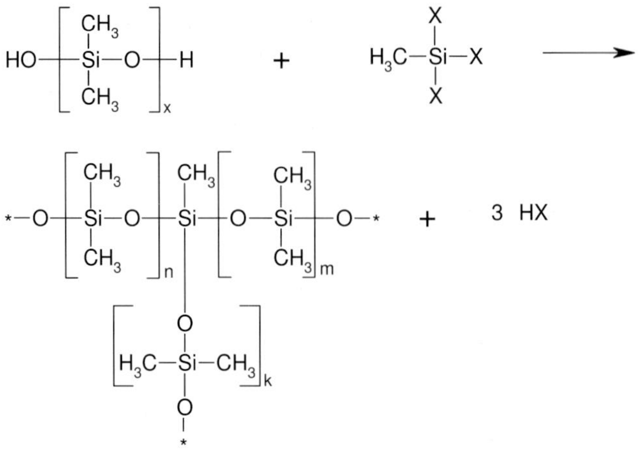

Figure 5.35 Reaction mechanism of a one-part silicone system.

the ambient moisture to generate silanol, which in turn reacts with the free –OH groups of the base polymer, resulting in a crosslink. The reaction is shown schematically in Figure 5.35. When methyl triacetoxy silane is used as the crosslinker, acetic acid is liberated during the reaction; indeed, the odor of this is familiar to many consumers who use one-part silicones at home or at work.

In RTV-1 systems, ambient water initiates the reaction, the rate of which is a function of the humidity of the air, the temperature and the bond-line thickness. Full crosslinking is obtained (depending on these parameters) within a period of few hours to several days. The addition of substances such as NaOH or $Ca(OH)_2$ enhances the crosslinking process. Indeed, products that are split off during condensation (e.g. acetic acid) are transformed into water by these substances. This water, in addition to that which penetrates by diffusion from the outside, enhances the curing process and prevents the development of acetic acid vapors.

5.8.2
Two-Part Silicone Adhesives

Two-part silicone adhesives are known as RTV-2 silicones, and are used for applications where curing takes too long with RTV-1 silicones due to the prevailing conditions of bond-line thickness, temperature and air humidity. Two types of silicone crosslinking have been identified, namely condensation and addition, but neither type requires the presence of moisture.

5.8.2.1 Condensation Crosslinking
Unlike addition crosslinking, condensation crosslinking of silicones takes place by release of the reaction products. Condensation silicones adhere well to a larger number of substrate surfaces, and the risk of the crosslinking reaction being inhibited is lower when they are used.

By splitting off the corresponding alcohol, branched silicones are generated using hydroxyl polysiloxane and an ester of silicic acid; organotin compounds are used as catalysts in this reaction. As the basic formulation is constant, the crosslinking rate is determined by temperature, pH value and catalyst concentration.

5.8.2.2 Addition Crosslinking

Addition silicones for bonding and sealing purposes were developed later than condensation silicones. These modern systems consist of a siloxane with a terminal vinyl group, and a siloxane with a hydrogen atom that is directly bound to the silicon. The exothermic addition reaction is platinum-catalyzed (Figure 5.36) and, when conducted at room temperature, full cure is generally obtained within 24 h. The reaction is significantly accelerated by an increase in temperature; raising the temperature to 150 °C leads to a full cure within 10 min. Special formulations, containing inhibitors, are stable at room temperature and cure at higher temperatures.

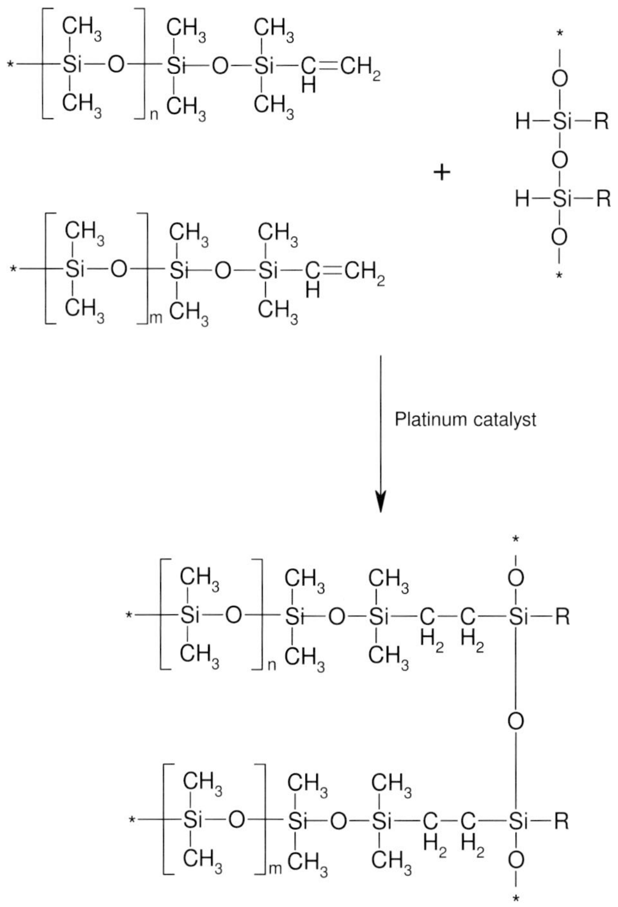

Figure 5.36 Polyaddition of a two-part silicone system.

5.9
Natural Adhesives

5.9.1
Introduction

For many years, up until the 1930s, natural inorganic or organic substances were the most important raw materials for the production of adhesives. However, with the emergence of polymer chemistry and its rapid development, they were increasingly replaced by synthetic products. In the history of mankind, there is abundant evidence of humans being inspired by the manifold manifestations of efficient bonding in nature. Indeed, bonding as a flexible joining technique was used a very long time ago, and appears to have been one of the most ancient joining techniques utilized by mankind. The most important original raw materials included pitch, bitumen and organic polymers obtained from bones and animal skins, later casein, and last – but not least – starch. The inorganic raw materials used to produce cement will not be discussed here as they have lost significance.

The performance of binders was dramatically improved when adhesives were first manufactured from synthetic raw materials. Over the years, the needs of industry have been increasingly demanding with regards to the strength and long-term durability of adhesives and, in many cases, could no longer be met by natural, raw material-based products. It was at this time when bonding was considered unusable for structural joining purposes. During World War I, for example, the commander of the German Naval Airship Department, Peter Strasser, complained angrily to the Admiralty that the wooden construction used for the Schütte-Lanz ships (the girders of which were made from aspen wood glued with cold-setting casein glues) delaminated in the damp environment of the airship hangars in Northern Germany. In fact, such airship girders would still be usable today had they been glued with the presently available phenolic resins used to manufacture plywood (see Chapter 2).

In some fields of application, however, natural material-based adhesives have survived to the present day. Today, approximately 6% of the total adhesives manufactured in Germany is based on natural raw materials, and this proportion may rise again within the context of economical or ecological aspects. The advantages and disadvantages of natural adhesives are listed in Table 5.10. One of the advantages listed – the avoidance of carbon dioxide discharge – is today much less important as

Table 5.10 Advantages and disadvantages of natural adhesives.

Advantages	Disadvantages
• Avoidance of carbon dioxide discharge (cycle)	• Quality of raw materials
• Often better physiological tolerance	• Shelf life (bacterial damage)
• Biodegradability	• Long-term durability of the joints
• Structural viscosity!	• Formulation limits
• (Enzymatic) debonding on demand	• Economic aspects (e.g. raw materials price)

<1% of organic raw materials is used worldwide in the manufacture of adhesives. Physiological tolerance, biodegradability – as well as the possibility to break bonds by enzymatic degradation – are all much more important features. It should be noted, however, that enzymatic degradation is very time-intensive and not well-suited to industrial applications in waste management.

One disadvantage of organic polymers is that they are not durable, largely because decomposition and decay serve as efficient recycling systems for natural materials, a fact rarely taken into account when considering their use. It has often been pointed out that the long-term durability of bonds created with natural adhesives is not good but, as will be discussed later, it is possible to improve this situation by adding hardeners that induce crosslinking of the polymer base materials. It may be surprising to find the property 'economic' listed as a disadvantage; however, while natural materials may be cheaper than synthetic, their quality may vary depending on the conditions of origin and growth. If such fluctuations occur, and/or the commodity price on the world market rises, economic problems may ensue and lead to difficulties in the costing of adhesives.

5.9.2
Natural Raw Materials

With regards to their chemical structure, the higher-molecular-weight, organic base materials used in natural adhesives can be classified as:

- *Proteins*: elastin, fibronectin, keratin, collagen
- *Polysaccharides*: cellulose, starch, gum arabic
- *Polyphenols*: lignin, urushiol
- *Lipids*: terpenes, terpene resins

Proteins represent the most important structural component of living cells in both the animal and plant kingdoms. They are high-molecular-weight, colloidal natural substances composed of a large number of different alpha-amino acids as building blocks, with long chains of amino acids being linked via peptide bonds. Proteins – or *macropeptides* – are composed of more than a hundred amino acids. *Collagen*, which is an important protein for the manufacture of adhesives, is the main component of the supporting connective tissue of tendons, ligaments, cartilages and organic bone substance. When boiled with water to which acid has been added the collagen is converted into 'bone glue' that disperses as a colloid. Further purification of the bone glue leads to the production of colorless gelatin, which has strong swelling capacity in aqueous media and jellifies upon cooling. *Casein*, another protein used in the manufacture of natural adhesives, exists in milk as its calcium salt. In this type of protein, phosphoric acid is bound to a free OH-group like an ester. A third protein, albumin, is found in the serum of the blood.

Polysaccharides form the second group of natural base materials for the manufacture of adhesives. Besides cellulose, which is important in the manufacture of natural adhesives, starch in particular is used to manufacture the dextrin adhesives that are used for many different applications (see Section 8.5). *Gum arabic*, which is also known as 'acacia gum' or 'Sudan gum', is an interesting historical example of a polysaccharide used in adhesive manufacture. Gum Arabic, which is the alkaline salt of arabic acid obtained from the barks of *Acacia arabica*, is soluble in warm water but insoluble in alcohol. Gum arabic is applied as an aqueous solution and then allowed to dry; it can then be reactivated (i.e. reconferred with its bonding capacity) by the re-addition of water. In former times, gum arabic was used as a moisture-activated adhesive for stamps; in fact, its slightly sweet taste was easily recognized is still appreciated by philatelists today. Although gum arabic has lost its commercial importance as an adhesive, it is still available as paper glue, with an odd trading pattern. One interesting use of gum arabic is as a binder in the manufacture of pharmaceutical tablets, with an annual world consumption of 50 000 tonnes.

During recent years, natural *phenols* such as lignin (obtained during the course of wood processing) have become less important in the manufacture of adhesives. Another important group is that of *lipids* – or rather *lipoids*, a sub-group of lipids – especially in the form of terpenes and terpene resins. In a broader sense, rosin belongs to this group, as does natural rubber, which is a polyprene. Lipids are the glycerin esters of even-numbered fatty acids. The so-called 'waxes' also belong to this group, and are still often used as additives in modern adhesives. Waxes are fatty acids esterified with higher monovalent alcohols. In a mixture with natural rubber, rosin was the first really suitable PSA for the manufacture of adhesive tapes. In a broader sense, shellac belongs to the group of terpenes; this is secreted by the Indian lacquer sign louse, and is essentially composed of aliphatic and aromatic hydrocarbon acids. Shellac can be dissolved in alcohol, or used as hot-melt adhesive with melting temperatures ranging from 65 to 85 °C. In former times, shellac was used for fastening ruby bearings to watch movement plates.

Among the combinations of the four groups of raw materials listed above is *celluloid*, a mixture of nitrile-cellulose and camphor (a terpene), and a material with which we are still familiar today. Celluloid is composed of subsequently nitrated natural cellulose plastified with camphor, and today is no longer used as adhesive but rather as a precious raw material for combs. As celluloid derivatives are pleasant to touch, they are also often used to make the cases of high-quality fountain pens.

Marine glue was used for the sealing of ships made from wood (caulking), and for the bonding of different materials. The glue was a mixture of rubber, asphalt and kerosene, and sometimes contained shellac; a typical formulation comprised (by weight) 100 parts of rubber which was dissolved (for two weeks) in 120 parts of kerosene. Molten bitumen (20 parts) was then incorporated into this solution. A variation was to dissolve 50 g of fragmented rubber in 200 g carbon disulfide, and 130 g of organic shellac in 300 ml of high-proof alcohol, and to mix both solutions to obtain a glue. It was essential to store this marine glue in the absence of air, and to apply it hot. The glue adhered extremely well to wood, stone, glass and metal.

Sealing wax is another traditional formulation of a natural hot-melt adhesive. In former times, this was composed of rosin, shellac, turpentine and cinnabar, but today it is a mixture of shellac, cinnabar, chalk, gypsum and turpentine. It may be astonishing to find sealing wax listed here as a natural adhesive, but as a very good hot-melt adhesive with very pleasant flow characteristics it can be used for the bonding of glass, making possible the formation of vacuum-sealed joints (up to 10^{-6} Torr).

The adhesives that can be produced from natural raw materials include:
- solvent-containing contact adhesives, such as shellac and alcohol
- dispersion adhesives, from *Ficus benjamina*
- water-activated solid adhesives, such as rubber/dextrin, gum arabic
- hot-melt adhesives, such as rosin, amber, shellac
- aqueous structural adhesives, such as casein, blood albumin, glutine, but without chemical crosslinking (structural viscosity)
- aqueous structural adhesives, such as glutine/formalin, with crosslinking
- pressure-sensitive adhesives (PSAs), such as rubber-rosin
- detachable adhesives (barnacles)

At this point it becomes clear that virtually all types of adhesive discussed in Chapter 4 can be produced (if desired) from pure natural raw materials. For example, solvent-containing contact adhesives do not exist naturally but can be produced by dissolving shellac in alcohol; antique restorers use this type of adhesive to reassemble broken turtleshell. It is easily possible at home to obtain a natural dispersion adhesive by cutting a bough from the popular indoor plant, *Ficus benjamina*, and then removing from the scissors (with quite some effort) the rubber dispersion that has leaked from the bough, after any water has evaporated. Besides gum arabic (see above), rubber dextrin mixtures are water-reactivated solid adhesives that are still used today as fastening adhesives for envelopes (see Section 8.5). As mentioned above, natural hot-melt adhesives for the assembly of watches are composed of shellac; likewise, rosin can be used as a hot-melt adhesive. For many years, the optical industry used *Canada balsam* as a binder for lenses, because it is crystal-clear and does not impair the refractive properties of the interfaces of various glasses. Today, *pitch* or *bitumen* are still used in some sectors of the building industry to join roofing felts. *Waxes* that were once used to adhere gold to wooden statues in Ancient Egypt have now lost their importance as adhesives, but are still used as additives. *Amber*, when used as an adhesive is a curiosity; it is still mined on a large scale and is noted for its very high melting point of 380 °C.

The details of *detachable adhesives* are discussed in Section 8.15; an example is the adhesive that barnacles use to attach themselves to a ship's hull, the high strength of which makes them very unpopular with sailors [42–45].

5.9.3
Modern Adhesives Based on Natural Raw Materials

The examples discussed above may give the impression that the use of these natural adhesives is limited to specialized niche technologies. In the following paragraph however it will be shown, quite on the contrary, that natural adhesives are also suited to large-scale industrial applications, and that there are not always reasons to replace them with synthetic products. This applies to *glutine glues* that are processed hot, and to *casein and dextrin glues* which are processed in the cold state. It should be noted that the term 'glue' is used deliberately here as, in the typical fields of application (i.e. paper and wood processing), this term is still employed to the present day (see Sections 8.4 and 8.5).

5.9.3.1 Glutine Glue

As mentioned above, the base material of glutine glue is collagen, a high-molecular-weight protein that is insoluble in the unprocessed condition. Depending on its origin, collagen has different mechanical properties: skin collagen is highly elastic, while bone collagen is very hard. In order to obtain water-soluble glutine, calcium hydroxide is added to collagen, or water vapor is allowed to react, thereby breaking the crosslinks between the collagen chains and partly reducing the length of the chains. The solution is then mildly concentrated by evaporation in a vacuum at temperatures below 100 °C. The solidified glue broth is then cut into small plates, cubes or other small shapes, and these are dried to obtain a residual moisture content of 12–15%.

Shortly before application, the dry glue material (which may contain antibacterial agents) is remixed with water at a temperature of approximately 60 °C to generate a 35–50% solution with a viscosity of 500–2000 Pa·s. If the glue is slightly overheated (up to 100 °C) for a short period, its viscosity decreases and application is facilitated. The solution, which is usually applied at 60 °C, abruptly solidifies when it reaches 40 °C, after which the solid must undergo a final drying process. The addition of paraformaldehyde can generate a hot-setting system; here, crosslinking takes place and the final product has an increased resistance to water.

Owing to the rapid change in viscosity at 40 °C and low application temperature, it may be difficult to replace glutine glues in very fast-bonding processes such as bookbinding.

Compared to other natural adhesives, glutine glues processed in the hot condition are noted for their high plasticity in the cured state. This is not the least of reasons why they are used for back bondings in bookbinding, and for wood composite materials which must be reshaped after bonding (see also Sections 8.4 and 8.5).

5.9.3.2 Casein Glues

Casein glues are still used today, in the form of so-called 'cold glues', in some sectors of the wood-processing industry. As described above, casein is a phosphoprotein which is insoluble in the purified condition and, in the presence of water, merely swells. Only if alkaline substances such as caustic soda solution are added is casein sufficiently hydrated to be soluble in water.

A typical formulation of cold-setting casein glue is obtained when 100 g of casein is allowed to swell in 250 g of water, and 11 g NaOH is added. If 20 g of calcium hydroxide is then added the dry glue will have higher resistance to water than simple glutine glues. This mixture can be used for a period of 6 h. In this case, 'resistance to water' has a different meaning from that used in modern structural adhesives. Although wood bonded with casein glues cannot be used outdoors (i.e. under weathering conditions), the water-resistance can be increased to some extent by using the following formulation: 100 g casein is allowed to swell in 250 g water whilst at the same time, a calcium hydroxide solution (20–30 g in 100 g water) is produced. Both solutions are then mixed, and 70 g water-glass ($Na_2O \cdot n\ SiO_2$) is added, resulting in a significantly improved resistance to moisture. Unfortunately, this formulation does not have the same resistance to water as do current phenolic resin adhesives in wood joints, and therefore casein glues will not likely play a minor role in the future.

5.9.3.3 Dextrin Glues

Dextrin glues are made from starch swollen that has been (but not dissolved) in water. Starch is commercially available in different qualities. At a temperature of 50–70 °C, mixtures of starch and water, with a starch content of 10–20%, will 'gelatinize' with a 100-fold increase in volume and a drastic increase in viscosity. Hydration with calcium hydroxide, for example, reduces the so-called 'gelatinization temperature'. These mixtures have a distinct structural viscosity (i.e. with increasing deformation rate), their viscosity significantly falls, and they become very thixotropic. Dextrin glues can therefore be manufactured with very low concentrations in starch and with a low initial viscosity that provides for rapid processing. When starch is heated to 110 °C and small amounts of nitric acid are added, water-soluble dextrins (and saccharides) are created; however, they have a less thickening effect and may therefore be dissolved at higher concentrations (up to 60%). Owing to their unusual viscosity properties, starch or dextrin adhesives are still used (in part besides casein glues) for the production of corrugated cardboard, and in the labeling sector. It was mentioned above that dextrin–rubber mixtures or dextrin–PVA mixtures generate moisture-activated adhesive systems for envelopes.

5.9.3.4 Natural Material-Based Pressure-Sensitive Adhesives

It was noted above that, during the second half of the nineteenth century, mixtures of terpene resins (e.g. rosin) and natural rubber were introduced as PSAs, and used initially for the production of adhesive plasters and simple adhesive tapes (see Section 5.1). The mixtures are produced in suitable organic solvents which evaporate after application of the adhesive, and this results in a self-adhesive mass characterized by a non-homogeneous dissolution of terpene resin in natural rubber. The terpene-rich phase forms low-rubber domains that are responsible for tack (the property that enables the adhesive to form a bond immediately on contact with another surface). Although the remaining rubber is an elastomer that may transmit forces, it is still sufficiently flowable to fill the surface roughness to a certain extent, even after the application of a second adherent. The molecular fundamentals underlying the phenomenon of tack is not yet been fully elucidated.

In the medical sector, these mixtures of rubber and terpene are still used today, whereas in other industrial sectors natural rubber has been replaced by synthetic rubber copolymerized with styrene; this results, upon solidification, in the formation of loadable domains.

Self-adhesive gummings that are used to fasten envelopes are based on natural rubber lattices that are used to coat both parts of the envelope closure. Bonding takes place by compressing both coatings, resulting in a diffusion process (see Section 8.5).

5.9.4
Future Trends

In recent years it has become apparent that natural material-based adhesives still have industrial importance. However, *gene technology* may make it possible to further develop these adhesive systems by skillfully influencing natural materials in such a way that advantage can be taken of adhesive systems already existing in nature. For example, barnacles are able to attach themselves to the surfaces of ships with a great durability that is not affected by water; yet, to change location, they simply debond themselves (probably by an enzymatic reaction). The chemistry of the barnacle adhesive system is only partly known, and obviously extremely complex.

It should also be noted that popular ecological aspects – that is, concerns about raw materials and recyclability – must not be overestimated as far as adhesives are concerned, since in quantitative terms, none of the products contain significant amounts of adhesives. As an example, around 100 kg of structural adhesive may be found in a large aircraft, while 7–15 kg is present in a bonded car body; hence, recovery is usually not profitable. It is important, however, to prevent those adhesive residues which remain on originally bonded components and are then recycled, from causing a significant impairment of the characteristics of any subsequent products. As mentioned above, organic natural materials are not suited to recycling where the aim is to recover the original material, because their durability is such that, for subsequent processing, the product characteristics will doubtless deteriorate in an unacceptable manner. On the 'plus' side, however, those products which are split off during the recycling of natural materials, as well as their residues, would generally not be expected to be ecologically harmful. Hence, in the field of natural, material-based adhesives, interesting specialties may be developed in future for which large-scale industrial applications will most likely be confined to the areas described above.

5.10
Chemistry of Primers and Adhesion Promoters

The effect of primers on manufacturing processes and adhesion, as well as their basic chemistry, was described in Section 4.4. It has also been mentioned that primers may contain adhesion promoters because it does not always make sense – even if it is possible – to directly add specific adhesion promoters to the adhesive. It is not easy to make a clear distinction between primers and adhesion promoters, since adhesion promoters contained in solutions used for the bonding of glass are sometimes also called primers. Yet, by confining us to adhesive technology and the primers and adhesion promoters used in this field, they may be fairly well defined:

- *Primers* are thin, polymer-bound layers which are located on the adherent to be bonded and contain components that may be similar or dissimilar to the adhesive.

- *Adhesion promoters* are mostly low-molecular-weight organic substances with special reactive groups that may form chemical bonds with the adherent surface on the one side, and with the adhesive on the other side.

These will be illustrated later, with examples.

5.10.1
Primers

The polymer chemistry of primers used in adhesive bonding technology is basically the same as the polymer chemistry of adhesives. A primer may simply be an adhesive solution that improves wetting and protects the surface in the dry or cured condition. Strictly speaking, the primer transforms the adherent surface into a polymer surface that, with regard to adhesive bonding technology, is more convenient than inorganic surfaces, particularly with regard to long-term durability, provided that the binding polymer contains at least polar groups or residual reactivities for good adherence of the adhesive. A standard example of this type of primer is a primer solution that has been used in aircraft manufacture for many years, containing resols, polyvinyl formal and ethanol and ethylene dichloride, for example, as solvents. In the dried condition, there is no chemical difference between this primer and the adhesive, except for the mixture of the components that is more homogeneous in the case of the primer (see Section 5.4); in this way the primer can be more brittle.

However, as the primer layer has only a minor thickness this is not critical. It is worthy of mention that, in the cured state, this phenolic resin primer also represents a surface with excellent adhesiveness for epoxy resin adhesives. Fokker advanced this concept for many years in the area of longitudinal joints in the fuselage of aircraft, where components primed with phenolic resin and bonded with cold-setting, two-part adhesives were riveted together to extend the service life; the adhesive was then post-cured at 60 °C using heating pads.

As phenolic resin adhesives were difficult to process and rather brittle, they were replaced – at least in aircraft manufacture – by epoxy resin systems with a lower curing temperature and higher plasticity (see Section 8.2.1). Unfortunately, however, these adhesives did not meet the high expectations of aircraft manufacturers, neither alone nor in combination with primers of a similar type. Therefore, anticorrosive primers were developed which had other formulations than the adhesives, and which are still in use today. These primers are composed of monuron-accelerated epoxy resins, resols once again, sometimes special hardening agents for both, and strontium chromate, for example, as functional component. Resols were added because experience had taught that resol-type phenolic resins, without other aids, provided good adhesion on aluminum surfaces. At present, this is explained with their capacity to form chelate complexes with the aluminum surfaces, and their weak acidity in the polymer condition (see Chapter 3). Strontium-chromate accelerates the curing process and improves the resistance to corrosion whenever the bonded joint comes into contact with water, for example at the edges or in the case of damage. It is dissolved and, put simply, stabilizes the aluminum oxides against electrolytic alterations which might lead to bond-line corrosion (see Section 7.5).

As mentioned in Section 4.4, dissimilar-type primers, based on epoxy resins, for example, are also used with polyurethane adhesives to improve durability. Today, they contain chromate-free, anticorrosive pigments.

Primers may contain components which, if added directly to the adhesive, would interfere with its characteristics. This opens up a wide array of possibilities to optimize adhesion properties, even in the case of surfaces that are difficult to bond, such as nonpolar plastics. The primers used for this type of material contain components that swell or etch the adherent, and this results in molecular mixing – and thus good adhesion – without any special oxidative pretreatment. These components, for example in the form of reactive solvents, may also be added to the adhesive, but this is difficult to formulate and would involve poor adherence of the adhesive on other substrates.

Today, the adhesive manufacturers keep the chemistry of the primers as trade secrets, just as they do the chemistry of the adhesives. While further details will not be discussed at this point, it has become clear that the primer system offers many opportunities for development, especially if it becomes possible to eliminate those solvents that are still needed at present, and to overcome the problems associated with chromates.

5.10.2
Adhesion Promoters

The existence of adhesion promoters has long been known. For example, as far back as the eighteenth century, Henry Duhamel du Monceau (see Chapter 2) explained that the adherence of glue to wood could be improved by rubbing the wood with garlic before the glue was applied. Whilst this was mere speculative experience, reactive organosilanes [42, 46] have become standard adhesive promoters on the basis of a detailed knowledge and description of their chemistry and mode of action. In the following, the relevant information will be confined to adhesive bonding technology.

Organosilanes are bifunctional silicon compounds wherein the Si atom is substituted three organic groups that can be transformed into OH-groups by hydrolysis, as well as an organic, mostly chain-like molecular fragment that, at the end facing away from the Si atom, carries a reactive group capable of forming bonds with adhesive molecules. The reaction scheme and the adhesive mechanism are illustrated in Figure 5.37.

As shown in Figure 5.37, the hydroxyl groups required for the addition reaction with the substrate are also capable of a condensation reaction with another silane molecule to yield a silicone. This reaction starts without any catalyst, and must be prevented before using these compounds as adhesion promoters. Therefore, the silane is blocked with the organic groups (–OR) shown at the top of the reaction scheme, and it becomes storage-stable. Water is added only shortly before application, and hydrolytic activation takes place. Catalysts may accelerate this process. The proneness of the promoter to polycondensation at application, or in the applied layer,

Organosilane

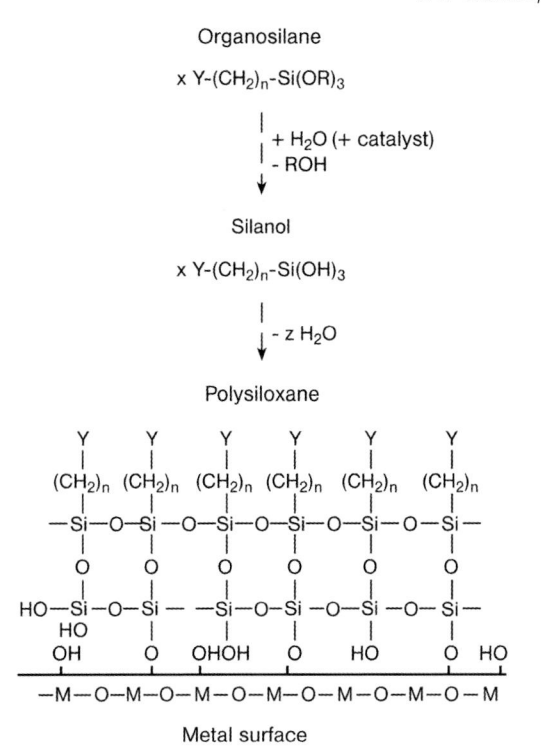

x Y-(CH$_2$)$_n$-Si(OR)$_3$

$\downarrow$ + H$_2$O (+ catalyst)
 − ROH

Silanol

x Y-(CH$_2$)$_n$-Si(OH)$_3$

$\downarrow$ − z H$_2$O

Polysiloxane

Figure 5.37 Hydrolysis and condensation of an alkoxysilane at the surface of an inorganic substrate.

cannot be prevented completely, so that a silicone-like intermediate layer is formed within the assembly that may impair its load capacity if too large a quantity is applied.

In silane adhesion promoters, the spacer groups between the reactive groups can be of a different nature. Indeed, their chemical structure and length have been shown to be primordial for the reactivity and functionality of the silane.

These adhesion promoters are mainly used for the bonding of glass as they provide high durability. However, when applied to the surfaces of other inorganic substances, they often also enhance the long-term durability of the bond, particularly on metal surfaces where hydroxyl groups provide for chemical reactions [47]. However, silane adhesion promoters are not capable of protecting the bond against bond-line corrosion, which is a potential risk factor in the case of metal bondings (see Section 7.5).

There has, therefore, been good cause for giving some thought to other types of adhesion promoter, without foregoing the principle of specific bifunctionality. An important approach here is to replace the silanol configuration by molecular groups capable of forming chelate compounds with metal surfaces, or rather with their oxides and hydroxides. These chelate compounds may have a high resistance to water; an example is shown in Figure 5.38.

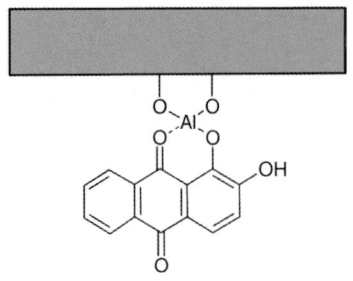

Figure 5.38 Chelate bond between alizarine (dihydroxyanthraquinone) and aluminum oxide.

As early as the 1970s, it was shown that simple chelate-complexing agents such as alizarin (dihydroxyanthraquinone) (Figure 5.38) or morin, once adsorbed onto aluminum surfaces from alcoholic dilutions, promote adhesion in phenolic resin adhesive bonds [48]. At that time, it was supposed that the adhesive, with its reactive methylol groups, was added to the free OH-groups of the chelate complexing agents. It was also shown that the aluminum oxides have an increased resistance to hydration when chelate complexing agents are used.

Later, it became possible to add spacers and, for example in the case of epoxy resins, to add compatible epoxy or amine groups to those substances [49, 50], generating adhesion promoters (Figure 5.39, left) that considerably improved the age resistance of aluminum bonded with these adhesives.

As noted in Chapter 3 (where polyhydroxy-benzoic acid was referred to as a standard example), chelate-complexing agents may also be added to other metals, for

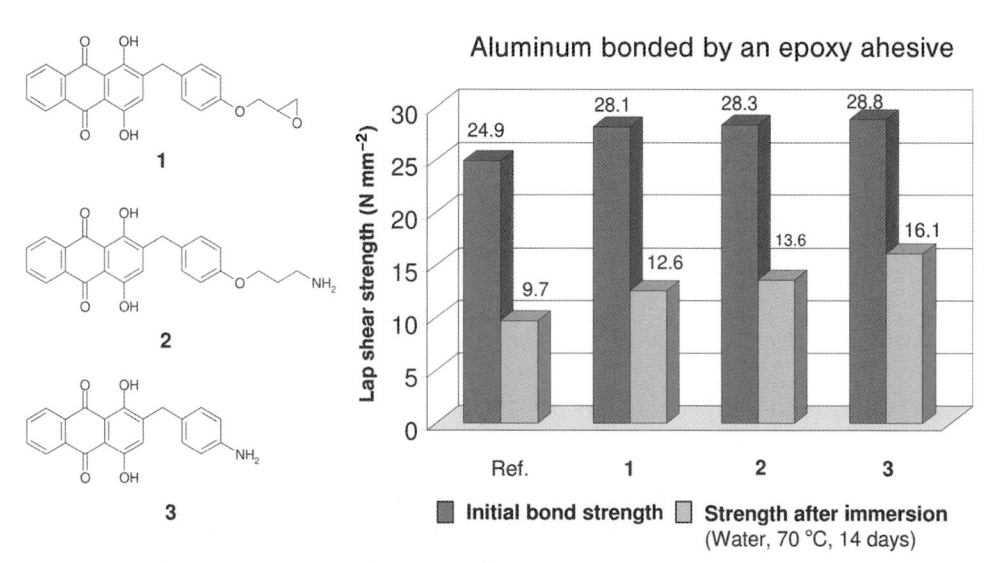

Figure 5.39 Derivatives of quinizarin (dihydroxyanthraquinone) and their efficiency in aluminum bondings.

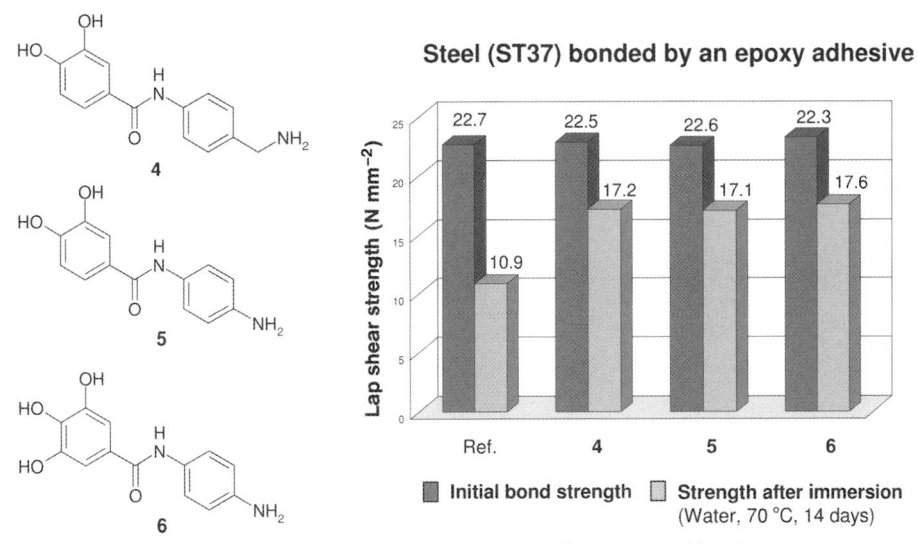

Figure 5.40 Derivatives of polyhydroxy-benzoic acid and their efficiency in steel bondings [50].

example iron alloys, with high resistance to water. Also, when bonding steel surfaces with amine-curing epoxy resin adhesives, adhesion is significantly enhanced if spacers and groups that are reactive with regard to epoxy resin adhesives are added to the chelate-complexing agents of the adhesion promoter (Figure 5.40).

Chelate-complexing agents are a good example of the immense development potential inherent to adhesion promoters. Adhesion promoters represent a matter of special importance because they can replace costly surface preparation methods.

5.11
Fillers

Fillers are fine solid materials that are added to adhesives and are, at best, adhesively bound to the polymer. Generally, they have an inorganic nature, although organic material (e.g. rubber) is used in special cases. The importance of fillers is also often underestimated, because they are believed to be used only to reduce the content of the costly polymer constituent of the adhesive.

Fillers offer more, however. If their chemical nature is similar to that of the adherents, then both the thermal conductivity and the coefficient of thermal expansion of the adhesive can be adapted to those of the adherents. Fillers in the form of flakes provide for electrical conductivity (see Section 8.7.2). The viscosity and thixotropy of uncured adhesives can also be significantly increased by using special fillers (e.g. Aerosil/silica), at only a very low volumetric content.

Finally, it must be noted that fillers can influence the kinetics of adhesive curing mechanisms and the structure of the polymer matrix if they have active surfaces. This is one of the reasons why fillers are also used in adhesives to improve resistance to heat. The surface characteristics of nanoparticles seem to be primordial for their effect when they are used as fillers. Nanoparticles have a very low volume, but large surfaces, and consequently the addition of minor amounts of nanoparticles has a major effect on the polymer matrix; alternatively, it may induce the formation of domains with different polymer structures within the adhesive layer. An inhomogeneous adhesive layer is an advantage with regards to fracture mechanics, because the inhomogeneities can act as 'crack stoppers'. Clearly, nanoparticles will undoubtedly play an important role in the future, particularly with regard to impact resistance, which is of utmost importance in the car manufacturing industry.

It has become clear that fillers play a very important role in the formulation of adhesives.

6
Design, Production and Quality Assurance of Adhesive Bonded Joints

When designing adhesively bonded joints, one should be aware that many of the mechanical properties of the polymer in the bond-line are different from those of metallic and other inorganic materials. Indeed, its strength is relatively low, its deformation properties are nonlinear and time-dependent (viscoelastic–plastic behavior), and irreversible material damages may occur when exceeding a deformation of about 1%.

6.1
Design and Dimensioning

6.1.1
Interaction of Polymer Behavior and State of Stress

For the design of structural adhesive bonded joints, it is necessary to take the time-, deformation- and temperature-dependent behavior of the adhesive polymeric materials into account.

Modern finite element analysis methods and tools allow a prediction of the stress distribution and deformations to be expected in an adhesively bonded joint, provided that the relevant material parameters are available. Only then, is it possible to exploit the full potential of adhesive bonded joints for lightweight construction and for the realization of economically optimized joints. As an example, Figure 6.1 shows the typical dynamic mechanical analysis (DMA) plot of a cold-setting, two-part epoxy

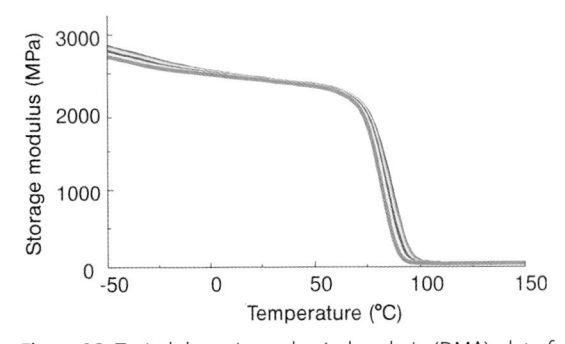

Figure 6.1 Typical dynamic mechanical analysis (DMA) plot of a two-part epoxy adhesive.

Adhesive Bonding: Materials, Applications and Technology
Walter Brockmann, Paul Ludwig Geiß, Jürgen Klingen, and Bernhard Schröder
Copyright © 2009 WILEY-VCH Verlag GmbH & Co. KGaA, Weinheim
ISBN: 978-3-527-31898-8

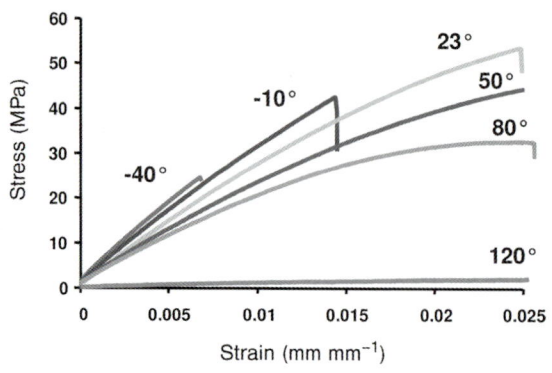

Figure 6.2 Temperature-dependent stress–strain behavior.

resin adhesive, illustrating the decrease of the storage modulus as a material characteristic of the elastic behavior in a temperature range between −50 °C and 150 °C, which is relevant in most applications.

Tensile tests performed on polymer specimens at different temperatures provide an insight into the stress–strain behavior within this temperature range (Figure 6.2). Several single curves add to a three-dimensional surface indicating the stress and strain to be expected as a function of temperature (Figures. 6.2 and 6.3). The further materials data needed are the Poisson's ratio and the coefficient of thermal expansion, which again both depend on temperature. The course of the storage and loss moduli is illustrated in Figure 6.4. If a bonded joint is intended for long-term exposure in a humid environment, the polymer will usually absorb water, thus modifying its strength and deformation properties which, in turn, may also be influenced by the surface condition of the adherents (see Chapter 3). A precise localized prediction of the stresses in a nonequilibrium state of water uptake is difficult in this case. It is recommended, therefore, to determine all of the above-described material characteristics on dry adhesive specimens and again after conditioning in water until saturation to obtain an indication of the properties range, which needs to be considered.

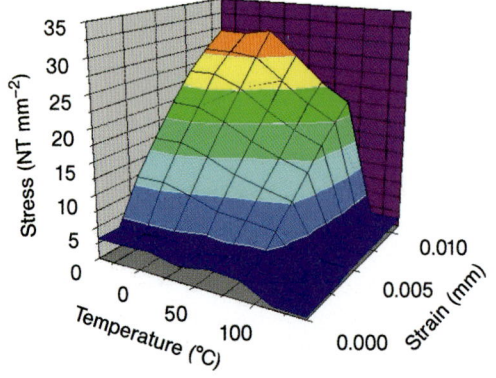

Figure 6.3 Stress–strain behavior at different temperatures, visualized as a three-dimensional surface.

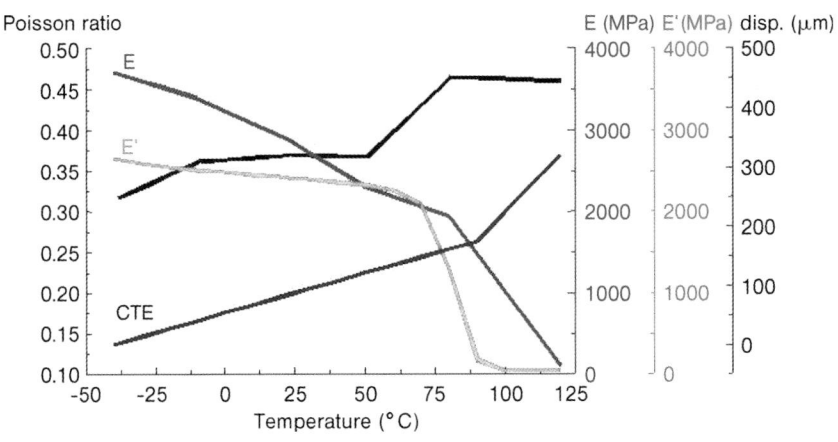

Figure 6.4 Temperature-dependence of the coefficient of thermal expansion (CTE), Poisson's ratio and modulus of elasticity.

In order to include the failure behavior in the simulation and to apply a multiaxial failure model, more sophisticated design tools are required. Commonly, the von Mises yield criterion is used for this purpose but, in the case of polymers, this often does not adequately describe the maximum admissible strain in the case of a multiaxial state of stress. The polymer behavior is better described by the Drucker–Prager model, which describes the multiaxial yield stresses by the surface of a cone.

To ensure that the maximum admissible strains and failure mechanisms obtained from simulation properly describe the individual case of application, calibration tests must be performed on standard samples, and the results verified on the components in order to assess whether the data obtained from the finite element analysis, which are based on simplified material parameters and strength hypotheses, are reliable and valid.

6.1.2
Design of Adhesive Bonded Joints

The design considerations for optimizing a bonded joint should mainly include:

- avoiding peel forces
- generating compressive stress rather than tensile stress
- avoiding stress concentrations in the bond-line [1]

The stress concentrations present at the ends of the overlapped joints are due to the difference in axial strain occurring in the adherents as a result of the stress transfer in the adhesive joint (Figure 6.5).

An equal stress distribution in adhesive joints can be improved by several approaches. Some alternatives for the creation of lap shear joints are illustrated in Figure 6.6; other possibilities to improve the stress distribution in lap shear joints include the use of adhesives with different moduli of elasticity (Figure 6.7) as described by Schliekelmann [2].

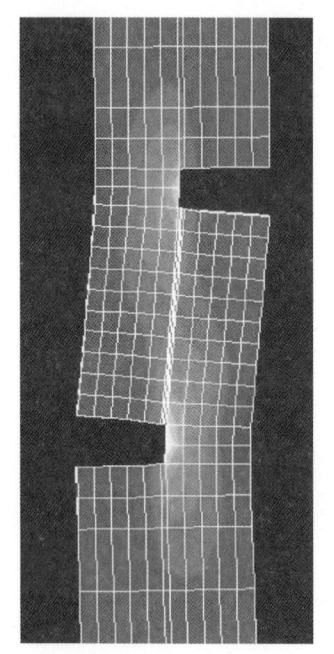

Figure 6.5 Deformation and stress in a loaded single overlapped joint (FE simulation).

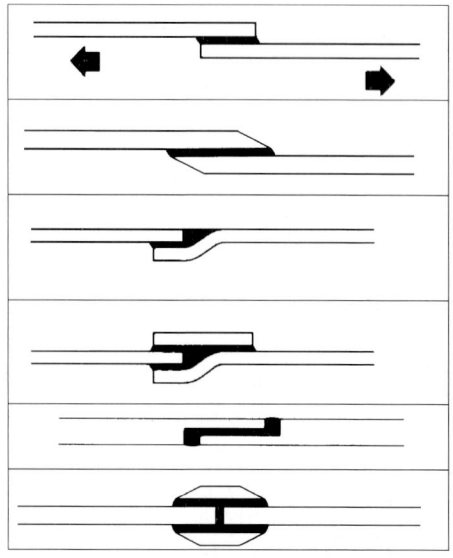

Figure 6.6 Reducing stress concentrations in overlapped joints. The stress distribution is seen to improve from the top to the bottom of the figure [1].

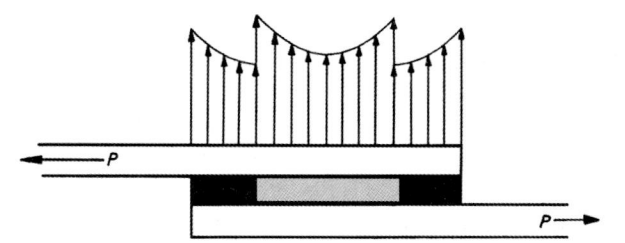

Figure 6.7 Combination of adhesives with different moduli to reduce stress concentrations in a single lap shear joint.

Some simple design rules to avoid peel forces at the ends of bonded joints are shown schematically in Figure 6.8

In many cases tensile stresses can be changed into compressive stress by using a different assembly of the parts to be bonded (Figure 6.9).

6.1.3
Dimensioning of Adhesive Bonded Joints

Many factors influence the load capacity of adhesive bonded joints [3–7]. In a first approach, simplified design rules may be used to estimate the average stress in basic joint layouts [3] (Figure 6.10).

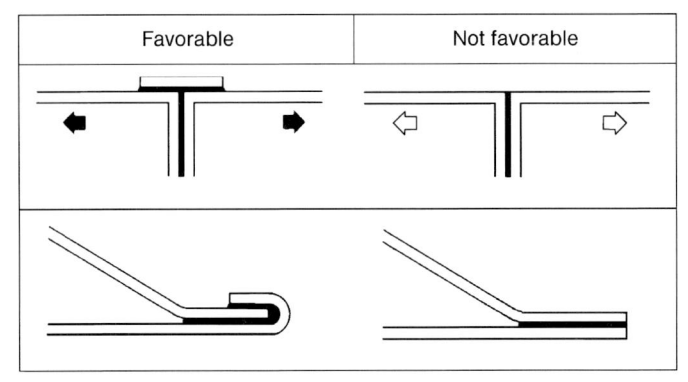

Figure 6.8 Avoiding peeling in bonded joints [1].

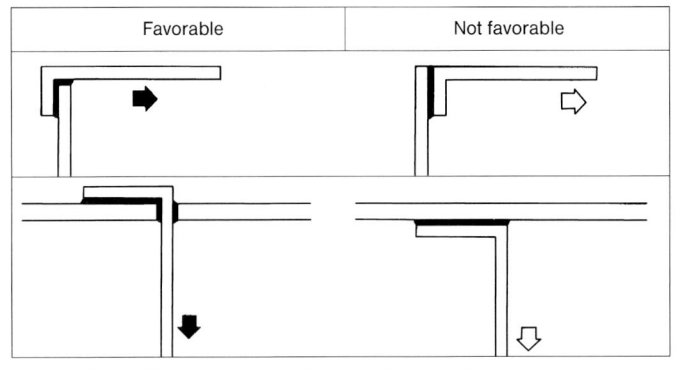

Figure 6.9 Avoiding tensile stress in bonded joints [1].

When creating adhesively bonded joints with different adherent materials, thermal internal stresses will occur that superimpose on the external mechanical loads and thus reduce the load capacity of the joint. Elastic adhesives are superior in this type of application as they reduce the level of thermal stress, especially when applied in thick bond-lines (Figure 6.11).

In a single lap shear joint, in addition to shear stress, bending moments occur due to the eccentricity of the load introduced. This bending moment causes additional normal stresses within the bond-line (Figure 6.12).

Matting [8] calculated the distribution of the shear stresses along the center axis in single lap shear joints, with and without taking into account the bending moments. The stress distribution, for three different overlap lengths, is illustrated graphically in Figure 6.13. Note that the distribution becomes increasingly uniform – that is, the stress concentration factor $C = \tau_{a,max}/\tau_{am}$ decreases with decreasing overlap length, L_{ov}.

When using brittle adhesives the calculations and representations of the shear stress distribution in single overlapped shear joints can be based on the simplified assumption that the adhesives had a quasi-elastic deformation behavior. Present-day adhesives, however, are usually plasticized and can therefore reduce stress concentration by plastic deformation (Figure 6.14).

Type of joint	Load	Bonding area	Average shear stress	Load capacity
A) single lap joint 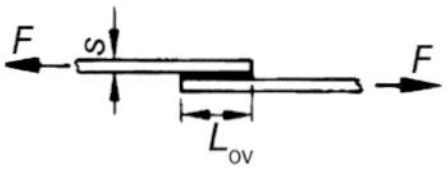				
B) folded lap joint 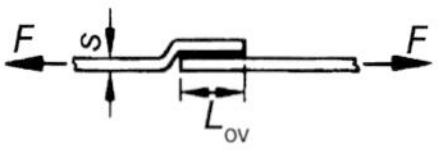				
C) sloped lap joint	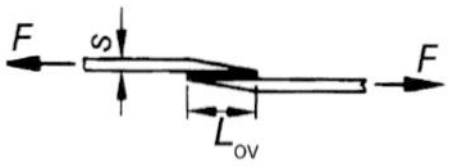 tension / compression	$A = L_{ov} \cdot b$	$\tau_s = \dfrac{F}{L_{ov} \cdot b}$	$F = \tau_{kl} \cdot A$
D) single sided strap 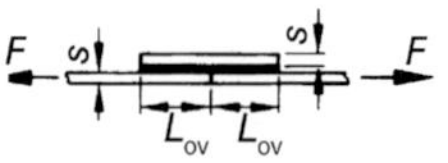				
E) double sided strap	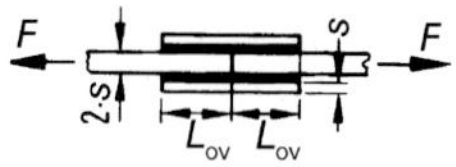 tension / compression	$A = 2 \cdot L_{ov} \cdot b$	$\tau_s = \dfrac{F}{2 \cdot L_{ov} \cdot b}$	
F) slanted joint	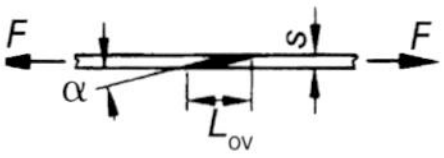 tension / compression	$A = \dfrac{L_{ov} \cdot b}{\cos \alpha}$ (for $L_u/s = 10$)	$\tau_s = \dfrac{F \cdot \cos \alpha}{L_{ov} \cdot b}$	$F = \tau_{kl} \cdot A$
G) tube joint	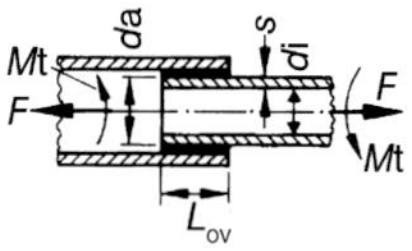 tension / compression	$A = L_{ov} \cdot \pi \cdot d_a$	$\tau_s = \dfrac{F}{\pi \cdot L_{ov} \cdot d_a}$	$F = \tau_{kl} \cdot A$
	torsion		$\tau_s = \dfrac{2 \cdot M_t}{\pi \cdot L_{ov} \cdot d_a^2}$	$M_t = \tau_{kl} \dfrac{\pi \cdot L_{ov} \cdot d_a^2}{2}$
H) pin and collar joint	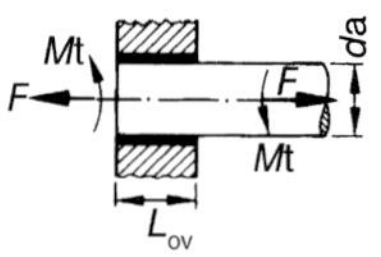 tension / compression	$A = L_{ov} \cdot \pi \cdot d_a$	$\tau_s = \dfrac{F}{\pi \cdot L_{ov} \cdot d_a}$	$F = \tau_{kl} \cdot A$
	torsion		$\tau_s = \dfrac{2 \cdot M_t}{\pi \cdot L_{ov} \cdot d_a^2}$	$M_t = \tau_{kl} \dfrac{\pi \cdot L_{ov} \cdot d_a^2}{2}$
I) cylinder flange with butt collar	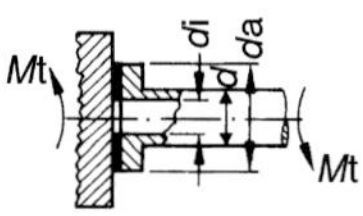 torsion	$A = \dfrac{\pi}{4}(d_a^2 - d_i^2)$	$\tau_s = \dfrac{16 \cdot M_t \cdot d_a}{\pi (d_a^4 - d_i^4)}$	$M_t = \tau_{kl} \dfrac{\pi(d_a^4 - d_i^4)}{16 \cdot d_a}$

Figure 6.10 Simplified calculation of average stresses in bonded joints.

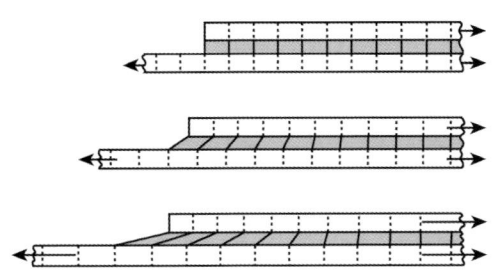

Figure 6.11 Deformations of adherents and adhesive in a single overlapped joint under increasing load (note: the illustration shows only the left half of an overlapped joint).

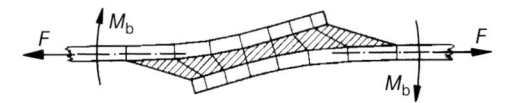

F = force
M_b = bending moment

Figure 6.12 Bending moments and deformations in a single lap shear joint with thin adherents.

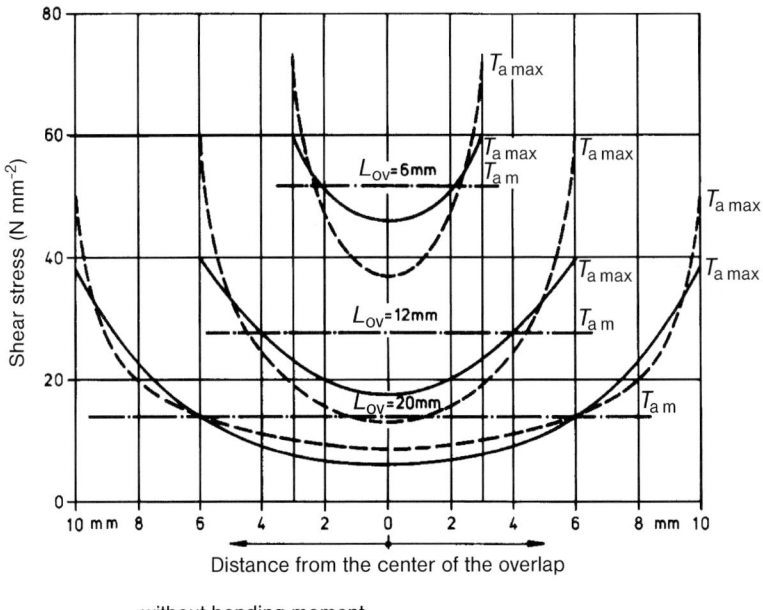

——— without bending moment
- - - - - with bending moment

Figure 6.13 Stress distribution, according to Matting [8], in a single lap shear joint with different overlap lengths.

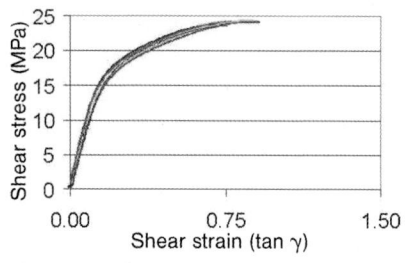

Figure 6.14 Shear stress–strain behavior of a plastified epoxy adhesive.

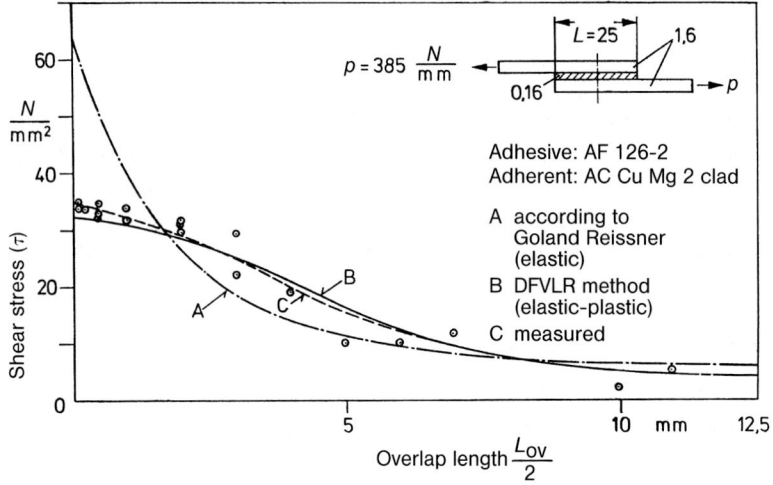

Figure 6.15 Calculated (A, B) and measured (C) stress distribution in a single lap shear joint (half overlap length).

Figure 6.15 compares the theoretical stress distribution calculated with elastic deformation properties of the adhesive (according to Goland–Reissner) to a stress distribution calculated with real (elastic–plastic) deformation of the adhesive, and to experimental data [9].

The experimental data show a more uniform stress characteristic than expected due to viscoelastic deformation occurring under the effect of repeated load, an increase in temperature, and by a slight softening under the influence of moisture.

6.2
Surface Preparation

The surface condition of the adherents is one factor which has the greatest impact on the quality and durability of adhesively bonded joints. Therefore, both in the literature and in adhesive manufacturers data sheets it is strongly recommended that the adherent surfaces are cleaned and carefully prepared prior to bonding. Only in few cases where the type and amount of contamination are predefined and the adhesive is designed to tolerate this specific degree of contamination, can bonding be carried out without further cleaning procedures. This applies especially to the bonding of steel sheets in body-in-white construction. Here, hot-setting adhesives with a special formulation are used that are capable of building up high-strength adhesion with long-term durability on oil-protected surfaces – 'against all the rules'. This requires the adhesive manufacturers and users, in cooperation with the steelmakers and oil producers, to take appropriate measures to ensure the compatibility of lubricants with the adhesive systems. In the majority of other fields of application a thorough surface preparation, including cleaning and degreasing, must be performed before adhesive bonding.

The purpose of cleaning is to remove dust, grease, oil, rust, scale and miscellaneous dirt from the surface of the adherents. A variety of cleansing agents such as organic solvents and water-miscible detergents is commonly available. Polar hydrocarbons (e.g. esters, ketones, alcohols) are the material of choice for degreasing, but in open facilities the safety and health of the workers, as well as environmental precautions, must be considered when handling these hazardous substances. Acetone is often used because it has a good balance of cleaning capability and comparatively low environmental concern, although fire-protection measures must be taken due to its liability to ignite. This applies especially when processing cannot be carried out in a hermetically closed circuit.

The following cleaning methods are used in industrial manufacture [10]:
- mechanical cleaning (e.g. sanding or grit blasting)
- solvent wipe
- immersion in an ultrasonic solvent bath
- vapor degreasing with solvents
- washing with aqueous solutions
- high-pressure water steam cleaning

The most common surface preparation methods are discussed briefly in the following sections.

6.2.1
Mechanical Surface Preparation Methods

6.2.1.1 Grit Blasting
Grit blasting is carried out by aiming abrasive materials (e.g. corundum) by means of a jet of air onto the surface to be treated. The degree of abrasive surface removal and deformation caused by the particle impact is determined by the form and hardness of the blast medium, the applied pressure, and the distance of the blast nozzle from the surface. When the blast medium impacts the surface, very high, localized energy concentrations are created. Thus, it is assumed that during a short duration, a plasma state is created in the impact area. When the plasma collapses, condensed residues may recontaminate the adherent surface. Hence,

the material properties of the blast medium may influence the adhesion properties of the blasted surface (see Sections 7.4 and 7.5). The two blast systems generally used are:

- dry blast processes, such as flywheel-blasting, grit-blasting and vacuum-blasting
- blast processes using fluids, such as water-jet blasting, hot-water and steam blasting

The *blast media* comprise metallic, organic and mineral materials. It is important that the blast medium (e.g. compressed air) is free from contaminants such as oil and silicone. In the case of plated substrate materials, the plating may be damaged and a mechanical deformation of sensitive materials may occur under the influence of compressive internal stresses at the superficial areas zone.

In the special blasting method known as SACO (SAndblasting and COating), corundum coated with reactive silicate is used as the blast medium. When the grit hits the surface and the plasma collapses, part of the silicate coating is assumed to transfer to the adherent. In a next step, a silane adhesion promoter is applied to the silicate-containing layer (see Section 5.10) to yield a surface with good adhesive properties. Precautions must be taken when using this process, however, as harmful silicate dust may be produced.

6.2.1.2 Grinding

As the adhesion properties of surfaces after grinding with abrasives are poorer than those of blasted surfaces, it is recommended that the adherents should be ground at right-angles to the loading direction, or crosswise. Before grinding, the surfaces must be carefully degreased since, if the contaminations are not removed they will be distributed uniformly over the entire surface and no improvement in bond quality will be achieved. Typically, belt sanders, angle grinders and random orbital grinders are used for this type of surface preparation method. Although the localized energy concentrations during grinding are lower than during blasting, it is still possible that the components of the binding material of the abrasive grains may be transferred to the adherent surface. However, such contaminants have been shown as nondetrimental in a variety of tests as they adhere firmly to the surface.

6.2.1.3 Brushing

Brushing also acts as a special type of abrasive surface preparation. For a given size of grit, the surface roughness obtained by brushing is less than that achieved with grinding, due mainly to the fact that, compared to the grinding wheel, the brushes are relatively elastic. It is again essential that the surfaces are degreased before brushing (as with grinding), and that the working direction again be at right angles. Since with brushing, the compressive pressure applied to the adherent surface is lower than with grinding or blasting, this surface preparation method is also suitable for sensitive parts, without any risk of distortion. Residual binding agents may be transferred to the adherent surface, as with grinding.

6.2.2
Chemical Surface Preparation Techniques

Chemical surface preparation can have two different effects on the surface of the adherent:

- Non-oxidizing acids (e.g. hydrochloric acid, dilute sulfuric acid) cause an electrochemical oxide/acid or metal/acid reaction which results in removal of the oxide layer or, if etching times are long, of the deeper metallic interfaces. In this method – called 'acid degreasing' – the surface becomes not only chemically clean but also rough in its submicroscopic structure. At the same time, chemically active sites are created on the surface, which are a prerequisite for the formation of adhesive bonds.

- Oxidizing acids (e.g. nitric acid, phosphoric acid, chromic acid), with or without the addition of oxidizing salts such as sodium or potassium dichromate, induce the formation of an additional firmly adhering conversion coating (e.g. phosphate or chromate oxide layer). This method is known as 'pickling' or 'phosphatizing'. The application of an electric current to the adherent can enhance the chemical reaction. Since the parts to be treated are generally used as the anode, this process is called 'anodizing'. On aluminum alloys, zinc and stainless steel, the process yields an oxide layer of up to several hundred microns thickness which, in the majority of cases, exhibits filigree nanomorphological structures with very good adhesion properties (see Chapter 3 and Section 7.5).

6.2.3
Physical Surface Preparation Techniques

Physical surface pretreatment methods are mainly used for polymer surfaces to increase their polar free surface energy.

6.2.3.1 Flame Treatment
During flame treatment, the polymer materials are gently touched by an open flame with a surplus of oxygen. This induces an oxidative effect that improves the adhesion properties, especially in the case of nonpolar plastics, depending on the type of adhesive used. Flame treatment is an environmentally safe, continuous method with short treatment periods.

6.2.3.2 Microplasma Treatment
In a microplasma burner, the plasma arc between the tungsten electrode and the workpiece is initiated by means of a pilot arc. The process gas between the electrode and the workpiece becomes ionized and the flow of gas directs the plasma to the surface to be treated. Microplasma pretreatment is easy to handle, but the adhesion strength is only slightly increased.

6.2.3.3 Atmospheric Plasma Treatment

The plasma generator consists of a ceramic tube surrounded by an electric conductor made from sealed metal powder. In the center, an electrically conductive rod with good thermal conductivity (inner electrode) acts as a counter-electrode. An ionizable gas (e.g. air, oxygen) is introduced into the ionization gap between the inner wall of the ceramic tube and the inner electrode, and a high-frequency, high-voltage field is applied to the electrodes. The gas that passes through the gap is ionized by the alternating electrical field and then escapes at the end of the nozzle.

Atmospheric plasma pretreatment provides significant improvement to the adhesion properties of many polymers, and is easy to handle. It has also been successfully used for the pretreatment of metal parts if only a small degree of superficial contamination is to be removed.

6.2.3.4 Low-Pressure Plasma Treatment

In low-pressure plasma treatment the adherents are placed in a process chamber that is then evacuated. After introducing a process gas, the plasma is ignited by means of a high-frequency voltage or microwave excitation. The color, lightness and distribution of the plasma arc depend on the pressure in the chamber and the type of process gas used.

The plasma modifies the surface of the adherents, considerably improving the adhesion properties. Low-pressure plasma pretreatment is environmentally safe and allows components with a complex shape to be prepared for bonding.

6.2.3.5 Corona Treatment

In corona treatment, a high-frequency voltage is applied between two electrodes or between an electrode and the substrate material. This causes the atmosphere between the electrodes to become ionized, and a corona discharge is ignited. When introducing an insulator between the electrodes, its surface is targeted with ions and radicals, causing a modification of the adhesion properties.

6.2.4
The Preconditions of Adhesion

To gain insight into the adhesive–surface interaction on a molecular level, and thus compare the effectiveness of different surface preparation procedures, contact angle measurements have frequently been discussed. Prediction of the adhesive properties of a surface based on its interfacial tension can be possible because adhesion will only take place if the adhesive wets the adherent surface as completely as possible (see Chapter 3).

The shape of a liquid drop (e.g. a liquid adhesive) placed on a solid surface depends on the interfacial surface–liquid interaction. In the case of complete wetting, the liquid spreads on the surface (contact angle $\theta = 0°$). If the wetting is incomplete, then a finite contact angle is established and a three-phase contact area appears. At this point, three phases are in contact: the solid; the liquid; and the vapor (Figure 6.16). Here, Young's equation relates the contact angle to the interfacial

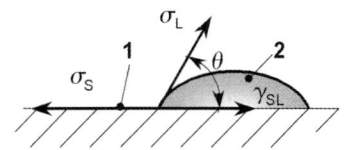

Figure 6.16 Surface and interfacial tensions.

tensions γ_{SL}, σ_L and σ_S:

$$\sigma_L \cdot \cos\theta = \sigma_S - \gamma_{SL} \tag{6.1}$$

If the interfacial tension of the bare solid surface σ_S is higher than that of the solid–liquid interface γ_{SL}, then the right-hand side of Young's equation is positive. Then, $\cos\theta$ must be positive, the contact angle is less than 90°, and the liquid partially wets the solid. If the solid–liquid interface is energetically less favorable than the bare solid surface, the contact angle will exceed 90°, allowing $\cos\theta$ to be negative. If the contact angle is greater than 90°, the liquid is said not to wet the solid.

The Dupré–Young equation relates the work of adhesion to the interfacial tensions:

$$W_{ad} = \sigma_S + \sigma_L - \gamma_{SL} \tag{6.2}$$

By introducing Young's equation into Dupré's equation, the Young–Dupré equation is obtained:

$$W_{ad} = \sigma_L \left(1 + \cos\theta\right)_L \tag{6.3}$$

Depending on the surface chemistry, the interfacial tension in the liquid–solid contact can be influenced by polar and nonpolar (dispersive) interaction. According to Owens, Wendt, Rabel and Kaelble, the interfacial tension γ_{SL} can be calculated by subtracting the polar and dispersive share of the interfacial tension in the following way:

$$\gamma_{SL} = \sigma_S + \sigma_L - 2\left(\sqrt{\sigma_S^D \cdot \sigma_L^D} + \sqrt{\sigma_S^P \cdot \sigma_L^P}\right) \tag{6.4}$$

After introducing into Young's Equation and linear regression according to $y = mx + b$:

$$\underbrace{\frac{(1+\cos\theta) \cdot \sigma_L}{2\sqrt{\sigma_L^D}}}_{y} = \underbrace{\sqrt{\sigma_S^P}}_{m} \underbrace{\sqrt{\frac{\sigma_L^P}{\sigma_L^D}}}_{x} + \underbrace{\sqrt{\sigma_S^D}}_{b} \tag{6.5}$$

As a general rule and precondition for adhesive bonding, the interfacial tension of the bare solid σ_S, which can be determined by contact angle measurements in the manner described above, must be larger than the surface tension of the liquid adhesive. Typical surface tension values for adhesives range from 35 to 50 mN m^{-1}, depending on the composition. When the surface tension of the surface to be bonded is too low (e.g. in the case of nonpolar polymer materials such as PTFE, PE or PP, or fatty contaminants), it is recommended to perform a suitable surface preparation that

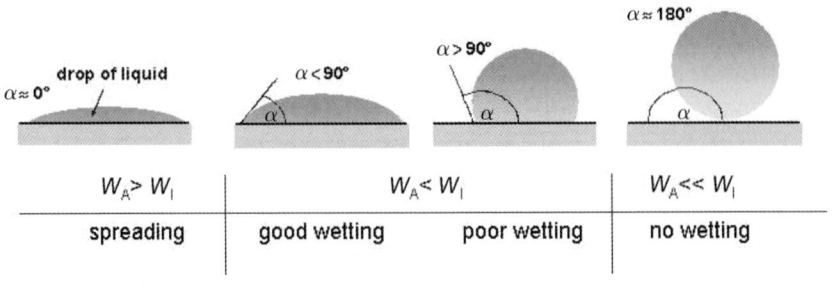

Figure 6.17 Different types of wetting.

increases the surface tension of the adherent surface, or thoroughly removes the contaminants. Figure 6.18 provides an overview of the current surface pretreatment techniques used in bonding technology. In order to obtain good adhesive properties, those surface preparation methods should be chosen which especially enhance the polar share of the substrates surface energy for it to be able to interact with polar groups in the adhesive and thus promote interfacial bonds with a high level of binding energy.

6.2.5
Protection of Prepared Surfaces

Following extensive surface pretreatment, adherents with high surface energies and good adhesion properties show a great attraction for any type of contaminant within the environment. Therefore, if bonding cannot be performed directly after pretreatment, the use of a protective primer should be considered to avoid uncontrolled recontamination or loss of wettability (see Section 5.10).

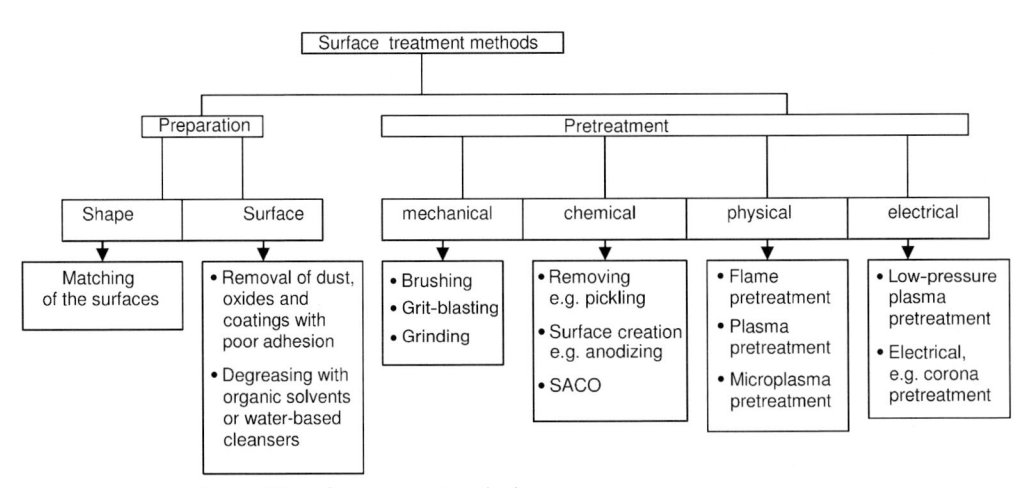

Figure 6.18 Surface treatment methods.

6.3
Application of the Adhesive

The integration of bonding into a process chain, or the replacement of a mechanical joining technique by bonding, is a decision that must be taken if the cost–benefit ratio is to be favorable. Indeed, improvements in product characteristics, as well as increases in manufacture process productivity and quality must outrank the investments effected in terms of production equipment, facilities and variable costs involved. Bonding must be compatible with existing process steps and facilities, as well as with manufacturing materials logistics and disposal concepts.

An example of adhesive process integration, which is employed not only in the automotive industry, is to use the drying equipment for thermal paint-curing processes also to cure adhesive-bonded joints with single-component, heat-curing adhesives.

A further potential benefit might also be to avoid labor-intensive wet-coating processes by using prefinished surfaces, thus taking advantage of the adhesive-bonding technologies and minimizing distortions of the finish in the area of joining.

The wide variety of conditions and parameters under which adhesives can be processed are presented in the following sections.

6.3.1
Processing of Adhesives

During the creation of an adhesive bonded joint, all types of adhesive must undergo transition from the liquid to the solid state, except for pressure-sensitive adhesives (PSAs), which comprise both liquid and solid properties and thus have a permanent tack and adhesive bonding ability. *Viscosity* is a prerequisite of the adhesive to bring it and the adherent into close proximity, and to allow adhesive forces to originate. The material properties that prevail after curing are responsible for the cohesive strength of the bonded joint.

Curing may take place according to a wide variety of chemical and physical mechanisms, all of which have been translated into adhesive systems with advantages and disadvantages for the manufacture process, depending on the materials to be joined and the structure of the paramount process chains (see Chapter 4).

Some combinations of application forms and processing/curing conditions of physically setting and reactive adhesive systems are summarized in Tables 6.1 and 6.2.

Specific examples will be discussed in the following sections in order to illustrate the possibilities and processing parameters available today.

6.3.1.1 Processing of Two-Part Adhesive Systems
The curing reaction of two-part adhesives takes place after mixing at room temperature, without any need to apply pressure. The adherents must be fixed to avoid movement of the adhesive joint before the cure is complete; this can be accomplished

Table 6.1 Processing of physically setting adhesive systems according to their application forms and setting mechanisms.

System	Temperature	Drying	Permanently tacky
Liquid/ viscous	Hot melts, e.g. based on polyurethane, polyamide block-copolymers or EVA, spraying of hot-melt PSAs	Solvent-containing adhesives, dispersions, water-soluble glues	Dispersion or solvent-containing PSAs
Film	Hot-melt films	–	PSA tapes
Granulate	Hot-melt powder	–	–

EVA = ethylene vinyl acetate; PSA = pressure-sensitive adhesive.

by mechanical fixtures and fasteners, or by supporting pressure-sensitive adhesive strips.

Depending on the chemical setting mechanisms of the adhesives, specific mixing qualities are required for proper processing and curing. A wide range of mixing ratios may be used for radical curing systems without any serious loss in strength, while polyaddition adhesives (e.g. epoxy resins) and polyaddition curing polyurethane adhesives require a precise mixing ratio to be met with a maximum allowance of a few percent. Otherwise, nonreacted monomers carry the risk of softening the polymeric network, leading to a poor mechanical performance of the bonded joint.

The quality of the adhesive mixture depends on the metering ratio and the homogeneity of mixing. The following methods are available for metering and mixing two-component adhesives:

- batch processing; metering by weighing and mixing with manual or power stirrers
- static mixers
- dynamic mixing heads

For *batch processing*, the components are dosed and mixed batchwise. During and after mixing, the pot life – that is, the period of time in which a multipart adhesive can be used after mixing the components – has already started to expire. If the whole mixture is not scheduled for immediate processing, freezing may quench the reaction almost completely. In the optical industry, this method is used in order to carry out extensive quality assurance measures on the adhesive mixture before releasing it for further manufacture.

A continuous processing of two-part adhesives is made possible by means of static mixers consisting of a cylindrical tube with mixing elements, applying the principles of fluid mechanics. The *Kenics helix* is the most popular device, and comprises a series of mixing elements with right-hand and left-hand twisted elements being arranged alternatively. This principle is applied both in micromixers and in the mixing heads of machine processing units (Figures 6.19 and 6.20, right-hand image).

Table 6.2 Processing of reactive adhesive systems according to their application forms and setting mechanisms.

System	2-P	1-P/temperature	1-P/radiation (UV)	1-P/moisture	1-P/anaerobic
Liquid/viscous	• Epoxy • Acrylate • Polyurethane • Polyester • 2-P silicone • Ambient temperature-setting phenolic and resorcin wood glues	• Hot-setting 1-P epoxy-dicyandiamide systems • Phenolic resin adhesives • Polyurethanes with blocked hardeners	• Acrylates • Cationic epoxy resins	• 1-P polyurethane adhesive sealants • RTV silicones • Silane-terminated polymers • Reactive hot melts • Cyanoacrylates	• Methacrylates
Film	• Reactive A/B adhesive tapes	• Epoxy films • Phenolic resin films • Adhesive tapes crosslinking under the effect of temperature	• UV-setting adhesive films		
Granulate	• Encapsulated epoxy resins and hardeners	• Solid epoxy granulate			• Encapsulated thread-locking adhesives

1-P = one-part adhesive system; 2-P = two-part adhesive system; RTV = one-component, room-temperature curing silicone (1C RT) silicones.

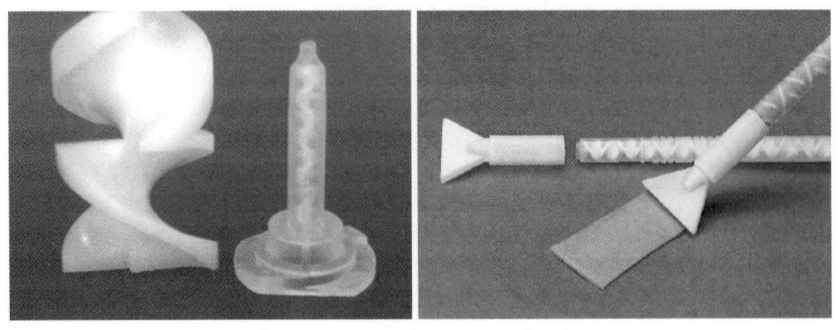

Figure 6.19 Different types of static mixer which use Kenics helices.

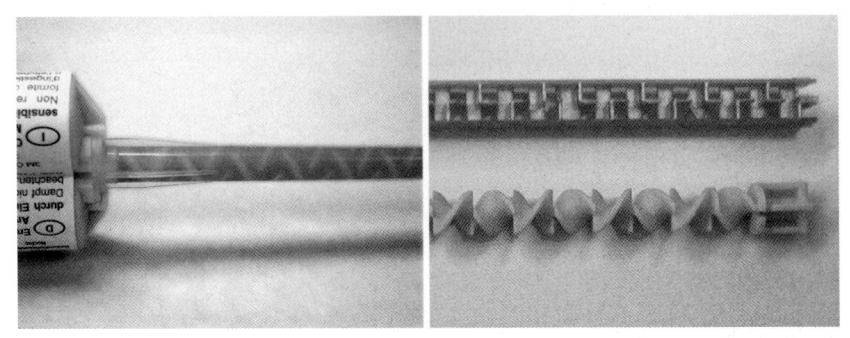

Figure 6.20 Left: The adhesive components mixing as they pass along the mixing elements. Right: Different mixing helices; upper: a Sulzer static mixer; lower: the Kenics helix.

Alternatively, static mixers with rectangular cross-section profiles according to Sulzer may be used (Figure 6.20, right-hand image). These provide for a compact length of the mixer.

The mixing quality obtained with this type of system depends on the number of mixing elements arranged in series, and on the viscosity properties of the adhesive. The driving energy force for the mixing process is the initial pressure of the adhesive components. Unless the manufacturer has already released the static mixers, it is of utmost importance to test the equipment settings by taking samples for controlling the metering and mixing quality. *Note*: A visual check of whether the material has been mixed homogeneously is usually not sufficient for the purpose of quality assurance.

Due to their simple layout and low unit costs, *disposable static mixers* are widely used. In any case, static mixers must be replaced whenever the pot life at the opening of the nozzle has been exceeded due to interruptions and delays in the processing chain, or whenever the viscosity of the adhesive material in the mixing tube has increased to such an extent that the dispensing pressure has exceeded limiting values.

More complex dynamic mixing heads (Figure 6.21) use a rotor in the mixing chamber of the application head to mix the adhesive components.

Figure 6.21 Metering and dynamic mixing of two-component adhesives. (Photo courtesy of Rampf, Dosiertechnik).

Whenever production comes to a standstill for a long period, the mixing chamber must be purged with an inert liquid to prevent the adhesive from curing and the rotor from blocking. The purging liquid is then removed using compressed air, and may be recycled after having removed the adhesive components in a closed circuit. Whenever there is a need to span short pauses, in order to save costs, the mixing chamber can also be purged by dispensing a small quantity of adhesive ('blind shots') or purging with just a single component of the adhesive system; this allows the mixing chamber to be refilled upon reuse, without the need to use a special purging agent. Importantly, any incorrectly mixed adhesive material must be disposed of according to the manufacturer's health and environmental safety recommendations [11].

6.3.1.2 Processing of One-Part Adhesive Systems

In particular, elastic polyurethane adhesives that cure under the effect of moisture are widely used in the automotive and building industries. The introduction of adhesives in automotive glazing (e.g. adhesive fitting of windshields) has provided a significant increase in the rigidity of the bodywork compared to the use of nonadhesive, rubber seals. Today, the bonding process in this field of application is highly automated, and includes robotic application of the primer, adhesive bead application and accurate positioning and joining of the windshield element (Figures 6.22 and 6.23).

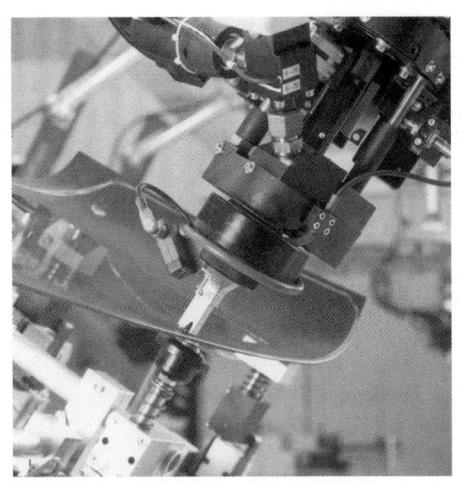

Figure 6.22 Adhesive application in the bonding of a car windshield. (Photo courtesy of KUKA Schweissanlagen GmbH).

For the bonding of smaller components in medical and electronic applications, one-part adhesives that are cured by UV radiation are widely used (Figure 6.24). The specific interaction between the photoinitiator used and the high-energy UV radiation allows a very rapid curing process, down to fractions of seconds of exposure time (see Section 8.14).

UV light sources for radiation curing range from light-emitting diodes (LEDs) to small mobile mercury pressure lamps with optical conductors for medical applications in dentistry, to large systems for use on assembly lines. The most popular light

Figure 6.23 Handling of car windshields by robots during adhesive application. (Photo courtesy of KUKA Schweissanlagen GmbH).

Figure 6.24 UV-curing of adhesive using a fiber optic conductor. (Photo courtesy of Ellsworth Adhesives).

sources are electrode-excited or microwave-excited mercury lamps, and microwave-excited excimer lamps. The following parameters influence the total dose and maximum intensity of UV radiation to be controlled:

- speed of the conveyor belt or throughput speed
- distance between the lamp and the conveyor belt
- initial performance and aging of the lamp

The UV dose available on the surface of the substrate consists of the direct lamp radiation and the reflected radiation. Commercial UV devices are available with different reflector types. Below a wavelength of 240 nm, short-wave UV-C radiation generates ozone that, on an industrial scale, may necessitate treatment of the exhaust air for reasons of health and environmental safety. For the in-line control of UV radiation intensity, UV sensors with a good long-term stability are available from different manufacturers.

6.3.1.3 Hot-Melt Adhesives

A wide variety of melting equipment is available for the processing of hot-melt adhesives, ranging from simple hand-held hot-melt stick dispensers to pneumatic cartridge dispensers or hot-melt drum- or tank-melters. If the molten adhesive is sensitive to oxidation, a shielding inert gas blanket can help to prevent decomposition. After processing, curing one-part hot melts react with moisture from the ambient air, resulting in a higher thermal resistance after curing. Sealed metal cartridges in storing and shipping are used to prevent these adhesives from curing until application. Hot-melt adhesive can be applied to large areas by means of spraying or spin-coating, such as in the packaging industry (Figure 6.25).

6.3.1.4 Film Adhesives

The processing of film adhesives is particularly simple, accurate and clean if die-cut shapes are used (Figure 6.26). Structural film adhesives are available as nontacky films made from modified hot-setting epoxy resins or phenolic resins, respectively.

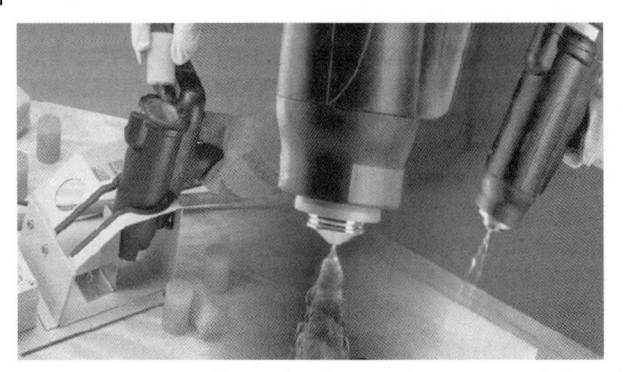

Figure 6.25 Spraying of hot-melt adhesive. (Photo courtesy of Ellsworth Adhesives).

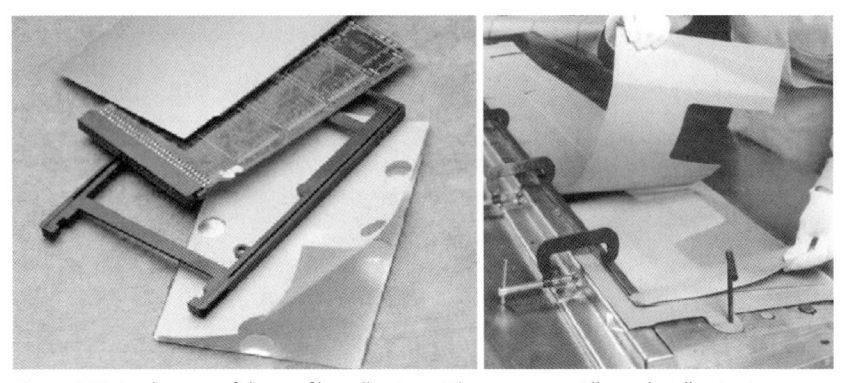

Figure 6.26 Application of die-cut film adhesives. (Photo courtesy Ellsworth Adhesives).

In order to prevent bonding faults, the adherents must be pressed thoroughly against each other. Compressed air cushions or vacuum foil bags are used for large-size adherents.

PSAs are physically setting systems and generally do not require any curing process. Common PSAs are permanently tacky, and are mostly processed in the form of converted adhesive tapes. Automatic application devices are available for industrial manufacture purposes.

Die-cut films (Figure 6.27) are particularly well suited for large surface bonding and lamination purposes. It is possible to create either permanent or detachable joints according to the properties of the different types of adhesive.

6.3.1.5 Hybrid Bonding

Hybrid bonding makes it possible to take advantage of the mechanical and processing characteristics of different joining techniques, to improve the quality of the joint, and to join specific materials and material combinations.

When high-strength adhesives are used, the adhesive bond usually takes the greatest share of the mechanical load under service conditions. The mechanical fastener or spot-weld then takes the role of an initial fixation aid, and may serve to

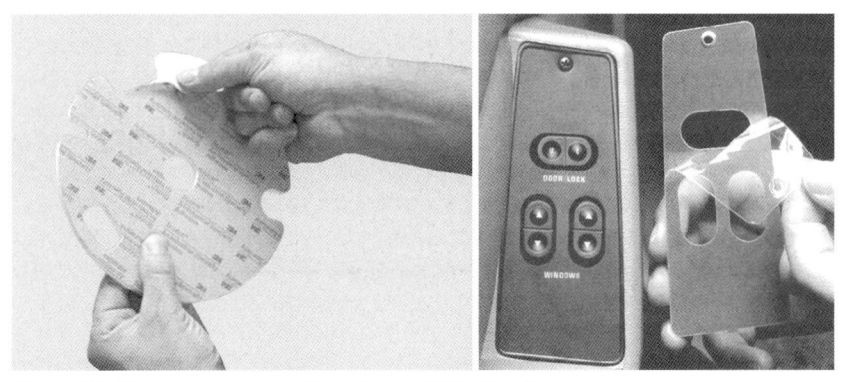

Figure 6.27 Die-cut pressure-sensitive adhesive (PSA) films for surface bonding. (Photo courtesy of Ellsworth Adhesives).

avoid peel stresses near the edge of a joint or long-term static stresses. Examples of successful hybrid bonding include the combination of bonding with spot-welding, riveting or clinching. Bonding combined with riveting is used primarily in the aircraft industry, where the joints are expected to have a high level of fatigue resistance and dynamic mechanical stress.

6.3.2
Quality Assurance

Besides the cost factor, one decisive parameter when assessing a manufacturing process is the process safety in terms of controllability and reproducibility. Hence, in order to select the correct technique it is crucial to know which parameters may influence the quality of the bonding.

Failure Mode and Effects Analysis (FMEA) is an approach used to ensure process safety by systematically examining potential failures, assessing the likelihood of occurrence of failures, and evaluating the effect of the failures on product quality. Furthermore, the likelihood of detection and the necessary actions to be taken when a failure has occurred are identified. Process FMEA assesses every step of a manufacturing process, from materials supply to shipment, while Design FMEA assesses the design and layout of a specific article. As a result of these analyses, the Risk Priority Number (RPN) for a specific fault is then created, representing a direct criterion of priority.

7
Adhesives and Adhesive Joints: Test Methods and Properties

Among different industrial branches and trades, a variety of test methods for the evaluation of adhesively bonded joints have been developed and established, including the ISO (International Organization of Standardization), EN (European Committee for Standardization) and/or ASTM (American Society for Testing and Materials) standards.

While some test procedures require sophisticated equipment to assess the intrinsic material properties of the adhesive material, other test protocols may represent simple means to judge the characteristic properties of an application specific joint geometry on a comparative basis.

The basic test procedures to assess the bulk properties of cured adhesives include:

- stress–strain analysis of the bulk specimen
- determination of the glass transition temperature (T_g) by dynamic mechanical thermoanalysis (DMTA) or dynamic scanning calorimetry (DSC)
- determination of the coefficient of thermal expansion (CTE)
- temperature resistance by thermogravimetry (TG)

In general, the spectra of mechanical tests for adhesive joints can be subdivided by the type of mechanical stress to be created in the adhesive joint:

- tensile stress (e.g. butt joints)
- shear stress (e.g. single lap shear specimen)
- cleavage (e.g. wedge test)
- peel (e.g. 180° peel test)

In terms of timescale and test duration, the following test conditions may apply:

- short-time tests
- long-term static tests (e.g. creep experiments under dead load)
- impact tests (e.g. impact wedge test to evaluate the crash performance)
- cyclic loading (e.g. fatigue tests)

Mechanical tests are often combined with preceding aging procedures to evaluate the durability of adhesive joints. Reasonable aging cycles depend on the expected

Adhesive Bonding: Materials, Applications and Technology
Walter Brockmann, Paul Ludwig Geiß, Jürgen Klingen, and Bernhard Schröder
Copyright © 2009 WILEY-VCH Verlag GmbH & Co. KGaA, Weinheim
ISBN: 978-3-527-31898-8

in-service conditions of an adhesive joint. Accelerated aging tests typically include one or a combination of the following conditions:

- moisture
- immersion in water or other liquids
- elevated temperature
- cyclic freeze–thaw aging
- corrosive environment
- UV-exposure (for transparent adherents)

Further standards on adhesives relate to aspects of health and safety, storage and handling. Examples include:

- determination of words and phrases relating to the product life
- open time and working life
- extrudability
- sagging
- flammability
- volatile organic compound (VOC) emission

The current ISO CEN and ASTM databases contain some hundreds of standards related to adhesives which can be searched and accessed online at: http://www.iso.org, http://www.astm.org or http://www.cen.eu.

The European Guideline CEN/TR 14548 *Adhesives – Guide to test methods and other standards for the general requirements, characterization and safety of structural adhesives*, provides a detailed and comprehensive summary of up-to-date test methods for adhesives in structural applications. For nonstructural adhesives, industry-specific standards may apply, such as those of the Pressure-Sensitive Tape Council (PSTC), the Association des Fabricants Européens de Rubans Auto-Adhésifs (AFERA) or the Fédération Internationale des Fabricants et Transformateurs d'Adhésives et Thermocollants sur Papier et autres Supports (FINAT). Therefore, the scope of this chapter is limited to the basic principles and concepts of adhesive testing, and does not aim to provide a full coverage of currently active standards.

7.1
Adhesive Bulk Properties

7.1.1
Stress–Strain Analysis of Bulk Specimens

Tension tests are performed on dog-bone-shaped test specimens (Figure 7.1). This geometry provides for a uniform stress and strain distribution in the central gauge section, where these values are at a maximum. Thus, the specimen extends and fails under well-defined conditions. Some flexible adhesives will sustain large extensions up to several hundred percent prior to failure, and this will place limitations on the sample geometry in order to obtain a failure within the travel range of the tensile

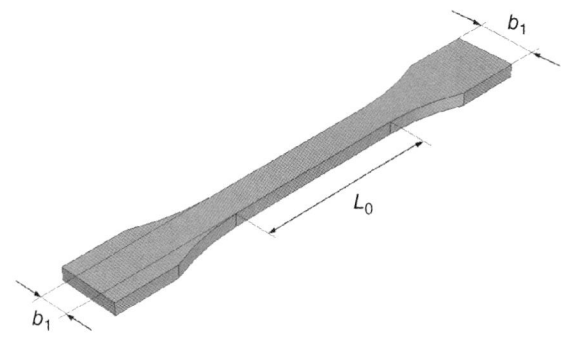

Figure 7.1 The basic geometry of a 'dog-bone'-shaped bulk tensile specimen.

testing machine. According to the standard ISO 3167 a multipurpose test specimen with a minimum length of 150 mm and a gauge width-to-gripped width ratio of 1 : 2 is recommended for structural adhesives. Smaller test specimens may be more appropriate for flexible adhesives, while the plastics tensile standard allows for a half-sized test-piece. The rubber test standards specify a sample with a smaller gauge width-to-gripped width ratio. Although, there is no general preference as to which type of specimen is to be tested, it should be noted that test results obtained from different geometries might not be identical.

7.1.2
Determination of T_g by Dynamic Mechanical Analysis or Differential Scanning Calorimetry

Thermal analysis provides a rapid method for measuring transitions due to morphological or chemical changes (or both) in a polymer as it is heated/cooled through a specified temperature range.

For dynamic mechanical thermoanalysis (DMTA; e.g. ASTM D4065), a specimen of known geometry is subjected to mechanical oscillations at fixed or natural resonant frequencies.

The elastic or loss moduli (or both) of the specimen are measured while varying the time, temperature of the specimen or frequency, or both, of the oscillation. Plots of the elastic (E', G') or loss moduli (E", G"), or both, are indicative of viscoelastic characteristics of the tested material. Rapid changes in viscoelastic properties at particular temperatures, times or frequency are normally referred to as 'transition regions'. A plot of the elastic modulus and loss modulus of material versus temperature provides a graphical representation of elasticity and damping as a function of temperature or frequency.

By using differential scanning calorimetry (DSC; e.g. ASTM D3418), changes in specific heat capacity, heat flow (in conjunction with Test Method D 3417) and temperature values can be determined for the polymer transitions mentioned above. The test methods consist of heating or cooling the material at a controlled rate in a specified purge gas at a controlled flow rate, and continuously monitoring the difference in temperature or the difference in heat input between a reference material and a test material due to energy changes in the material.

DSC can be used to assist in identifying specific polymers, polymer alloys and certain polymer additives, which exhibit thermal transitions. Chemical reactions which cause or affect certain transitions that can be measured using this technique include oxidation, crystallization, the curing of thermosetting resins, or thermal decompositions marked by absorption or release of energy by the specimen resulting in a corresponding endothermic or exothermic peak or baseline shift in the heating or cooling curve.

7.1.3
Coefficient of Thermal Expansion (CTE)

The CTE describes the temperature-dependent change in length and volume of a material. Various types of instrument can be used to measure the CTE of solid adhesives, including push-rod dilatometers or dilatometers with optical measuring devices. Within the glass transition range the CTE changes rapidly and causes additional internal stress in adhesively bonded joints, depending on the material and rigidity of the adherents and the elastic properties of the adhesive. Therefore, *coefficients of linear thermal expansion* are used, for example, for design purposes and to determine if failure by thermal stress may occur when a solid body composed of two different materials is subjected to various temperatures. The test method ASTM E831 (Standard Test Method for Linear Thermal Expansion of Solid Materials by Thermomechanical Analysis) uses a smaller specimen and substantially different apparatus than test methods E 228 and D 696.

7.1.4
Temperature Resistance by Thermogravimetry

The temperature resistance of polymer materials is often limited by thermal decomposition at elevated temperatures. However, by using a micro thermobalance the loss in weight of an adhesive sample can be monitored as a function of temperature. ASTM E2550, for example, covers a test method for the assessment of material thermal stability by determining the temperature at which the materials begin to decompose or react, and the extent of their mass change, using thermogravimetry.

7.2
Test Categories by Type of Mechanical Stress

7.2.1
Tensile Stress (e.g. Butt Joints)

A tensile test of an adhesive bond puts the adhesive in a state of stress in which the bondline is perpendicular to the direction of the force applied to the specimen. A typical sample for evaluating the tensile properties of an adhesive is shown in Figure 7.2.

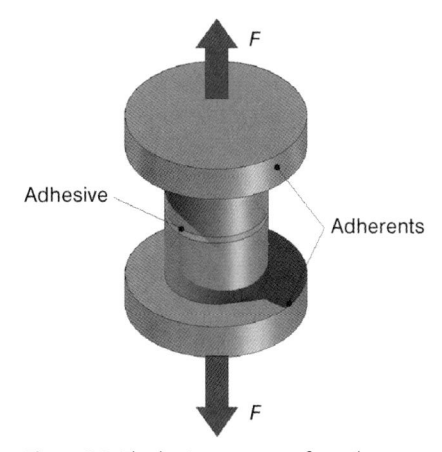

Figure 7.2 The basic geometry of tensile tests.

This specimen is similar to that used in ASTM Test Method D2095. Metal rods are generated according to the exacting specifications described in ASTM Standard Practice D2094. The ends of the metal rods must be machined so that the surfaces contain no burrs that might cross the adhesive gap, and that a parallel adjustment of the adherents is guaranteed during assembly. Any angular misalignment of the surfaces could force the test to be one of cleavage rather than tension. The metal rods are butted up to an adhesive which joins them – hence the term 'butt tensile' test. When the adhesive has cured or set, the specimen is loaded in tension up to failure of the joint, as depicted in Figure 7.2. Both, the tensile stress at break and the mode of failure are reported. The major drawback of this type of test is that, although an average stress at failure is reported, the actual stress distribution in the bond-line is not uniform throughout the adhesive, with higher stresses occurring at the center of the butt tensile specimen compared to the edges. In fact, the average stress at failure is more likely due to these edge effects rather than to the actual tensile strength of the adhesive. Butt tensile tests are rarely used to evaluate adhesives because the mode of loading is typically not used in adhesively bonded structures. Therefore, tension tests provide reasonably accurate information with regard to the tensile strength of adhesives. Tensile strength data may be suitable for specification acceptance, service evaluation, manufacturing control, and research and development purposes. However, tension tests are not considered significant for applications differing from the test in rate, direction and type of loading [1].

A test method similar to D2095 is described in ASTM D897 in which studs are glued together and pulled apart in tensile mode. This 'stud pull-off test', which is an interesting variation on the tensile test, involves a metal stud being adhered to a surface by means of a structural adhesive. The stud is machined in such a way that an actuator can be attached to its back, allowing a tensile force to be applied between the stud and the sample surface. This test can be used to determine a practical adhesion value to a particular surface. The method can also be modified so that the stud is bonded to a coated surface. Then, if adhesion between the adhesive used to bond the

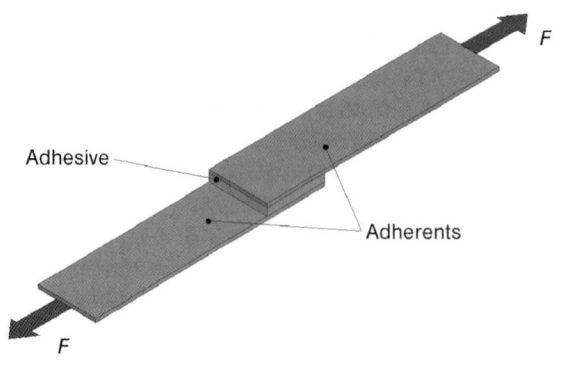

Figure 7.3 The basic geometry of a lap shear specimen.

stud and the coating is good, a 'figure of merit' can be obtained for the adhesion of the coating to the surface.

7.2.2
Shear Stress (e.g. Single Lap Shear Specimen)

The true shear strength of an adhesive can be determined only if normal stresses are entirely absent. These conditions can be approached under special conditions, but not in single-lap specimens made with thin adherents, which are normally used in manufacturing and in most standard test specimens (Figure 7.3).

Due to the specimen geometry, the application of force is eccentric and the adherents rotate as a result of bending moment, as shown in Figure 7.4.

This rotation introduces *adhesive peel stress* in the vicinity of the joint edges. Due to the eccentric sample geometry, the stress distribution within a lap joint is not uniform in the load direction. Furthermore, the shear stress nonuniformity in the through-thickness direction becomes appreciable if the bond-line thickness exceeds certain values. In most cases the tensile stress in the adhesive layer controls joint failure and, as a consequence, the strength value obtained by single lap shear specimens is unrelated to – and an unreliable measure of – the true shear strength

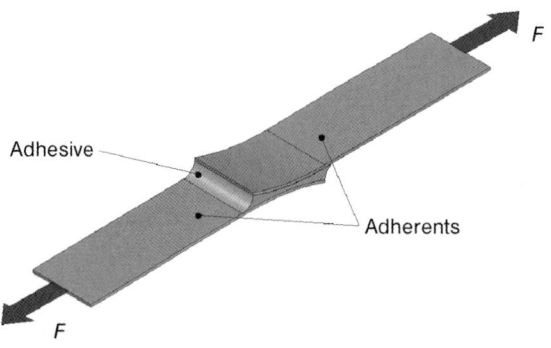

Figure 7.4 Rotation and bending in a lap shear specimen due to eccentricity.

of an adhesive. Other commonly used ASTM standard shear tests on adhesives include:

- ASTM E 229: Standard Method of Test for Shear Strength and Shear Modulus of Structural Adhesive.
- ASTM D1002: Standard Method of Test for Strength Properties of Adhesives in Shear by Tension Loadings (Metal-to-Metal).
- ASTM D 3165: Strength Properties of Adhesives in Shear by Tension Loading of Single-Lap-Joint Laminated Assemblies.
- ASTM D 5656: Thick-Adherend Metal Lap-Shear Joints for Determination of Stress-Strain Behaviour of Adhesives in Shear by Tension Loading.

The so-called 'thick adherent specimen' (TAS) as described in ISO 11003-2 or ASTM D3983 specifies a test method for determining the shear behavior of an adhesive in a single lap joint-bonded assembly when subjected to a tensile force. The test is performed on specimens consisting of thick, rigid adherents with a short length of overlap, in order to obtain the most uniform distribution of shear stresses possible and to minimize other stress states which may initiate failure (Figure 7.5).

The design of the test specimen is based upon the theoretical analysis by Goland and Reissner relating stress concentrations (i.e. nonuniformity) in single-lap joints to the geometry of the joint and the mechanical properties of the materials involved. The controlling factor in the Goland and Reissner equations is a composite of essentially two ratios which can be manipulated to improve the stress uniformity in the joint, and thereby control the accuracy of measurement. Stress uniformity is improved by: (i) increasing the adherent tensile modulus in relation to the shear modulus of the adhesive; and (ii) increasing the adherent and adhesive thickness while minimizing the overlap length. The test method is capable of providing shear modulus and shear strength values for adhesives with accuracy suitable for the use by design engineers in predicting the characteristics of building assemblies bonded with nonrigid adhesives. In general, the thick adherent lap-shear test has become a useful tool in research during studies of both short- and long-term load–deformation

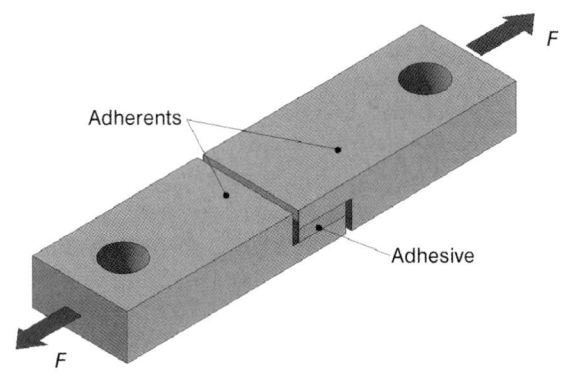

Figure 7.5 The basic geometry of a thick adherent shear specimen.

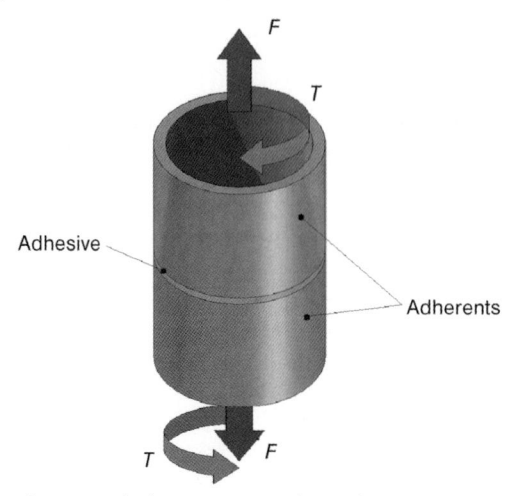

Figure 7.6 The basic geometry of a 'napkin ring' specimen.

properties of adhesives. However, it should be noted that pure shear strength cannot be monitored by this test method, because even in this joint geometry some tensile and compression stresses and stress concentrations are present at the edges. Hence, the estimation of shear strength using this test method will be conservative. If the measurement of pure shear strength is demanded, then Test Method E 229 should be applied [2].

Both, ISO 110003-1 and ASTM E 229 relate to the so-called 'napkin ring test', where specimens are loaded in torsion (T). An additional tensile stress (F) can be superimposed to assess the mechanical adhesive properties under multiaxial load conditions (Figure 7.6).

Torsional shear forces are applied to the adhesive through a circular specimen which produces a peripherally uniform stress distribution. The maximum stress in the adhesive at failure represents the shear strength of the adhesive. By measuring the adhesive strain as a function of load, a stress–strain curve can be established. Production cleaning and bonding processes should be used when applicable. The basic material properties obtained from this test method can be used in the control of the quality of adhesives, in the theoretical equations for designing bonded joints, and in the evaluation of new adhesives.

7.2.3
Cleavage (e.g. Wedge Test)

Wedge and cleavage tests are common test methods applied to assess the quality of surface treatments and provide an estimate of long-term durability. Variations of the test employ thicker and longer adherents to enable longer crack growth and, therefore, more accurate fracture energy calculation and avoid plastic deformation of the adherents.

7.2.4
Fracture Resistance

Determination of the fracture resistance of structural adhesive joints is typically tested under an applied Mode I opening load, using double cantilever beam (DCB) and tapered double cantilever beam (TDCB) specimens.

The resistance to *crack initiation* can be determined from both a nonadhesive insert placed in the adhesive layer and from a mode I precrack. The resistance to *crack propagation* is then determined from the mode I precrack. The adhesive fracture energy G_{IC} (also termed the critical strain energy release rate) for applied Mode I loading can be calculated and a resistance (R)-curve – that is, a plot of the value of the adhesive fracture energy G_{IC} versus crack length – can be deduced [3]. The TDCB shown in Figure 7.7 is designed so that, over a large range of values of crack length, the rate of change of compliance with crack length is constant and thus independent of the value of crack length. This is useful as it means that: (i) relatively tough adhesives may be tested without plastic deformation of the arms occurring; (ii) the substrates may possess a relatively low yield stress, but again no plastic deformation of the arms may be incurred during the test; and (iii) the measurement of G_{IC} is independent of the crack length, a. To develop a linear change of compliance with crack length, the height of the specimen is varied by contouring the substrate beam so that the quantity:

$$\frac{3a^2}{h^3} + \frac{1}{h} = m$$

is a constant, where m is the specimen geometry factor.

7.2.5
Wedge Test

The so-called 'wedge test' according to D 3762, for example, simulates – in qualitative manner – the forces and effects on an adhesive bonded joint at the metal–adhesive/

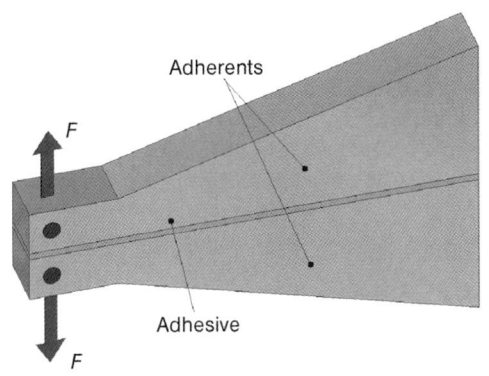

Figure 7.7 The basic geometry of a tapered double cantilever beam (TDCB) specimen.

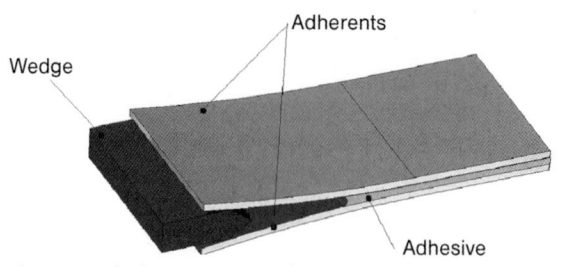

Figure 7.8 The basic geometry of a wedge test specimen.

primer interface. This test has proven to be highly reliable for determining and predicting the environmental durability of adherent surface preparations. The results obtained may be well correlated to the service performance of the adhesive in a way that is much more reliable than conventional lap shear or peel tests.

At the beginning of the test a wedge is driven into the bond-line of a flat-bonded specimen, thereby creating a tensile stress in the region of the resulting crack tip (Figure 7.8). The stressed specimen is exposed to an aqueous environment, usually at an elevated temperature, or to an appropriate environment relative to the use of the bonded structure in the later application. The resulting crack growth with time and failure modes are evaluated later. Variations in the adherent surface quality are easily observable when the specimens are forcibly, if necessary, opened at the end of the test.

7.2.6
Peel Test

Peel tests are quite similar to cleavage tests, except that at least one of the adherents is prepared from a flexible material which could be plastically deformed during the measurement. A typical peel test is illustrated in Figure 7.9; this is known as the 'T-peel test', and is described in ASTM D1876. Two adherents of equal thickness are

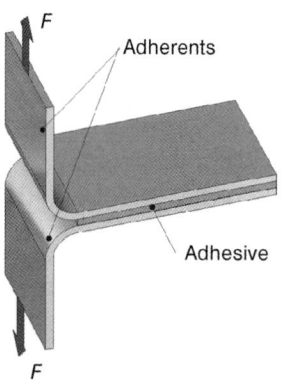

Figure 7.9 A T-peel specimen.

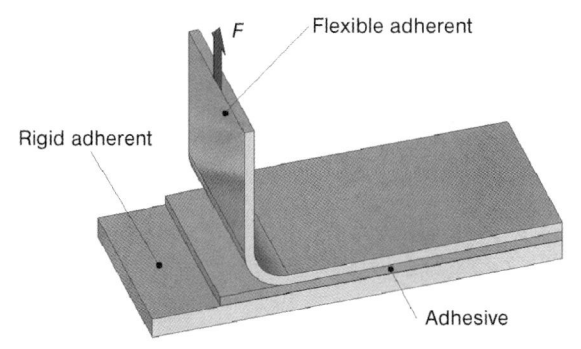

Figure 7.10 The 90° peel test.

bonded with an adhesive, after which the ends or 'tabs' of the specimens are gripped by jaws of a tensile testing machine and subsequently separated at a defined deformation rate. The test can also be carried out below and above room temperature. If the adherents have the same thickness and bending modulus and yield strength, the resultant peel is symmetrical, while the crack front usually propagates down the center of the adhesive bond-line. If one of the adherents has thickness or bending properties substantially different from the other, the adhesive bond rotates so that the bond end bends towards the test fixture which is grasping the thinner adherent. The locus of failure also shifts towards the thinner adherent.

In the *90°-peel test* a flexible adherent is adhesively bonded to a rigid adherent. The flexible adherent is peeled off the rigid adherent at a fixed rate so that the angle between the tab and the rigid adherent is maintained at 90° throughout the test. The fixture to which the rigid adherent is clamped must be movable so as to allow the peel front to stay in a constant position under the testing machine crosshead as the bond is peeled.

The *180°-peel test* (Figure 7.11) is described in ASTM D903. As with the 90°-peel test, the flexible adherent is bonded to a rigid adherent, after which the specimen is placed in a tensile testing machine so that the tab is pulled off parallel to the rigid adherent. The flexible adherent undergoes substantial bending to conform to the stress to which it is being subjected. Thus, this adherent must be flexible enough not

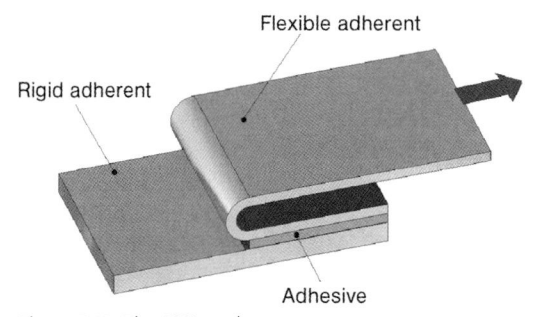

Figure 7.11 The 180° peel test.

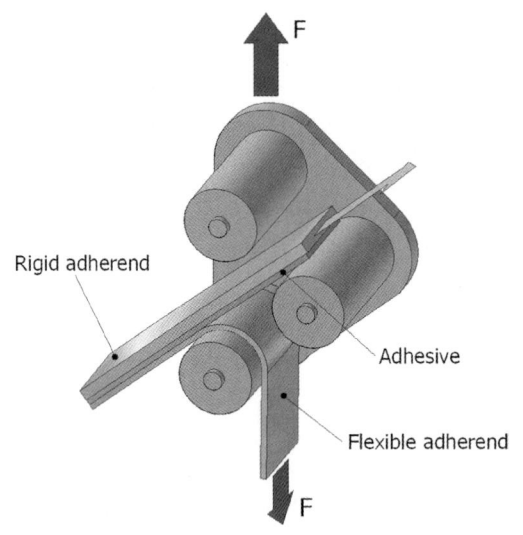

Figure 7.12 The floating roller peel test.

to yield to failure by such a bend. This test is often applied to investigate the adhesion of thin films or sheets to an adhesive. It is also used to examine the adhesion of very soft adhesives, but these must be supported by a canvas during the test. In this case the sealant or rubber-based material is applied uniformly to the rigid adherent. Before the adhesive is allowed to cure, or the solvent allowed to evaporate, the canvas is pressed into the adhesive. When the adhesive has cured, the canvas forms an ideal flexible adherent for this test.

Another variation of peel tests is shown in Figure 7.12. In the 90°- and 180°-peel tests the bending radius of the adherent close to the peel front is mainly controlled by the bending stiffness of the adherent material. Depending on the type of adhesive and the type of adherent, the crack could propagate either through the adhesive or close to the adherent's interface.

In the *floating roller peel test* the radius of curvature near the peel front is guided by the radius of the mandrel over which the flexible adherent is peeled. This system will also control the locus of the crack in the adhesive layer. For example, if a flexible adherent is peeled over a mandrel of small radius, the locus of crack propagation shifts towards the thin adherent. This method is useful if the aim is to examine the effect of surface preparations on adhesion.

A modification of this peel test takes into account the sensitivity of the interfacial adhesion to moisture. The European standard EN 1967, *Structural adhesives – Evaluation of the effectiveness of surface treatment techniques for aluminum using a wet peel test in association with the floating roller method*, has been developed to assess pretreatments for aluminum substrates, and is intended for laboratory evaluations. The objective of this standard was to develop a rapid test procedure with a high sensitivity, which could differentiate between the pretreatment processes. The application of water containing a wetting agent leads to a change of the failure mode from cohesive failure – which usually is found in dry joints – to interfacial failure when a wetting agent was injected into the peel zone. The change in failure mode corresponded to a reduction in peel resistance if the surfaces were not pretreated according to the complete CAA or PAA (C̲hromic A̲cid A̲nodizing, P̲hosphoric A̲cid A̲nodizing) (see Section 8.2.1.1) process. The failure mechanisms in the boundary layer zone initiated by this test are mainly explained by the diffusion of water, as the wetting agent essentially improves surface wetting. The main objective of this method is evaluation of the quality of a surface pretreatment used in the preparation of aluminum or its alloys.

7.3
Test Categories by Duration and Load Rate

7.3.1
Long-Term Static Tests (e.g. Creep Experiments Under Dead Load)

Long-term life prediction for adhesives based on laboratory tests requiring only days, weeks or months remains a demanding challenge. Testing is carried out with the intention of simulating and often accelerating the time-dependent aging effects that may happen to a joint during its lifetime. Due to the specific polymer network structure of most adhesives, their viscoelastic–plastic deformation behavior is strongly time- and temperature-dependent. For long periods of static loading the creep properties of adhesives are dominated by viscous flow. Adhesive creep can be investigated in tensile mode with a polymer bulk specimen (e.g. according to ISO 527-2) or in shear mode with, for example, single lap shear joints according to ISO 11003-2.

Various tests and standards, such as ASTM D2294-96, ASTM D1780-99, ISO 15109 or EN 1943, have been developed to evaluate the creep resistance of adhesives, by monitoring the time- and load-dependent displacement of an adhesive specimen under shear load, or by recording the time to failure.

In the engineering process of bonded structures the limits for maximum tolerable deformations in the structure to maintain functionality and appearance are important for the design of the adhesive joint. The shear strain of adhesive joints is generally expressed as $\tan \gamma = s/d$, where s is the displacement of the adherents and d is the thickness of the adhesive layer (Figure 7.13).

In static load creep experiments, a dead load is applied to the adhesive joint and the initial displacement, as well as the increase in strain over time during the test, are recorded. Standard creep experiments provide basic results for the assessment of time-dependent deformation of adhesives with high plasticity.

Another common experimental procedure is the *relaxation test*. While creep experiments apply a constant load to the sample and follow the increase of strain over time, the relaxation technique applies a fixed strain to the specimen at the start of the test and then records the decrease in stress over time caused by plastic deformation of the adhesive polymer. The difference in creep testing versus relaxation testing is illustrated in Figure 7.14.

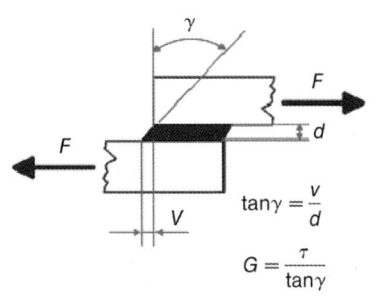

Figure 7.13 Deformation of an adhesive joint under shear stress.

7.3.2
Impact Tests (e.g. Impact Wedge Test to Evaluate Crash Performance)

In some cases adhesive joints are required to maintain the structural integrity of bonded parts even under high dynamic loading conditions, for example in a car body during a crash situation. ISO 11343 [4] describes a procedure for the 'Determination of dynamic resistance to cleavage of high strength adhesive bonds under impact conditions – Wedge impact method', while ISO 9653 relates to a 'Test method for shear impact strength of adhesive bonds'. According to the standard, ISO 9653 determines the average cleavage resistance, expressed as force or energy, of a structurally bonded metal joint. The cleavage corresponds to the separation of the adherents by a wedge, moving at high speed, the displacement of which is initiated by the impact of, for example, a pendulum hammer (Figure 7.15).

In practice, tests with application-specific parts and geometries are necessary to assess the impact resistance of a bonded structure. Since, for example, automotive crash tests are cost-intensive significant effort has been applied to seek ways in which to simulate and predict crash performance by using numerical tools such as finite element simulation.

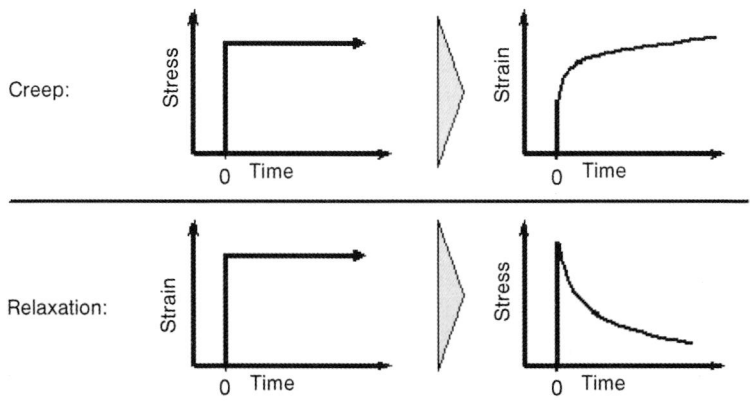

Figure 7.14 Time-dependent stress–strain behavior in creep and relaxation experiments.

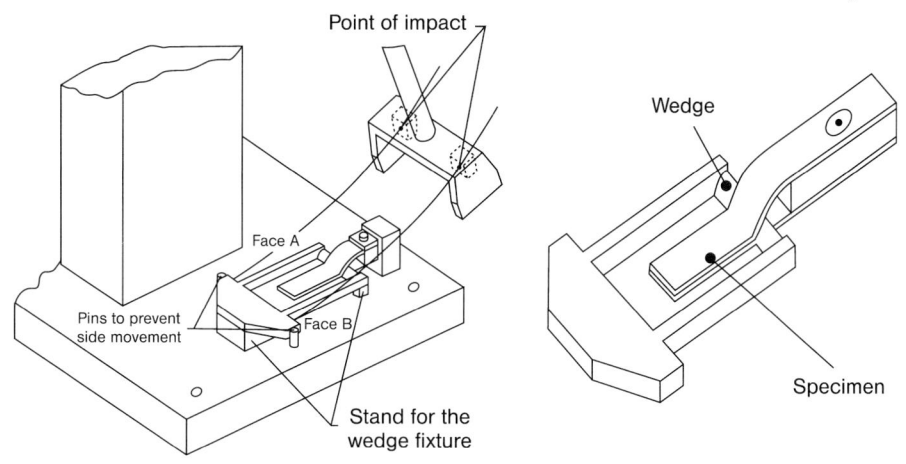

Figure 7.15 The test principle of the impact wedge test.

7.3.3
Cyclic Loading (e.g. Fatigue Tests)

Test methods such as ASTM D3166, *Standard Test Method for Fatigue Properties of Adhesives in Shear by Tension Loading (Metal/Metal)* and ISO 9664 *Adhesives – Test methods for fatigue properties of structural adhesives in tensile shear*, describe ways to assess the durability of adhesive joints under cyclic loading conditions. In both cases a testing machine, capable of applying a sinusoidal cyclic axial load is required. Depending on the experimental parameters (testing frequency load or displacement-controlled tests) the results can vary, and therefore these parameters should be exactly specified beforehand in the material specification. The testing frequency must be carefully chosen so as to avoid significant heating of the specimen due to its viscoelastic energy dissipation. In ASTM D3166 a maximum frequency of 30 Hz is specified, with specimens being tested at five or more different maximum load amplitudes selected such that failures occur with regular spacing over a range varying from at least 10 000 000 cycles to not less than 2000 cycles. (As a guide, the initial maximum load may be 50% of the strength of the adhesive determined by shear tests beforehand). On completion of the test the location of failure (adhesive or joint material), as well as the failure mode (cohesive, adhesive or mixed fracture) are relevant when assessing the fatigue performance of the adhesive joint.

7.4
Assessment of Durability and Service Life of Adhesive Joints

The durability (long-term performance) of a bonded joint depends on the properties of the adhesive, of the materials being joined together, and their surface pretreatment prior to bonding.

The true *service life* of an adhesive joint is influenced by a variety of factors such as climate, environment and mechanical stresses, which in common practice occur with a limited amount of predictability. Service conditions can vary widely, depending on the final application of the joint. In order to carry out an individual service life assessment it is necessary to make assumptions as to the 'normal' and 'worst' conditions that the adhesively bonded joint will be subjected to during its application. These conditions are generally chosen so that the majority of adhesive joints will be at or below these conditions [5], although in some fields of application the basis for the assumption of load and service conditions is regulated by standards and perhaps building codes. As many of the joints will be located under conditions deemed less severe than 'normal', it is obvious that many products will achieve service lives greater than predicted.

The next step in assessing and predicting the service life of an adhesive joint is a *sensitivity analysis*, taking into account possible degradation mechanisms related to the chemical and physical properties of the adhesive and the adherent. The most common mechanisms that cause changes in the mechanical properties or irreversible degradation of adhesive joints include:

- physical or chemical (e.g. thermal or UV) degradation of the bulk adhesive
- desorption or hydrolysis of superficial adhesive bonds under the influence of moisture
- loss of adhesion due to an interfacial weak boundary layer caused by, for example, a poorly cured adhesive (e.g. caused by acidity of a surface after chemical pretreatment)
- dissolution and deterioration of superficial oxides or conversion coatings promoted by the chemical conditions inside the adhesive joint under climate exposure
- corrosion of the metal adherent in contact with corrosion-promoting media; examples are marine environments (high chloride levels) or industrial sites involving exposure to SO_2, H_2S or NOx, typically starting at the bond-line (bond-line corrosion)
- debonding under thermal cycling (e.g. freeze–thaw) due perhaps to mechanical thermal stress caused by mismatch in thermal expansion of the adhesive and the adherents

The standard experimental procedure in the assessment of service life of adhesive joints is based on the comparison of characteristic mechanical and physical properties before and after a certain period of aging.

If the aging conditions to assess the service life correspond directly to the expected service conditions, the designation 'natural exposure' has been widely accepted in case of adhesive joints subjected to outside weathering. In the absence of confirmed data on the relation of accelerated aging procedures to the performance in use, natural aging remains a major source for useful information on degradation mechanisms and service life projection. However, when climatic conditions of exposure are similar to those of the intended use, the period of natural exposure should be at least one-tenth of the anticipated service life in order to create useful data, and could be considerably longer. If the product is to be used in climatic conditions more severe than the natural exposure test site (e.g. higher ambient temperatures and/or radiation), then even significantly longer periods of natural exposure will be required to obtain equivalence. The analysis of aging tests should not only be limited to the acquisition of data from mechanical tests but also include a careful examination and documentation of any changes in outer appearance, such as cracking or crazing, distortion, or change in failure patterns after testing according to ISO 10365 (Adhesives – Designation of the main failure patterns).

In spite of the fundamental importance of test results from natural exposure either under service conditions or under defined exposure conditions, there is a strong demand for accelerating aging tests to cut down the duration of tests and approvals, and to allow for a lifetime prediction beyond the actual time of test exposure. Therefore, testing following accelerated aging will generally form the basis of predictions of service lives.

The guidance document of the European Organization of Technical Approvals for the assessment of the service life of products differentiates three main groups of testing after accelerated aging: direct, indirect and torture tests.

- *Direct testing* is performance- or use-related, and links directly to measurement of the characteristic in question. Direct testing is often accomplished by using the actual joint geometry instead of a standardized specimen.

- *Indirect testing* relates to the measurement of properties which have a known relationship to performance in use. When using indirect tests it is critical that a proven correlation between the property measured and long-term performance has been established (i.e. the property measured must be significant in terms of performance).

- So-called '*torture tests*' are short-term tests where the conditions are significantly more severe than the service conditions of the product; such tests may be used to negate the need for long-term aging. If the product *passes* the severe test, no further work needs to be done on that particular factor. If the product *fails* the test, it does not necessarily mean that it will not perform well, but additional testing over a longer period will be required under conditions closer to the service conditions (i.e. simulatory aging conditions) in order to establish the product's credentials. Examples of additional tests include water boil tests or salt spray tests with increased acidity.

Torture tests should be used with extreme caution – the analogy that 'boiling cannot accelerate the brooding of an egg' has often been cited. The application of torture tests assumes a profound knowledge of the performance of the material in question under the conditions proposed. Any aging conditions applied and subsequent testing should be questioned concerning their relation to the 'basic' phenomena observed on site.

Although accelerated aging may be carried out in a number of different ways, the most appropriate method will depend on the type of adhesive and adherent material, and the intended use of the adhesive joint. In general, simulatory aging methods are used in which the aging conditions attempt to simulate natural conditions, usually with only a moderate acceleration of the factors. Some examples of simulatory aging methods include:

- heat aging
- water resistance

- freeze–thaw
- corrosion and chemical resistance
- artificial weathering

The international standard EN/ISO 9142 (*Adhesives – Guide to the selection of standard laboratory ageing conditions for testing bonded joints*), contains a comprehensive overview of applicable accelerated aging conditions.

Because the presence of mechanical stress during aging can have a significantly accelerating effect, standards such as ISO 14615 (*Adhesives – Durability of structural adhesive joints – Exposure to humidity and temperature under load*) take the interaction of climate exposure of adhesive joints under load into consideration. Susceptibility to fatigue crack growth under hot humid conditions is one of the major concerns for the durability assessment of adhesively bonded metallic joints. The recently published standard EN 15190 (*Structural adhesives – Test methods for assessing long-term durability of bonded metallic structures*) therefore specifies test procedures for determining the long-term durability of an adhesive system subjected to environmental and fatigue loads. The procedures are based on measurement of the crack growth rate and resistance to crack propagation through the adhesive layer in DCB-type specimens under an applied mode I opening cycling loading.

In specific application areas, such as the automotive industry, accelerated aging procedures have been established on the basis of experience. The standardized test procedure VW PV 1200 consists of cycles between $-40\,°C$ and $80\,°C$ at a relative humidity of 95%, and promotes damage mechanisms related to moisture and temperature gradients. The VDA test procedure 621-415, named after the German Association of Automotive Manufacturers, consists of several test cycles with exposure to a corrosive chlorine environment, humidity and phases where the specimens are being stored at room temperature.

For general applications, the following overview may serve to suggest accelerating aging cycles as being both applicable and relevant to the major mechanisms of degradation:

- *Exposure to moisture*: Aging in climate chambers at 95% relative humidity (no condensation on the surface of specimen) and $60\,°C$ or $80\,°C$ for 300–1000 h, depending on the intended application.

- *Exposure to water or other liquids*: Immersion at $40\,°C$, $60\,°C$ or $80\,°C$ for typically 500–1500 h, depending on the intended application. Note that this type of exposure stimulates degradation under elevated temperature and high humidity, but does not promote corrosive attack due to the lack of oxygen in the immersion bath unless constantly aerated.

- *Corrosion in a chlorine environment*: Standard cabinets for corrosion testing create a salt spray environment at $35\,°C$. Exposure times typically range from 300 to 500 h, depending on the intended application.

- *Aging at elevated temperature*: Aging at elevated temperature in the absence of moisture especially promotes oxidation, embrittlement and thermal degradation.

Typical temperatures depend on the intended application. For adhesive bonds under outside weathering temperatures of 80–120 °C are common, depending on the geographical area of application.

- *Thermal cycling*: Accelerated aging tests under cyclic change of temperature include the freeze–thaw transition to consider the detrimental effect of water penetration and icing. Temperatures typically span from –20 °C or –40 °C to 60 °C or 80 °C with 95% relative humidity during the warm period of the cycles. Typical numbers of cycles range from 100 to 300. It should be considered that rapid transitions between high and low temperature limits may induce thermal stress beyond levels representative of outside weather conditions, for example.

- *UV-exposure*: Sophisticated test cabinets such as the Xenotest or Weatherometer (WOM) are available to simulate aging conditions, combining moisture, temperature and UV radiation. The test duration depends on the irradiation doses to be expected in a specific application, and typically ranges from 500 to 3000 h. Clearly, such tests should mainly be considered if one or both adherents is/are transparent to the applied spectra of radiation.

7.5
Tests Related to Storage and Handling of Adhesives

7.5.1
Determination of Words and Phrases Relating to the Product Life

The European standard EN 12701 *Structural adhesives – Storage – Determination of words and phrases relating to the product life of structural adhesives and related materials*, contains the main designations relating to limits of the usability of adhesives after manufacturing.

The ASTM D 1337 standard covers means by which the storage life of an adhesive can be measured using rheology and adhesive performance testing. The test method is applicable to all adhesives having a relatively short storage life, and is intended to determine whether the storage life conforms to the minimum specified storage life required of an adhesive by consistency tests (Procedure A) or by bond strength tests (Procedure B), or by both.

7.5.2
Open Time and Working Life

ASTM D 1338 deals with the working life of an adhesive – that is, the time elapsing between the moment an adhesive is ready for use and the time when it is no longer usable. This practice covers two procedures applicable to all adhesives having a relatively short working life. It is intended to determine whether the working life conforms to the minimum specified working life of an adhesive required by consistency tests or by bond strength tests, or by both.

Further test methods are described in ISO 10364 *Adhesives – Determination of working life (pot life) of multi-component adhesives*.

7.5.3
Extrudability

This test method measures the amount of force necessary to extrude an adhesive from a cartridge at a given temperature. Cartridge adhesives are typically used at job sites over a wide variety of temperature ranges. These products may be exposed to cold temperatures, where one of the limiting application factors is whether the product can be extruded from the cartridge. This provides the manufacturer with results that can assist in formulation development, and the end user with information for use in selecting a product for general usage. By measuring (the ease of) extrusion from the cartridge, the test method can also be used as a quality control tool to measure the shelf-life stability of a product.

7.5.4
Sagging

Methods to determine the resistance of adhesives to flow after application (sagging) are described e.g. in EN ISO 14678. In practice a rope of adhesive is applied to a test panel which is then placed in a vertical position. The deflection of the adhesive rope after a defined period of time gives an indication of the sagging properties.

7.5.5
Volatile Organic Compound (VOC) Emissions

The release of VOCs into the environment has become a major concern since health issues such as 'sick building syndrome' have been assigned to emissions from adhesives used, for example, in flooring.

Although today a still wider variety of mostly solvent-based adhesives are employed, since 2005 the European VOC Solvents Directive has made solvent usage more difficult and more expensive. Therefore, manufacturers are seeking solvent-

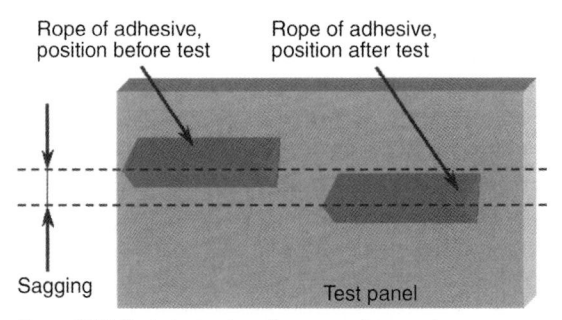

Figure 7.16 Basic test set-up for measuring sagging.

free alternatives to stay abreast of changes in the regulations stipulated by law, to consolidate their reputation as environmentally friendly companies, to save cost, and to optimize their processes.

7.6 Nondestructive Testing Methods

For quality assurance purposes in the aircraft and car manufacturing industries, it has been a matter of particular concern to develop methods for the nondestructive testing of adhesive bonds. However, these tests do not all provide significant results and are relatively easy to use in practice. Yet, at the same time, due to increased product quality expectations, they are expected to be more and more performing.

In both the literature and in industry, there is a body of different quality assurance methods. The National Materials Advisory Board (NMAB) Ad Hoc Committee on Nondestructive Evaluation adopted a system that has classified the different methods into six major categories: visual; penetrating radiation; mechanical vibration; thermal; magnetic-electrical; and chemical-electrochemical (Figure 7.17).

Each method can in turn be characterized in terms of five principal factors: energy source or energy medium; nature of the signals; image or signature resulting from

Nondestructive Test Methods for Quality Assurance

Visual	Penetrating radiation	Mechanical vibration	Thermal	Magnetic-electrical	Chemical-electrochemical
CCD camera	Radiography	Vibrometry	Impuls video thermography	Eddy current	Intrusion
Light section	Laminography	Pulse echo	Reflection impulse thermography	AC field measurement	
Speckle interferometry	Tomography	Through-transmission	Induction thermography		
Moiré technique	Microwaves	Air-coupled ultrasound	Eddy Therm		
Laser scanning microscopy	Penetration of radionuclides	Nonlinear ultrasound	Lockin thermography		
FTIR		Laser ultrasound	Ultrasound burst phase thermography		
Shearography		Sound emmision analysis	Imaging vibrometrie		
Holography			Stress analysis		

Figure 7.17 Nondestructive test methods [7].

interaction with the test object; means of detecting or sensing resulting signals; method of indicating or recording signals; and basis for interpreting the results [6].

7.6.1
Visual Techniques

Although, visual or optical techniques are the best technology available for nondestructive testing, their usefulness is rather limited for the testing of adhesive bonds. For example, when used for the control of bead application these tests do not yield any information about the quality of the bond, since delaminations or adhesion failure are not detected. Visual techniques are relatively inexpensive; whilst the behavior of a component under load can be tested by means of speckle interferometry, detailed information about the physical properties of the bond cannot be obtained from these tests.

Another available optical method is *shearography*, which is based on the differences between the deformational states of an unloaded and a loaded component. The load is applied by different means depending on the component to be tested, for example, by deformation through vacuum or by twisting of the component. Until now it has not been possible to detect adhesion flaws satisfactorily with shearography, whereas it is possible to detect a position error of the bead. But even here, a high degree of practice and experience are needed to correctly interpret the results [7].

7.6.2
Radiographic Techniques

As a technique for the nondestructive testing of adhesive joints, penetrating radiation is still at the research stage. It is not possible to use X-radiography because, as the adhesives have extremely low absorption coefficients, there would be no significant differences detectable by radiography. However, in some specialized cases it is possible to visualize the adhesives by adding special pigments.

Unlike X-radiography, neutron radiography is very well suited to the testing of adhesive bonds. Neutrons react with hydrogen, thereby inducing a very low absorption in metals, and a high absorption in organic adhesives. In this way thin layers of organic substances or adhesives can be visualized with great contrast, and differences in transmission due to nonuniform adhesive layers, pores, moisture and extraneous substances can be detected. Unfortunately, this method also requires a high degree of technical and economic provisions, such that it will most likely not be implemented in a series production.

7.6.3
Magnetic, Electrical and Chemical Techniques

Magnetic, electrical or chemical techniques are not well suited to the nondestructive testing of adhesive joints, due to the fact that the adhesive layer acts as an insulator when these techniques are used. Consequently, the measurement of capacitance or induction for the testing of adhesive joints is used only to a limited extent. It is not possible to obtain information about the quality of the adhesive joint by using these methods.

7.6.4
Mechanical Vibration

7.6.4.1 Impulse-Induced Resonance

This technique was developed to test assemblies made from sheet metals and nonmetallic backing elements.

Pulses that are created by broadband ultrasonic transducers and long enough to produce stationary waves are transferred to the metal sheet via an appropriate medium. The metal sheet has a thickness resonance spectrum of frequencies within the band of the pulse. The signals detected are not corrected. The first echo observed is the reflected pulse from the surface of the specimen, followed by a second signal produced by the stationary wave in the metal. This signal is partly captured and transferred to a spectroscope that displays the frequencies of the fundamental and harmonic vibrations of the thickness resonance of the half-wave. When a nonmetallic plate is bonded to the metal as a backing element, the resonances are damped to an extent that corresponds to the degree of bonding. Consequently, a high resonance amplitude corresponds to 'no bonding', while that of a partly bonded assembly will be average, and a perfect bonding will have a low amplitude.

During the early 1950s Fokker first introduced a bondtester based upon a well-known impulse-induced resonance technique, and indeed the ultrasonic resonance–impedance-based Fokker Bondtester is still used today, primarily for manufacture or maintenance purposes in the aircraft industry. The procedure involves a piezo-electric transducer being excited with a high frequency and then coupled with the sample such that changes in resonance frequency and impedance can be detected, depending on the characteristics of the sample and the physical properties of the adhesive joint. This allows conclusions to be drawn with regards to the quality of the joint.

The sensor consists of a disk-shaped, spring-loaded transducer coupled with to the upper end of the sample by a viscous substance, for example a fatty substance or other highly viscous medium. The transducer must be supported so as to allow axial and radial vibration and to forward the elastic waves at given resonance frequencies. A commercially available Fokker bondtester is shown in Figure 7.18.

The Fokker bondtester is especially suited to the testing of metal-to-metal joints, and sheet-to-honeycomb epoxy resin-bonded metal assemblies to detect delaminations in the outer sheets.

Unfortunately, the method is very time-consuming and so is particularly well suited to small sample quantities. Another disadvantage is the extremely high spread of the measured values, so that only large-sized defects such as large-area delaminations are detected.

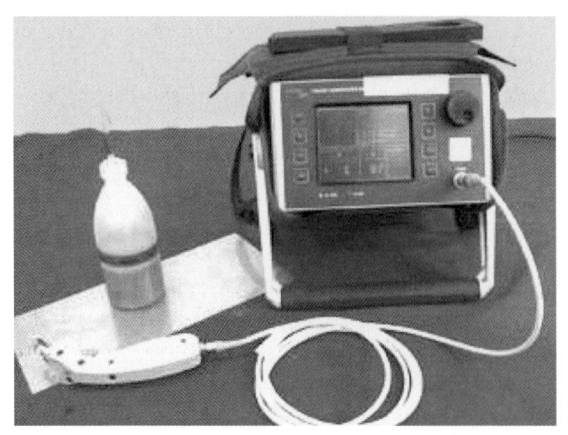

Figure 7.18 The Fokker bond tester.

7.6.4.2 Pulse Echo

The pulse echo technique is the simplest method for detecting defects in an adhesive joint. The sample is placed into a water bath, as is the probe that supplies ultrasound to the sample; the system is shown schematically in Figure 7.19. The amplitude of the measured signal allows conclusions to be drawn with regard to the sound transmission of the wetting surfaces and the adhesive. Good wetting and sound transmission within the adhesive produce a high amplitude, whereas the ultrasound wave of the back-wall echo is attenuated or totally shadowed in the case of defects within the adhesive bond. Consequently, there is higher amplitude of the reflector in flawless regions than in areas with defects. This can also be detected by means of shorter echo times.

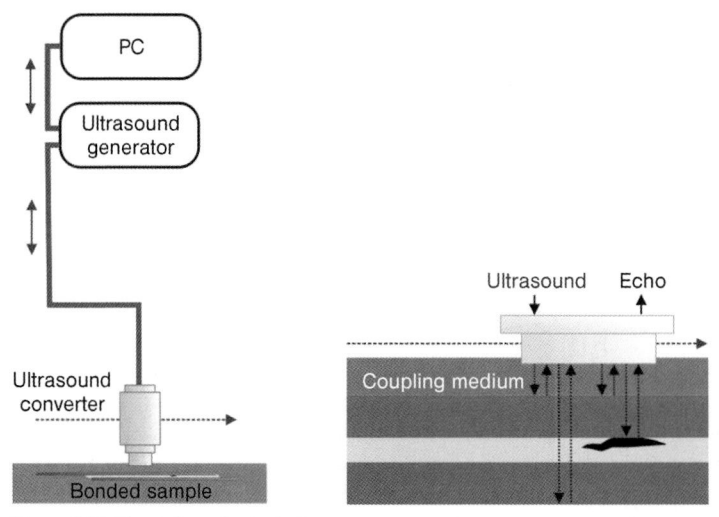

Figure 7.19 The pulse echo technique.

Low testing frequencies (at 5 MHz) reduce the lateral resolution. However, due to the integral evaluation of the total structure it is not possible to map the defect along the thickness of the adhesive joint. The depth of a defect can be estimated on the basis of the signal attenuation. With C-scans, the defects appear in color, and can be detected by means of any color inhomogeneities, although neither the size nor the form of the defect can be determined precisely. The defects are displayed with a larger size than reality due to the funnel-shaped geometry of the sound beam; the higher the parallel alignment, the better the resolution.

As the test results may be impaired by background noise, this method is not an all-purpose technique. In relation to the wavelength, the thickness of the components to be bonded must be relatively large in order to obtain a good resolution of the echo of each interface. This method, among others, is very suitable for the testing of metal-to-metal bonds. Less satisfactory results have been obtained when testing bonded nonmetals due to the higher attenuation coefficient.

The major disadvantage of this method is that it is possible to evaluate only the first interface – the detection of cohesive defects within the adhesive or detachment on the opposite interface is not possible. In addition, each component must be placed into a water bath for testing. Finally, as this is a scanning technique it is relatively time-consuming.

7.6.4.3 Through-Transmission Sound Testing

In this method an ultrasound pulse is propagated through the entire structure and received by a second probe, making it possible to evaluate the adhesive joint and to detect delaminations in glass-fiber reinforced plastics.

The amplitude of the signal received is compared to a reference value, for example the amplitude of the initial signal. The frequencies required to detect flaws in honeycomb structures are relatively low (0.5 MHz).

In 1972, Schliekelmann reported the development of a scanning device in the form of a rotating sensor to test metal-to-metal bondings. A hub filled with water is fitted with sensors, and several hubs are fixed with a rubber tire at each side of the sample. During scanning the tires rotate, while the sensors remain fixed and centered. Water jets are used as couplants by directing them towards the areas between the tires and the sample, where contact is needed. Among its disadvantages, this method does not permit the mapping of defects, and a couplant must be placed between the probe and sample to provide proper propagation of the applied ultrasound. For these reasons this procedure is generally excluded from use during manufacturing processes.

7.6.5
Thermal Methods

7.6.5.1 Optical and Ultrasonic Lockin Thermography

In optical lockin thermography (OLT), a thermal wave is produced on the entire surface of a sample by periodic irradiation with a infrared heat source, the intensity of which can be modulated (see Figure 7.20).

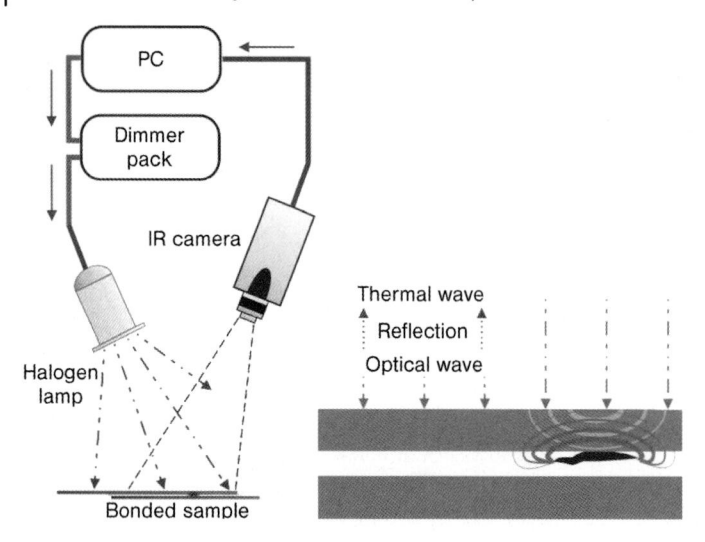

Figure 7.20 Optical lockin thermography.

For each picture element, a thermographic camera records the thermal response of the surface that follows the periodic application of heat. The measurement is noncontact, and a PC is used to calculate the magnitude and phase of the temperature modulation induced in parallel for each pixel of the detector during the whole measurement period. The magnitudes and phases calculated for a given initial frequency, respectively, are visualized in color in the form of a magnitude image and a phase image (see schematic in Figure 7.21).

As this method only reacts to a temperature modulation at a given excitation frequency, it is may be referred to as 'excitation thermography', 'lockin thermography', 'phase-sensitive modulation thermography' or 'multiplex photothermal imaging'.

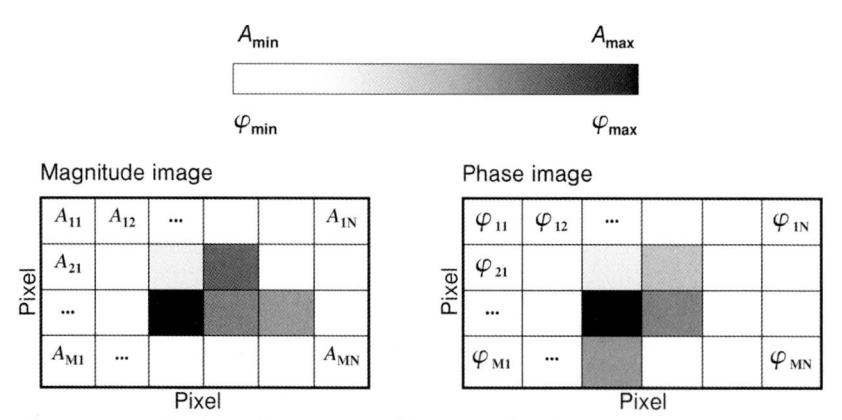

Figure 7.21 Two-dimensional image matrix of the magnitude and phase measurements, visualized in color (magnitude image, phase image).

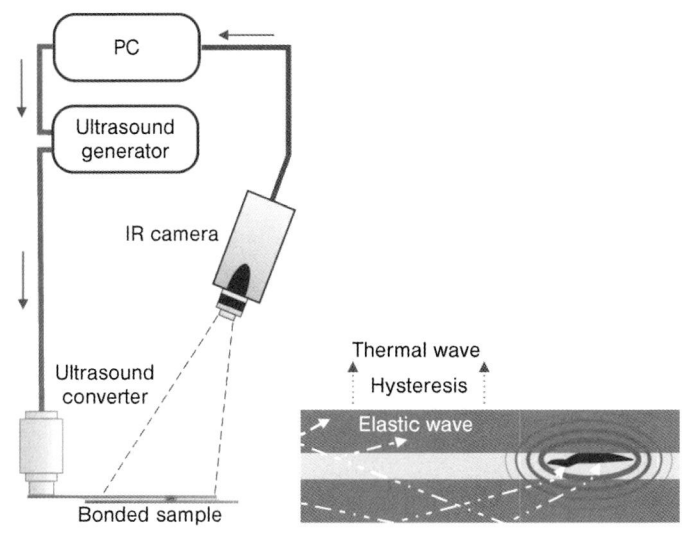

Figure 7.22 Ultrasonic lockin thermography.

With lockin thermography, the surface is subjected to a periodic (ideally sinusoidal) thermal stimulation. The temperature modulation is induced either *on the surface of the sample* and propagates as a thermal wave (external excitation, e.g. optically or by heated air), or *within the sample* (internal excitation, e.g. by means of ultrasound, microwaves, or electrically).

While external excitation makes it possible to detect interfaces that reflect the thermal wave, ultrasonic lockin thermography is based on internal excitation that requires a mechanical coupling of the ultrasound source with the sample (Figure 7.22).

With internal excitation, any flaws present produce a thermal wave; hence, this method is referred to as 'defect-selective imaging', a dark-field method that reacts very sensitively to areas with enhanced mechanical loss angle (Figure 7.23).

Lockin thermography provides for an interpretation of the magnitude and phase of the measured periodic temperature modulation at the surface in relation to the excitation. The phase signal is hardly disturbed by the emissivity distribution of the sample surface. When there is proper averaging – that is, sufficiently long periods of measurement – the method has an enhanced sensitivity for deep defects. A laboratory unit for lockin thermography, as performed at the Institute of Joining and Welding Technique, Technical University of Braunschweig (Germany), is shown in Figure 7.24.

7.6.5.2 Ultrasound Burst Phase Thermography

Ultrasound burst phase thermography is a new nondestructive defect-selective technique originating from ultrasonic lockin thermography. It provides thermal wave images of defects by ultrasonic stimulation.

In contrast to the sinusoidal-modulated excitation used by lockin thermography, conventional pulse thermography is based upon short optical or acoustical excitation

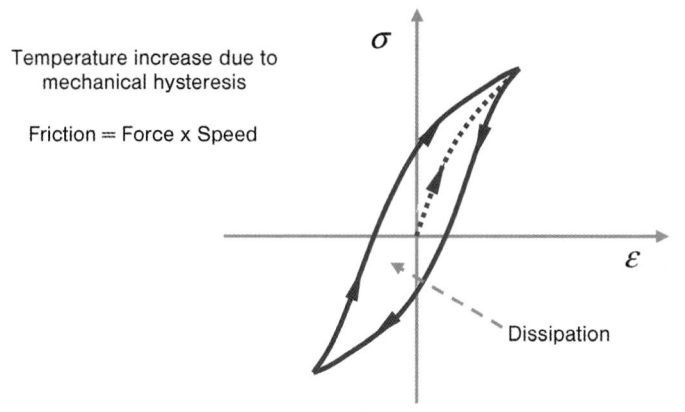

Figure 7.23 The transformation of vibration energy into heat by hysteresis [8].

Temperature increase due to mechanical hysteresis

Friction = Force x Speed

Dissipation

pulses, allowing the measuring period to be significantly reduced compared to the lockin method. However, the thermography images obtained are impaired by the inhomogeneities of the sample surface and the temperature gradients. One way of overcoming this problem is to sue *ultrasound burst phase thermography*, which combines the advantages of both methods. The sample is excited by short ultrasound bursts, and the temperature modulation is recorded with a thermography camera and the data Fourier-transformed. As with lockin thermography, phase images can be obtained using this method. Compared to the monofrequent sinusoidal excitation, the response spectrum is larger – that is, several modulation spectra can be interpreted with one single measurement [8].

Ultrasound burst phase thermography allows conclusions to be drawn from the phase image directly to the depth position of the defect because the cooling behavior – and thus the range of the thermal wave – depends on the modulation frequency. High

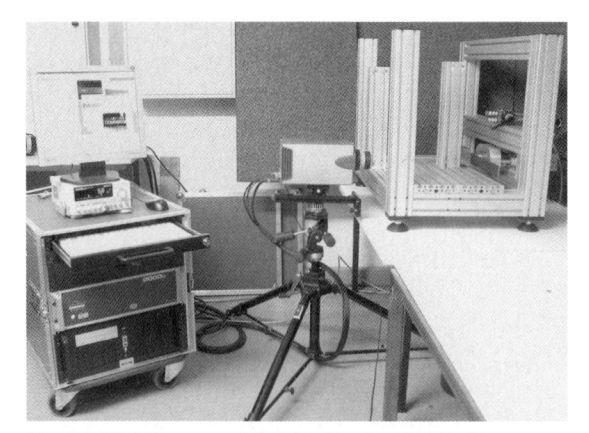

Figure 7.24 Laboratory equipment for ultrasonic lockin thermography at the Institute of Joining and Welding Technique, Technical University of Braunschweig (Germany).

frequencies display the near-surface area, while low frequencies with a higher range supply information about defects within the component. For interpretation, it is therefore recommended to relate the temperature sequence recorded to the frequencies used to determine the depth position of a defect.

7.6.5.3 Excitation Mechanisms for Active Thermography

In active thermography, a proper energy source must supply or transfer sufficient energy to the sample to generate an adequate temperature contrast, and this is the first objective of these testing methods. The amount of energy must also be capable of generating a contrast, without damaging the component [6].

Pulse thermography and lockin thermography use different heat sources. Lockin thermography requires a periodic or sinusoidal heat source, with the penetration depth of the thermal waves being altered by frequency modulation, good conductors being excited by higher frequencies, and materials with a low thermal conductivity being excited by low frequencies. In the case of external excitation of thermal waves, a temperature modulation must be generated on the surface of the sample that is then conveyed to inner regions of the sample. The following mechanisms provide for a transfer of heat:

- heat conductivity
- radiative heat transfer
- convection

For pulse thermography, a short heat pulse is supplied to the sample, with the duration of the pulse depending on the thermal conductivity of the material (Figure 7.25): the higher the thermal conductivity, the shorter the pulse duration. The following excitation mechanisms are used:

- *Laser*: When a sample is irradiated with laser light, the electromagnetic irradiation is absorbed by the sample, and heat is produced. Irradiation energy is propagated through electromagnetic waves (photons), so that it can be conveyed to a component without any contact. Optical irradiation sources are therefore all suited to the noncontact generation of thermal waves. In conventional photothermal imaging, laser light was used owing to its point excitation capacity. The radiation from a laser source is easy to focus and to modulate, even in the case of very high frequencies of up to several MHz. One advantage of this method is that, for excitation, a wavelength outside the sensitivity range of the detector can be used. However, provisions must be taken to prevent unwanted radiation from falling on the detector, and this can be achieved by using a filter. Lockin thermography, on the other hand, is based on the heating of an area, and the low divergence of a laser beam does not provide any benefit here. Furthermore, the necessary safety precautions required are a fundamental disadvantage of all high-performance laser sources, particularly in industry.

- *Light-emitting diodes (LEDs)*: These are considerably less expensive and serve as an alternative to laser diodes for use as an excitation source. Power LEDs, such as those used in car manufacture, have a luminance of up to 6000 candela. However,

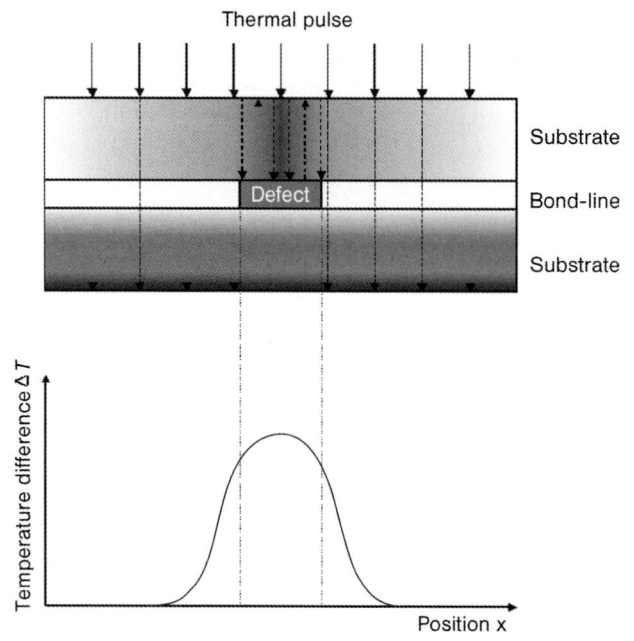

Figure 7.25 Pulse thermography: heat builds up above a defect.

their optical performance is low, which is a disadvantage. Currently, LEDs are undergoing rapid development and, if their performance continues to grow, they will represent an interesting alternative for conventional flash lamps and radiators as they provide flash sequences at any desired rate for lockin thermography.

- *Halogen lamps*: These are particularly well suited to the illumination of large areas. As their spectrum comprises both the visible and infrared (IR) ranges, filters must be used when halogen lamps are employed as the excitation source. According to the Wien radiation law, halogen lamps have their maximum radiant excitation at a wavelength of 1 µm. Consequently, the long-wave portion of their emission spectrum is within the sensitivity range of thermography cameras. This coherent noise must be filtered out, for example by using water or polycarbonate filters, because medium-wave infrared cameras (MWIR) function in the adjacent range of 3–5 µm. With current-controlled temperature modulation, the maximum frequency is limited to 1.85 Hz, and computational correction is required due to nonsinusoidal optical output. Computer-controlled lamps with wattages of 1 to 6 kW are well-suited heat sources for all types of low-frequency application, provided that the test object absorbs their radiation. If the surface is metallic bright, it is generally painted black for this reason.

- *Flash lamps*: These are only suitable for pulse thermography. With energies of up to 12 kJ, the flash lamps used frequently in the photographic industry make it possible to achieve an energy absorption of $1.5 \, \text{J cm}^{-2}$ on a black surface measuring $20 \times 20 \, \text{cm}^2$.

- *Hot air*: A hot-air blow-dryer provides heat via a modulated stream of air. In the wavelength range of 8–12 µm the atmosphere is virtually transparent, and consequently the thermography camera will not record the hot air. The optical properties of the sample are also irrelevant for excitation. Although the thermal waves are easily generated in practice, the upper frequency limit is 0.04 Hz due to the inertia of the hot wires.

- *Microwaves*: When microwaves are absorbed by the sample, the molecular rotation vibrations induced will result in a heat transfer. Microwaves work in a range from 1 GHz to 3,000 GHz; this method is confined, however, to electrical non-conductors.

- *Electrical heating*: Direct modulation of the voltage applied is well suited to the testing of electrical conductors, with defects being visualized as 'hot spots'. An example of application is the detection of leakage current in solar cells.

- *Induction*: An elegant solution for noncontact thermal wave excitation in electrical conductors is the generation of an oscillating electromagnetic field using an induction coil and a suitable high-frequency generator. In electrical conductors, eddy currents are highly attenuated and only exist within a thin layer near the surface (the 'skin effect'). The depth of penetration of the eddy current (d) is described by:

$$d = \sqrt{\frac{2c_0^2}{\Gamma \omega}}$$

where c_0 is the speed of light, ω the frequency, and Γ a material constant which is proportional to the electrical conductivity. The d-value for copper, for example, which determines the heating zone within the material, is 50 µm at 1 MHz. Consequently, with induction heating the thickness of the layer in which heat can be generated is variable, which is in fact an overall advantage. The heating response of flawless regions of a sample differs from that of defective regions, thus making it possible easily to detect flaws in contact joints of composite laminar materials (adhesive bond, soldering, welding), or near-surface fatigue cracks. Intensity modulation is achieved by varying the distance between the inductor and the sample surface.

- *Ultrasound*: A relatively new method for the detection of cracks, for example, is excitation by power ultrasound. Vibration is excited in a sample by means of an ultrasound generator with an electrical power of up to 5 kW. As a result, crack edges may rub against each other, or plastic deformations may arise in the defective region, and the heat induced as a consequence can be detected on the surface. This method is very rapid, and very sensitive, although there is a risk that the sample will be damaged.

7.6.6 Assessment of Nondestructive Testing Methods

The establishment of special requirements for the nondestructive testing of bonded assemblies helps to define suitable criteria for nondestructive testing methods in general. These refer to the applicability of a specific method for the nondestructive testing of bonded joints, the test equipment required, the time needed to test the component, the safety at work, the ease of use, and the provisions to be taken when preparing the samples. At present, with regards to general applicability, the following four methods have particular potential for the detection of defects in adhesive-bonded joints [7]:

- neutron radiography
- pulse echo
- shearography
- lockin thermography

Investigations have shown that, when using *neutron radiography*, accurate information can be obtained with regards to the position, the nature, and the depth of defects. However, as extensive equipment is needed for these measurements, and extensive provisions must be taken, for example with regards to sample geometry or to the nature or accuracy of the adhesive layer, an industrial application is out of the question. In the same way, when using the pulse echo method, one exclusion criterion for industrial application is the fact that a couplant is needed. The applicability of the method for serial production depends on the development of high-performance, air-coupled probes.

Shearography, on the other hand, could be perfectly suited because it does not require any extensive technical equipment, although the existing sensors were not adapted to the testing of adhesive-bonded joints. To date, no satisfactory results have been obtained with regards to the detection of flaws.

Lockin thermography, and particularly ultrasonic lockin thermography, requires only moderately technical equipment and provisions, yet yields very good results with regards to the detection of flaws. Furthermore, it is considered a robust and very rapid testing method that is perfectly suited to integration into serial production [8].

An example of ultrasonic lockin thermography is shown in Figure 7.26, where two sections of the flange joint, into which defects were introduced when the door was manufactured, are shown in selective enlargements. Within these areas, modified signals are detected. With regards to bonded joints, this technique is still at the research stage; however, this complex example illustrates a potential future application of ultrasound thermography as a nondestructive testing method for bonded components.

7.7
Applications of Test Methods in Structural Adhesive Bonding

The engineer describes the mechanical behavior of materials and components by means of quantitative parameters that can be determined by measurement under defined conditions, such as strength, deformability and weight. In the case of adhesive bonds, it is advisable to distinguish between the mechanical behavior as a function of time and temperature on the one hand, and the durability as a function

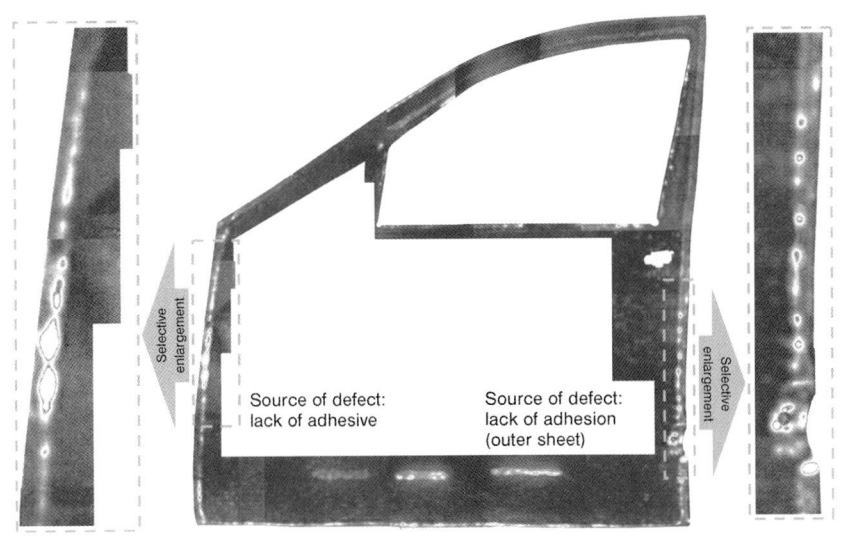

Figure 7.26 Nondestructive testing by ultrasound thermography (magnitude image of a car door).

of the environmental conditions on the other hand (see Section 7.5). In the following sections, some examples will be presented to illustrate the major strengths and deformability characteristics of adhesive bonds as a function of the time period during which they are submitted to a load, and the temperature. Until now, with regards to the properties of adhesive bonds, neither a systematic catalogue nor standardization is available. The reasons for this are provided in the following sections.

7.7.1
Tensile Shear Strength Under Short-Term Loading

As with any other technical component or material, the most frequently asked question about adhesive bonds concerns its strength, defined as the maximum stress or strain under which the bond will fail or be destroyed. The ultimate tensile strength of adhesive joints is determined almost exclusively in tensile shear tests using single-lap shear specimens that are easily manufactured (see Section 7.2.2). *Butt joints* are sensibly not used because the adhesives have a relatively low strength under tensile load, particularly with high-strength materials. The mean breaking strength, defined as the ratio of the failure load to the bonded area, can be determined by means of the tensile shear test.

In the examples shown in Figure 7.27, mild steel (St 37) was alternatively treated with the following surface preparation methods: degreasing with acetone in an ultrasonic bath (US); blasting with shot blast glass (SBG) or grit blast glass (GBG); shot blast ceramic (SBC); grit blast corundum (GBK); shot blast steel (SBS); or grit blast steel (GBS). It is easily recognized that different variants of a surface preparation method alone can have an influence on the strength behavior, at least with certain

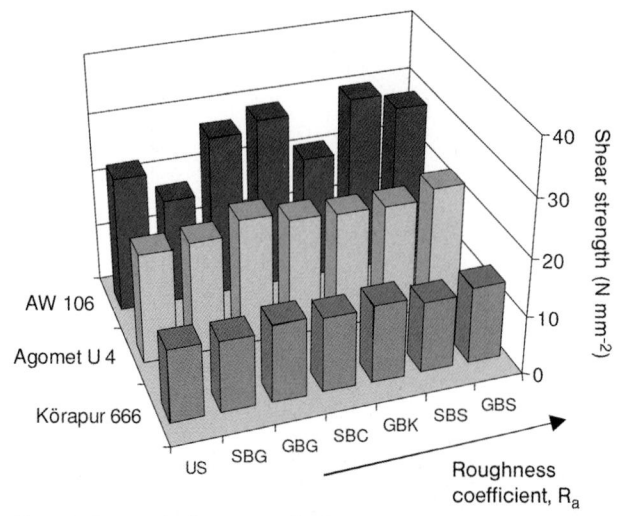

Figure 7.27 Tensile shear strength of unaged steel bonded with different adhesives as a function of different surface preparations.

adhesives, although the grain size of all blast compounds was within a range of 100 to 250 μm.

The adhesives used in this investigation were Ciba-Geigy AW 106/HV953U, Degussa Agomet U 4, and Kömmerling Körapur 666. Although these adhesives each have a different chemical base, they have comparable viscosity (so that wetting failure will not occur, despite the surfaces being very rough). After curing, however, they have different strength and failure patterns, respectively. In the unaged condition, the epoxy system AW 106 had tensile shear strength values >20 N mm^2 and displayed a cohesive failure pattern near the adhesive–substrate interface (surface near cohesive failure; SCF). Purely cohesive failure was not expected to occur. Agomet U 4 (acrylate adhesive) and Körapur 666 (polyurethane adhesive), which were used for comparison, each had a considerably lower inherent strength. In the unaged condition, due to a high content of fillers, these adhesives generally had cohesive failure patterns, as illustrated in Table 7.1. The effect of blasting with different compounds compared to reference values obtained with acetone degreasing (with ultrasound) is also indicated (as relative response).

It was also shown that the type of blast compound used influenced the tensile shear strength of the bond only in the case of the epoxy resin adhesive system. Due to a cohesive failure pattern near the adhesive–substrate interface, failure occurred within the zone of influence of the pretreated surface. The tensile shear strength values measured in this test run changed depending on the blast compound used – that is, the overall behavior of the bond was dominated by the behavior of the substrate–adhesive interface. The tensile shear strength was considerably impaired by the SBG surface preparation, but when using GBG, SBC, SBS or GBS it was considerably increased. Compressed air blasting with GBK slightly reduced the tensile shear strength. In the current literature [9], the mean arithmetic roughness R_a

7.7 Applications of Test Methods in Structural Adhesive Bonding

Table 7.1 Characteristics of fracture surfaces of steel bondings.

Adhesive	Characteristic	Treatment						
		US	SBG	GBG	SBC	GBK	SBS	GBS
AW 106 EP	Failure pattern	SCF	SCF	SCF	SCF	SCF	SCF	SCF
	Related to US (%)	–	78	120	128	93	133	121
	Response, related to US	–	–	+	++	o	++	+
Agometacrylate	Failure pattern	CF	CF	CF	CF	CF	CF	CF
	Related to US (%)	–	103	116	108	106	106	115
	Response, related to US	–	o	+	o	o	o	+
Körapur PUR	Failure pattern	CF	CF	CF	CF	CF	CF	CF
	Related to US (%)	–	98	107	103	107	98	106
	Response, related to US	–	o	o	o	o	o	o

CF = cohesive failure; PUR = polyurethane resin; SCF = particular cohesive failure near the adhesive/substrate interface (see Section 7.2.6); US = ultrasound.

(the surface preparation methods are classified according to the increase in roughness induced) was reported to have an influence on the tensile shear strength of steel samples bonded with epoxy resin adhesive. However, this effect was not recognized in the test run described in Table 7.1.

Within the standard deviation, those samples bonded with polyurethane and acrylate adhesives did not respond to the very intense surface preparation methods. These commercial adhesives have a high filler content, and therefore the polymer–substrate interface has a higher strength compared to the bulk; hence, cohesive failure will occur outside the zone of influence of the blasted surface. Consequently, whilst it is not possible to evaluate the effect of these different surface preparation methods, it should be noted that this behavior may drastically change as soon as these bondings are exposed to humidity (see Section 7.7.7).

It has become clear that whilst these simple tensile shear tests are adequate for determining certain parameters, they barely apply to different bond dimensions or surface conditions of other adherent materials. In general, values of ultimate tensile strength are not suitable for the dimensioning of components because, in simple terms, a component should not be evaluated by the maximum load under which it breaks. However, the results of tensile shear tests do provide useful indications for the evaluation of adhesive bonds, provided that only one influencing parameter is changed while all other parameters (adherent, surface condition, sample dimensions) are maintained.

This may be illustrated with a completely different adhesive system. The data in Figure 7.28 show the relatively low tensile shear strength of a pressure-sensitive adhesive (PSA) setting by UV irradiation, where the degree of crosslinking is a function of the exposure dose. As shown, an increase in crosslinking results in a considerable increase in shear strength, but this is limited by the chemical reactivity of the adhesive [10].

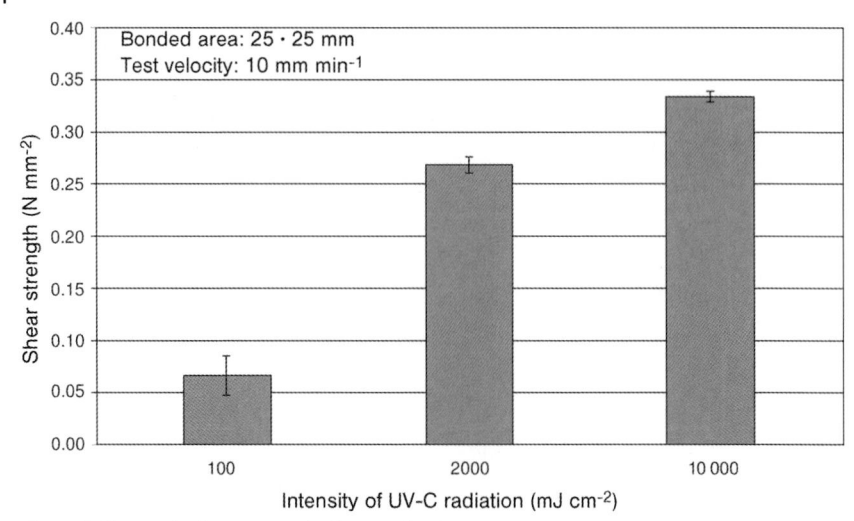

Figure 7.28 Tensile shear strength of an acrylate pressure-sensitive adhesive (PSA) on degreased AlMg3 as a function of the degree of crosslinking.

First, as can be seen in Figure 7.28, the absolute load-bearing capacity of this PSA (and others) is considerably less than that of structural adhesives. Furthermore, it becomes clear that the load capacity can be increased with a rise in UV irradiation intensity. In this adhesive system, as long as it is exposed to UV radiation, crosslinking reactions are induced in the formerly thermoplastic adhesive via chemically incorporated photoinitiators (see Section 5.7.2.6). However, the use of this effect to provide an overall assessment of the adhesive is limited because a higher degree of crosslinking may induce modifications of its other properties (see Section 7.7.3).

The tensile shear test is also suitable for the comparison of the temperature resistance of adhesives in defined assemblies (Figure 7.29) [11].

Figure 7.29 provides an overview of the general temperature behavior of structural adhesives. It is well known that the ultimate tensile strength values are also contingent upon the stress conditions prevailing in the bond-line, which in turn depend upon the plastic deformability of the adhesive. Possibilities for the plastification of high-temperature-resistant adhesives are limited, and therefore when performing tensile shear tests, low ultimate tensile strength values will be obtained.

As illustrated in Figure 7.29, a strength maximum occurred within a temperature range between 0 °C and 100 °C for almost all adhesives. This was followed by a decrease, with both events being recorded within the T_g ranges of the adhesives.

Tensile shear tests are also well suited to the evaluation of deformation behavior of adhesives. Displacement of the adherents during the test can be measured using fine-strain extensometers; the typical shear stress–shear strain curves shown in Figure 7.30 may be very useful in the evaluation of adhesives [12]. In this example,

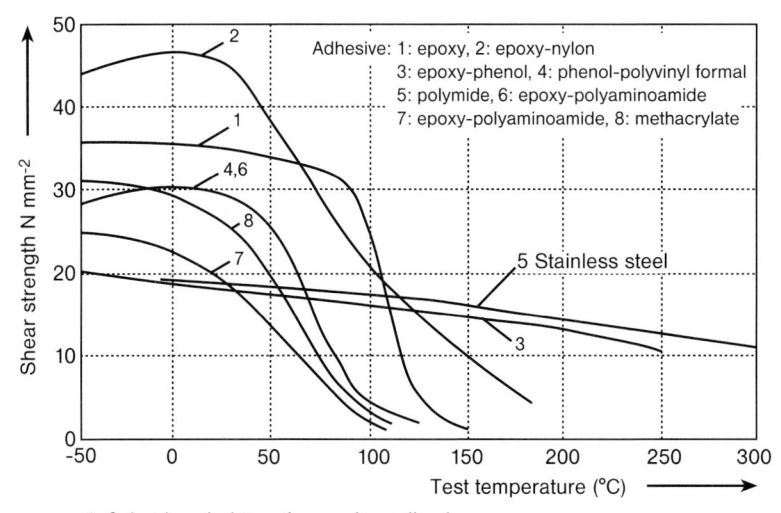

Figure 7.29 Tensile shear strength of metal bondings as a function of temperature.

1-6: hot bonded together curing adhesives
7, 8: cold bonded together curing adhesives
Material: Al Cu Mg 2 clad
Overlap length: 12 mm

Adhesive: 1: epoxy, 2: epoxy-nylon
3: epoxy-phenol, 4: phenol-polyvinyl formal
5: polymide, 6: epoxy-polyaminoamide
7: epoxy-polyaminoamide, 8: methacrylate

high-strength adhesives were used to join 6 mm-thick adherents with a single overlap length of 5 mm (to provide for a uniform stress distribution). The data in Figure 7.30 show that, up to a shearing of 0.1, there was a near-linear shear stress–shear strain behavior. At a higher shear deformation, the stress built up at a lower rate, which indicates that plastic deformation had occurred. Plastic deformation, however, may only be detected by means of the residual shear deformation following relief of the load (not shown here). This allows local peak stresses to be prevented, for example in single lap joints at the ends of the overlaps.

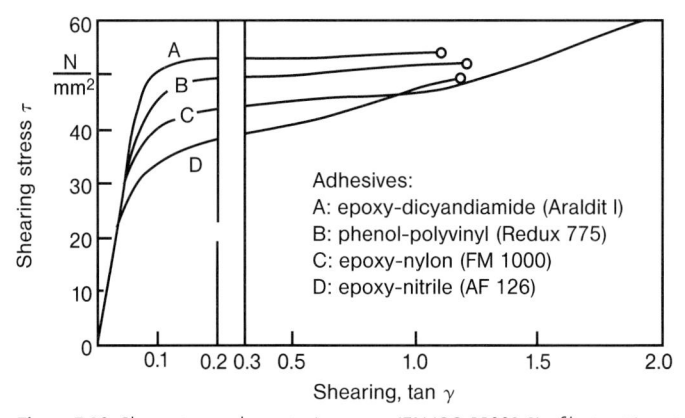

Adhesives:
A: epoxy-dicyandiamide (Araldit I)
B: phenol-polyvinyl (Redux 775)
C: epoxy-nylon (FM 1000)
D: epoxy-nitrile (AF 126)

Figure 7.30 Shear stress–shear strain curves (EN ISO 11003-2) of hot-setting structural adhesives.

It should be noted, however, that the deformation rules of organic polymers only apply to a maximum deformation of 1–2% of the total deformation capacity. Higher deformations may induce irreversible damage, the degree of which has not yet been determined. Therefore, the deformations should not be greater than the shearing tangent values displayed in Figure 7.30 (up to 0.1).

To summarize, simple tensile shear tests with short-term loading are an efficient, but rather limited – test method for the general evaluation of adhesives and adhesive bonds.

7.7.2
Tensile Shear Strength Under Long-Term Static and Alternating Loading

With few exceptions, all constructions, components and their joints are generally exposed to long-term static or alternating loads rather than short-term loads. For example, over the whole operating life of a car (around 12 years), the coil springs and their fastening elements must carry part of the car's weight without undue plastic deformation, and they must resist, without damage, dynamic states of stress produced during service over a total period of approximately 3000 h (corresponding to about 150 000 km distance covered). A reliable dimensioning of these components is possible by using material parameters determined under conditions at least similar to those met in service. With regards to adhesive bonds, the viscoelastic–plastic behavior of the adhesives must be translated into material parameters which, ideally, are reflected by isochronous shear stress–shear strain diagrams for given temperatures, and where each measuring point characterizes the shear stress–shear strain behavior of a given adhesive under long-term static or even dynamic load. With few exceptions, however, such diagrams are not available as they require an enormous testing effort. Another, easier testing method is therefore preferable. The simplest method would be to expose tensile lap shear specimens to different static or alternating mechanical loads, and to determine the time taken to rupture under static loading, or alternatively the number of loadings supported by the specimen until rupture of the joint (i.e. fatigue strength; see Section 7.3.3). Some typical results of these tests are shown in Figures 7.31 and 7.32. Up to a temperature of 40 °C, the phenolic resin adhesive supported up to 70% of its static failure load when exposed to long-term static loading. At 80 °C – that is, above its T_g-value – it was hardly loadable any more when exposed to long-term loading. The cold-setting epoxy resin could only be loaded with up to 50% of its short-term load capacity when exposed to long-term loading (up to a maximum of 60 °C; above this temperature, the time to rupture could no longer be measured).

Like the time to rupture, the fatigue strength of assemblies joined with a given adhesive depends on the adherent material (and its surface condition). Experience has taught us that numbers of stress cycles in excess of 10^6 induce a flattening of the Wöhler diagram, reflecting a form of fatigue limit. In this respect, the fatigue strength is approximately 15–20% of the short-term load capacity. This has to be put into perspective, however, as the fatigue strength of bonded joints also depends on the loading rate (Figure 7.33). This can be explained, on the one

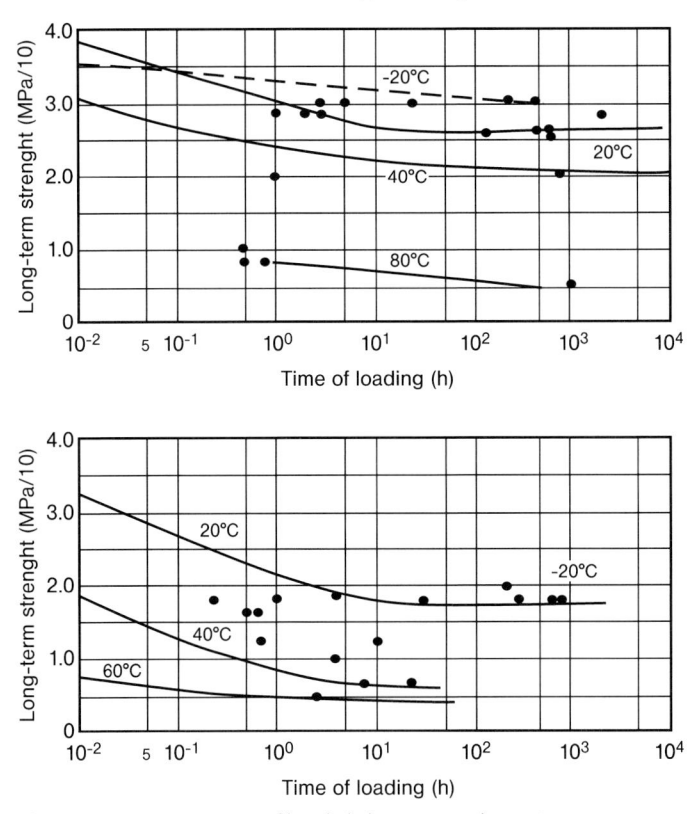

Figure 7.31 Time to rupture of bonded aluminum under static loading as a function of temperature. Adhesives: upper, phenolic resin; lower, cold-setting two-part epoxy resin.

hand, by the fact that at a low load frequency, testing until rupture takes considerably more time than at a high frequency. On the other hand, due to the viscoelastic–plastic behavior of the adhesives, the damage process is probably different when the specimen is exposed to short-term stress cycles (i.e. to a higher load frequency) as compared to a low frequency. Evidence suggests that, at very high loading frequencies, the loss in strength practically no longer depends on the number of stress cycles, although this behavior has not yet been systematically investigated.

In summary, the time to rupture of adhesive bonds under static load can cautiously be estimated to be 50–70% of its static failure load, and their fatigue strength can be estimated at around 15–20% of their short-term load capacity. As will be illustrated in Section 7.7.5 (the durability of adhesive bonds), in the majority of applications the mechanical load capacity of adhesive bonds is suitably described by the parameters determined under tensile shear loading.

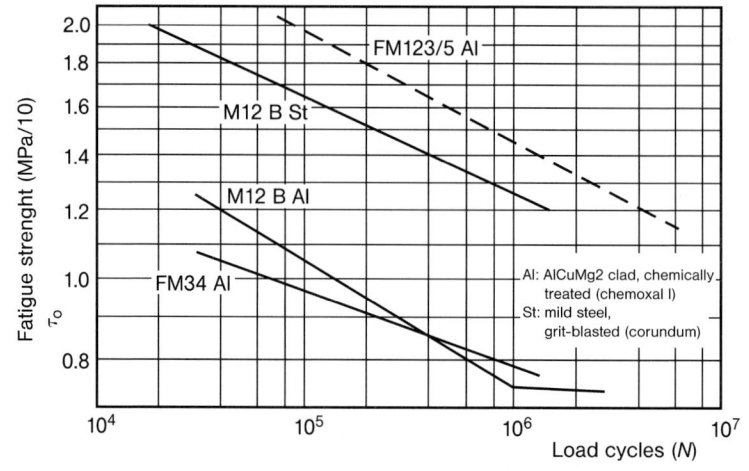

Figure 7.32 Fatigue strength (Wöhler diagrams) of bonded aluminum and steel. Adhesives: FM 123/5, hot-setting one-part epoxy-nitrile resin; M12 B, hot-setting phenolic resin; FM 34, polyimide.

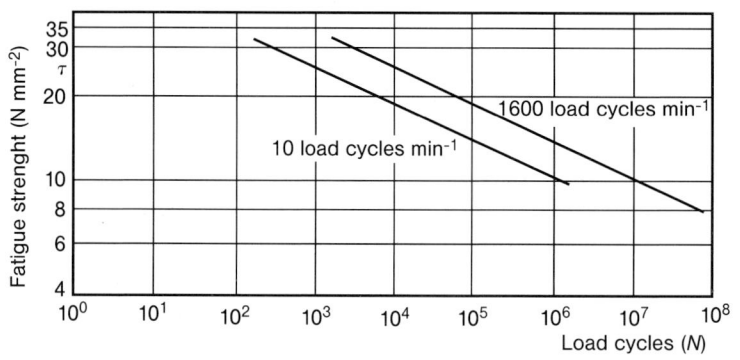

Figure 7.33 Fatigue strength of bonded aluminum as a function of the load frequency. Adhesive: hot-setting epoxy-nitrile resin.

7.7.3
Peel Strength

The peel strength of adhesive bonds can be determined using several test methods (see Section 7.2). This reflects the resistance of the bond-line to peel forces produced by a load that is applied eccentrically and perpendicular to the bond-line. Both, the crack initiation resistance and resistance to crack propagation within the bond-line are measured. The peel resistance depends not only on the properties of the adhesive used but also on the geometry, the surface conditions, and the material of the adherents, particularly on their deformation behavior, as well as on the rate of peeling. Therefore, a peel strength that was determined with a given adhesive bond configuration may not necessarily be applied to others. Rather, qualitative data are

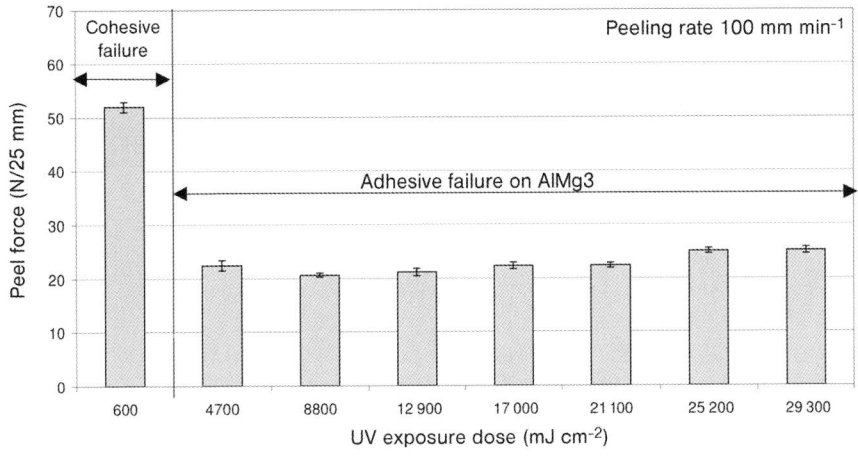

Figure 7.34 Peel strength of a UV-setting PSA on aluminum as a function of the crosslinking rate (180° peel test).

only provided for the purpose of comparison, for example in a test run where only one testing parameter is changed, and all other parameters are maintained.

The peel strength characteristics of a PSA are shown in Figure 7.34 (its tensile shear strength as a function of the crosslinking rate is illustrated in Figure 7.28). The tensile shear strength of this PSA increases with increasing crosslinking rate, whereas the peel resistance decreases. In peel tests, the deformation behavior of an adhesive is tested rather than its strength. The complicated peel behavior of adhesives, and its causes, will not be discussed in detail here [13]. In principle, peel tests are fracture–mechanical methods which involve testing conditions that are hardly suitable in terms of fracture mechanics; hence, they provide only qualitative data for comparative purposes. However, in this case the peel tests are perfectly valid, especially when the results are considered within the context of strength data, as in the above-mentioned example.

Well-plasticized structural adhesives can have a maximum peel strength (see Section 7.2.6) of approximately 20 N mm^{-1}, with similar values being obtained when using completely different systems and geometries.

In former times, peel strength was an important factor that had to be considered when deciding in favor of, or against, bonding. This was especially the case when formulating high-strength adhesives, as a high load capacity could not be combined satisfactorily with high peel strength. Today, however, this problem has been largely solved.

7.7.4
Impact Behavior

In the car manufacturing industry, and especially in bodywork construction, bonding has been adopted as an increasingly successful joining technique. Consequently, the

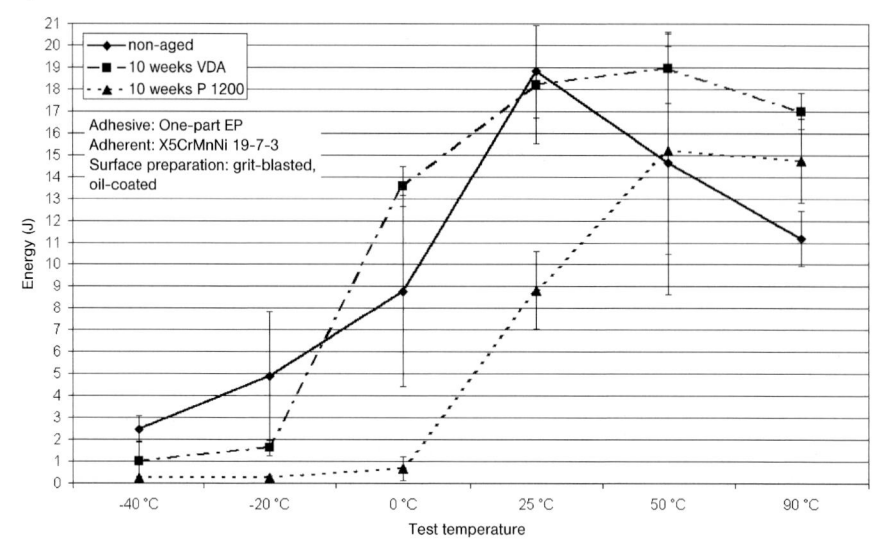

Figure 7.35 Energy absorption capability of bonded joints in the wedge impact test in accordance with ISO 11343 as a function of the temperature [14].

impact resistance of bonded joints has become a matter of growing particular interest (see Section 8.2.2). In this field of application, considerably more attention must be concentrated on the deformation capacity of adhesives, especially during a crash, than on the tensile shear strength or peel strength, for example. The crash behavior can be tested on component-like specimens such as double hat boxes exposed to an impact load, although this requires a considerable testing effort. Hence, simpler tests performed on smaller specimens have been accepted as tried and tested methods. The most important one – the wedge impact test – is described in Section 7.3.2; typical results are shown in Figure 7.35.

In this example (special steel bonded with an epoxy-resin based adhesive for bodyworks), it can be seen that the energy absorption capability depends heavily on the temperature. In anticipation of Section 7.7.5, it can also be seen that there was a considerable change in the behavior of bonded joints when they were exposed to a hostile environment prior to testing (see Section 7.4). Once again, the test results were seen to depend on the surface characteristics and surface conditions of the adherents. Whilst it is not possible to determine absolute parameters, these test results are useful for the comparison of different adhesives or, as in the above example, for the evaluation of crash behavior that changes under the influence of temperature and hostile environment.

7.7.5
Durability of Bonded Joints

In the non-loaded condition, the properties of bonded joints may change under environmental influences, as mentioned above at the end of Section 7.7.4 in connection with the wedge impact strength. In the loaded condition (i.e. in service) the effectiveness of environmental influences can be further increased under certain circumstances. The resistance of bonded joints to environmental influences is called 'age resistance', although this is not entirely correct because in other fields aging may also mean a change due to the influence of time alone, without further physical or chemical influences. A hostile environment generally causes an irreversible deterioration of the mechanical characteristics of a bonded joint. The deterioration depends on the type and intensity of the external influences, as well as on the quality of the manufacture process – that is, the conditions prevailing during the creation of the bonded joint. In the literature, with regards to the durability of bonded joints, water or humidity are referred to as the most hostile environmental influences. Hostile media enter the bond-line via diffusion processes or capillary attraction along existing microcracks in the polymer. This is considered to be the most frequent cause of poor durability of bonded joints, particularly if the adherent material is inorganic (e.g. metal or glass). The following, mainly humidity-induced, damage mechanisms are discussed:

- *Water uptake by the adhesive*: The adhesive polymer is plasticized by the uptake of water. This process is partly reversible upon redrying and generally results in a cohesive failure within the bond-line. The change in strength behavior of the bonded joint stops as soon as the bond-line is saturated with water. Irreversible impairment of the bond-line by moisture rarely occurs with modern adhesives used in industry, unless the adhesives are deliberately detachable on demand.

- *Detaching of the adhesive from the adherent surface*: A failure of adhesion between the adhesive and the substrate surface is explained by competitive adsorption of water instead of polar groups of the polymer molecules, or by hydrolysis of chemical bonds in the boundary layer.

- *Water that diffuses into the interface*: When this happens, the adherent surfaces are changed by solvolysis, for example in the form of a hydration of oxides on metals, and the bonding surface is weakened. Solvolytic reactions may generate byproducts in the interface that may further weaken the polymer near the interface.

- *The degradation process*: This is accelerated by an enhanced water uptake by the adhesive due to high temperatures and loading.

- *Electrochemical corrosion*: This starts at the unprotected edges of the adherents and is due, for example, to corrosive substances such as chloride or sulfate ions being dissociated in water. This results in bond-line corrosion, particularly in metal bondings.

With the exception of bond-line corrosion, these aging processes generally take place very slowly. Therefore, the durability of bonded joints is tested by means of most varied, more or less standardized aging simulations (see Section 7.3) in which several external influences are superposed and simultaneously affect the assembly, thereby accelerating the failure mechanisms. To achieve this acceleration effect, compared to a natural aging process with actual climatic conditions, almost all artificial aging tests share an increase in the intensity of the test parameters (temperature, salt exposure, etc.). This is one of the reasons why artificial aging cannot be exactly correlated with outdoor weathering of bonded joints.

In the following section, some examples are provided to illustrate the durability of different bonded joints as a function of the environmental conditions. The degradation mechanisms induced will be explained by means of fracture surface analyses. The durability of bonded aluminum, steel, glass and plastic materials, respectively, will be discussed separately. Whilst most of the available knowledge has been gained with bonded aluminum (owing to its importance in aircraft and automobile construction), many of the findings can be transferred to other adhesive systems [15–17].

7.7.6
Adhesive Bonding of Aluminum

To date, it has not been possible to create durable adhesive bonds with aluminum alloys using structural adhesives unless a special surface preparation is performed prior to bonding. This will be illustrated on EN-AW (AlCuMg) and EN-AW 5182 (AlMg4.5Mn0.4), which are noncurable wrought alloys. It is therefore of great interest to identify the surface conditions and surface preparation methods, respectively, which provide for good durability. An overview is provided in Table 7.2 of the surface preparations used for the above-mentioned aluminum alloys bonded with a one-part, hot-setting epoxy resin adhesive manufactured by Dow Automotive.

When used for strength tests of structural adhesive bonds, single lap-shear specimens produced according to DIN 1465 have well-known disadvantages. However, this simple sample geometry was employed here because a great number of parameters were tested. It seems as if these specimens are perfectly suitable and still widely used in practice to compare the aging characteristics of adhesives as a function of surface preparation and aging simulation. Following the aging process, the loss in strength is related to the initial strength, and the failure patterns are analyzed. Only in few cases were the so-called 'tensile shear specimens with thick adherents' used for the investigation of the shear stress–shear deformation behavior.

The durability tests were performed with the adhesives, substrates, and aging simulations that are mainly used in the automobile industry (Tables 7.3 and 7.4). The artificial aging procedures were divided roughly into categories according to the major failure mechanism. For example, in the salt spray test and the VDA 621-415 or VDA KKT test, failure is induced by corrosion, whereas in the climate change test and the immersion test, it is mainly induced by humidity or water.

For representation of the test results, conventional bar diagrams were hardly suited due to the great number of parameters investigated. Moreover, the test focused on the influence of the surface preparation and the aging simulation on the residual strength of the assemblies rather than on absolute strength values.

Therefore, the results were represented graphically so as to allow the percentaged residual strength after aging to be visualized, and to make it possible to differentiate between the major failure mechanisms occurring with different aging simulations – that is, 'humidity' or 'salt' (= corrosion induced by exposure to chloride ions).

Table 7.2 Surface preparation methods for aluminum alloys.

Surface preparation	Notes
Vapor-degreased with acetone	Considered as reference in terms of 'non-treated aluminum surface'. Generation of sufficiently reproducible starting conditions
Grit-blasted with white corundum (6 bar); degreased with acetone under ultrasound (15 min)	Mechanical surface preparation
Alkaline dip-etching in 1 M sodium hydroxide solution for 15 min; rinsing; etching in 20% nitric acid solution to remove copper enrichments for 3 min; rinse with demineralized water	Fresh, reactive oxide layers are created by chemical removal of covering layers
Acid dip-etching in 65% nitric acid solution for 15 min; rinse with demineralized water	The concentrated, highly oxidizing nitric acid hardly attacks aluminum oxide; there is no removal of covering layers by the etching process in the conventional sense
Gardoclean Bonder V 299/4M (Chemetall) Preparation: $50\,\mathrm{g\,l^{-1}}$ (demineralized water) $80\,°\mathrm{C}$; dip application for 15 min; rinse with demineralized water	Sodium metasilicate containing alkaline cleaner
Titanium-circonium mixed oxides (GB X 4711)	Chromate-free conversion coating
Dry lubricant on a titanium-circonium conversion coating	Drylub: Zeller + Gmelin Drylub C1

It becomes obvious that aging in corrosive media caused a relatively important loss in strength if the adherent surfaces were vapor-degreased with acetone (Figure 7.36) or blasted with corundum (Figure 7.37) prior to bonding. The failure mechanism that started from the adherent edges and progressed towards the inner region of the bonding (bond-line corrosion; Figure 7.38) can easily be detected by visual analysis of the fracture surfaces. In salt-free aging simulations with exposure to humidity, a relatively good durability was achieved when the adherent surfaces were grit-blasted with corundum prior to bonding. In all tensile lap-shear specimens there was cohesive failure (near the interface).

If the surfaces of aluminum adherents are pretreated with a mild alkaline (e.g. sodium metasilicate containing solution; 'Bonder') prior to bonding, the bonded joints have a near-opposite aging behavior. The surface preparation generates a silicate coating (conversion coat) that clearly protects the metal against corrosive attack by chloride ions. In the presence of humidity, however, the conversion coat

Table 7.3 Aging simulations with failure mainly induced by corrosion.

Aging test	Description	Notes
Salt spray test	• In accordance with DIN 50 021 • Salt spray fog from a 5% NaCl solution; temperature 35 °C; duration 1000 h; 2000 h	Exposure to corrosion and moisture
VDA change test 621-415	One cycle comprises: • 24 h salt spray test (DIN 50 021) • 96 h condensed water alternating climate (DIN 50 017 KFW) • 48 h RT storage (DIN 50 014) • Duration: 10 cycles	Exposure to corrosion and condensed moisture
VDA-KKT test	One cycle comprises: • 1 week VW P1200 (see below) • 3 weeks VDA 621-415 • Duration: 3 cycles	Alternating exposure to corrosion, moisture and temperature
Accelerated outdoor weathering	• Natural aging under real climatic conditions, accelerated by spraying a 5% NaCl solution on the samples every 14 days • Duration: 6 months	Alternating exposure to corrosion, moisture and temperature

Note: After removal from the aging simulation, all samples were conditioned at 40 °C for three days, and tested at room temperature.

seems to induce local alkalization of the damage medium within the interface (Figure 7.39).

Following artificial aging in standardized corrosion tests (e.g. salt spray test, VDA alternating climate test 621-415), there was only a relatively low percentage loss in strength. After four weeks of storage in a water bath at 70 °C, however, the diffusion of humidity into the interface, accelerated by the addition of a tenside, induced a dramatic loss in strength and correlated well with the results obtained following aging under natural climatic conditions. Neither of the other short-term aging tests (VW P1200, cataplasm test, etc.) with mainly humidity-induced failure mechanisms was able to simulate this damage pattern, with the exception of the three-months VDA KKT test which combined alternating exposure to humidity and temperature with corrosive exposure similar to outdoor weathering. There was also a significant loss in strength, similar to the above-mentioned damage pattern.

As indicated by the fracture surfaces (Figure 7.40), the adhesive was almost completely delaminated; this can be explained by a degradation of the aluminum oxide layer that is only stable in a pH range between 4.3 and 8.5.

The aging simulation tests were then performed with bonded joints that had undergone a chromate-free, no-rinse TiZrO conversion surface preparation. No

7.7 Applications of Test Methods in Structural Adhesive Bonding

Table 7.4 Aging simulations with failure mainly induced by humidity.

Aging test	Description	Notes
VW P1200 alternating climate test	One cycle comprises: • Constant climate at 80 °C/95% RH for 4 h • Cooling to −40 °C within 2 h • Isothermal storage at −40 °C for 4 h • Heating to 80 °C/95% RH within 2 h • Duration: 60 cycles; 120 cycles	Alternating exposure to humidity and temperature
Cataplasm test	• Each sample is enwrapped by a water-impregnated cotton-wool and aluminum foil; air-tight sealing in PE foil • Storage at 70 °C for 28 days, then at −20 °C for 48 h	Alternating exposure to humidity and temperature
Condensed water test	• Condensation on the specimens at 40 °C • Duration: 28 days	Exposure to humidity
Immersion test at 70°C	• Storage in demineralized water bath at 70 °C • Duration: 28 days	Exposure to humidity
Immersion test at 70°C + tenside	• Storage in demineralized water bath at 70 °C; • Addition of 0.1% commercial tenside in aging medium • Duration: 28 days	Exposure to humidity
Outdoor weathering	• Natural aging under real climatic conditions • Duration: 1 year	Alternating exposure to humidity and temperature Conditional exposure to corrosion

Note: After removal from the aging simulation, all samples were conditioned at 40 °C for three days, and tested at room temperature.

significant increase in durability was obtained compared to merely acetone-degreased reference bondings (Figure 7.41).

The aluminum alloy used here (AlMg4.5Mn0.4) was comparable to the AlMg3 substrates with regard to its technical and electrochemical characteristics according to the manufacturer's instructions.

Figure 7.42 shows the results obtained with a tenside-based dry lubricant coat applied as temporary anticorrosive agent and forming aid as used in pressing lines. Apparently, this was not sufficiently absorbed by the polymer, and caused an impaired adhesion. All aging simulations showed in common a rather important loss in

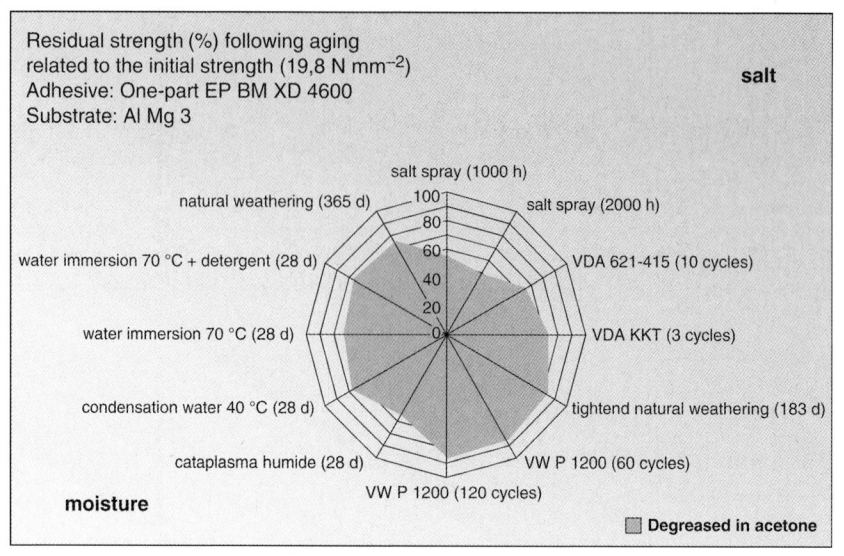

Figure 7.36 Aging characteristics of bonded aluminum after acetone degreasing.

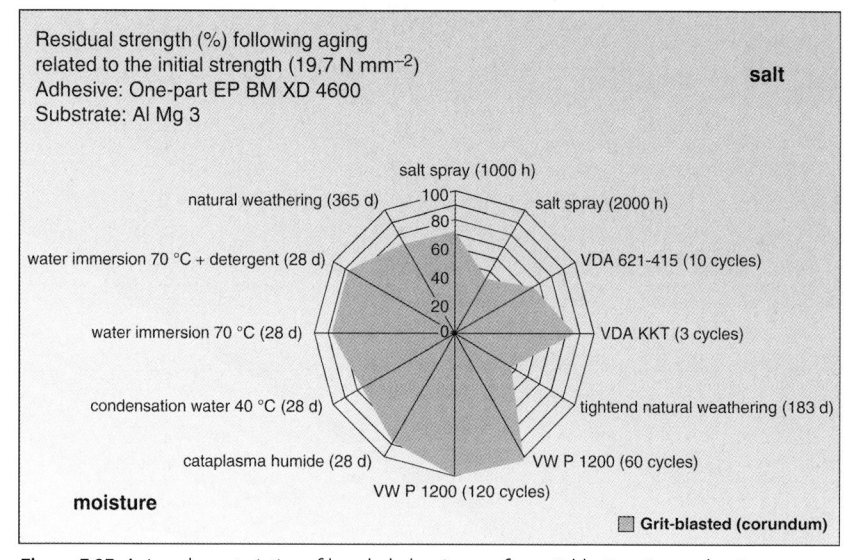

Figure 7.37 Aging characteristics of bonded aluminum after grit-blasting (corundum).

strength (Figure 7.42). It should be noted, however, that some aspects relevant to the manufacture processes used in the automobile industry were disregarded. For example, the dry lubricant coat beneath the noncrosslinked polymer must be resistant to being washed away during the washer processes prior to e-coating.

Specific tests on steel bondings are already in progress. Although representative only to a limited extent, the results obtained with the drylub-coated aluminum substrate used in these tests correlated well with the results of investigations

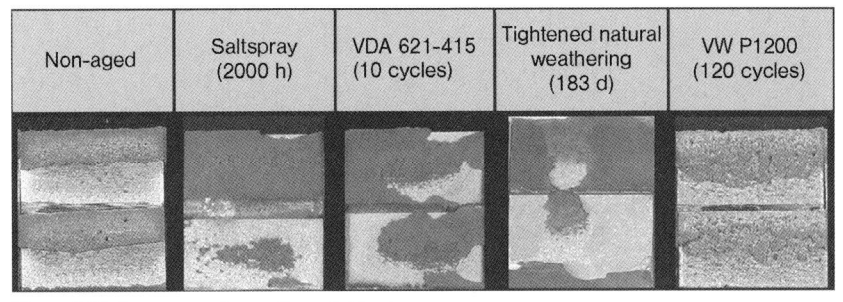

Figure 7.38 Fracture surfaces after grit-blasting.

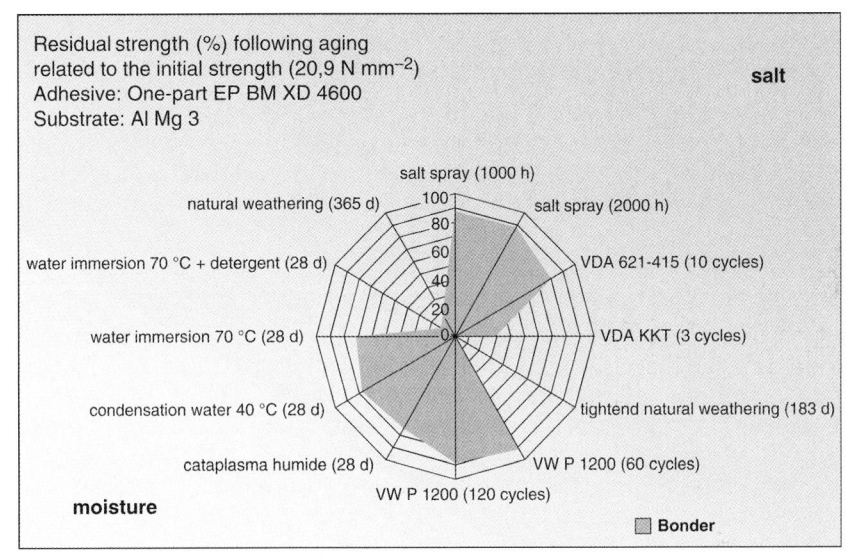

Figure 7.39 Aging characteristics after 'Bonder' treatment.

performed by Kleinert et al., who studied, among others, parameters that were comparable to those tested here [18].

The aging characteristics were most favorable following alkaline or acid etching. The chemical removal of contaminated covering layers by means of a sodium hydroxide solution (Figure 7.43) is an established etching method for aluminum, whereas the surface preparation with high-percentage nitric acid (Figure 7.44) is rather unconventional. Concentrated nitric acid barely attacks aluminum, and consequently there is no removal of the covering layers by the etching process in the conventional sense. In principle, this method is only suitable for closed systems, as nitrogen gases may develop. Following both surface preparations, the bonded joints showed a very good resistance to salt-containing aging media. Under exposure to humidity, their durability also was at least comparable to that of other surface preparations (Figures 7.43 and 7.44). Both, before and after aging (performed in accordance with EN ISO 10365), all fracture surfaces had a particular cohesive failure pattern in proximity to the substrate (SCF).

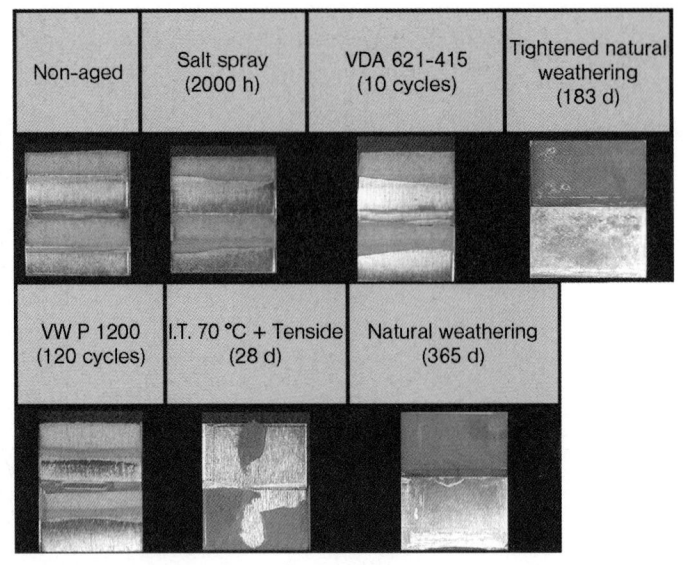

Figure 7.40 Fracture surfaces after 'Bonder' treatment.

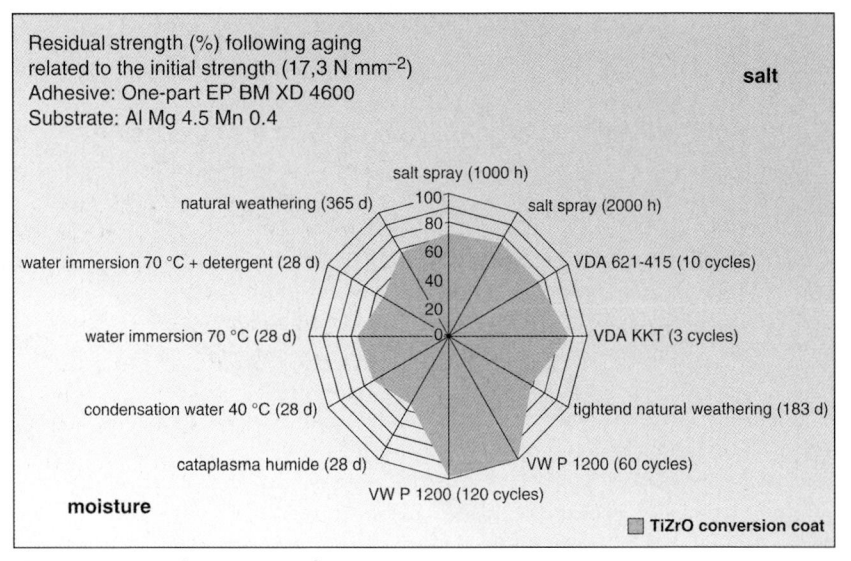

Figure 7.41 Aging characteristics after TiZrO conversion treatment.

It has become obvious that, when it comes to evaluating the durability of epoxy-resin bonded aluminum following different surface preparations, the significance of standardized aging tests may not be satisfactory. The influence of the simulated aging processes on the technical characteristics of bonded aluminum depends both on the type, duration and intensity of the major failure mechanism, and on the substrate surface conditions generated by the specific surface preparation method. Hence, the

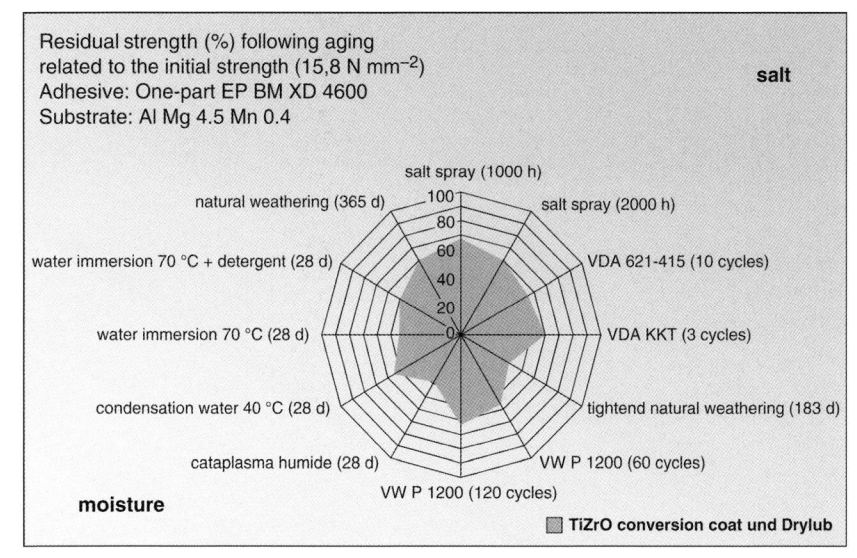

Figure 7.42 Aging characteristics after TiZrO + Drylub.

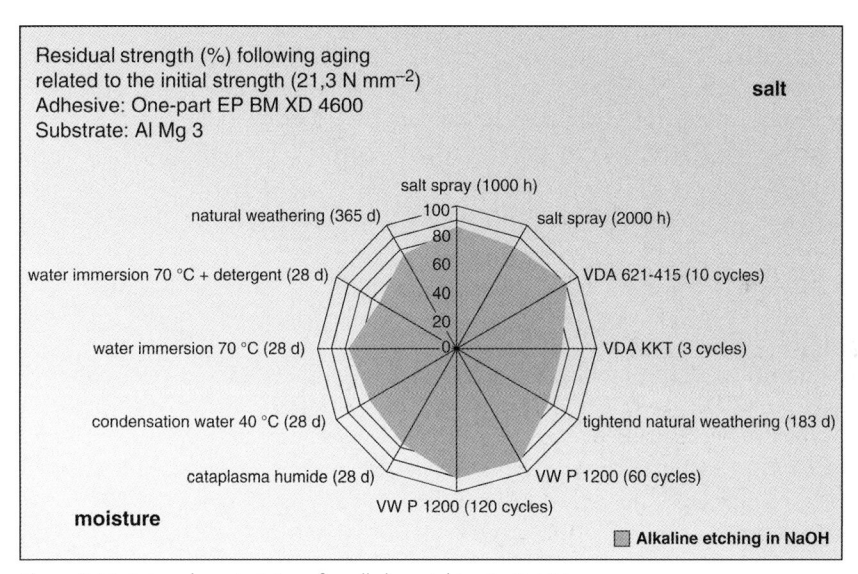

Figure 7.43 Aging characteristics after alkaline etching in NaOH.

surface preparation has a major effect on the sensitivity of the bonding to different aging processes.

The intensification of the individual parameters evaluated in short-term tests (e.g. temperature, humidity, concentration of chloride ions) does not necessarily induce the desired acceleration effect. Therefore, a correlation with natural aging processes is possible only to a very limited extent.

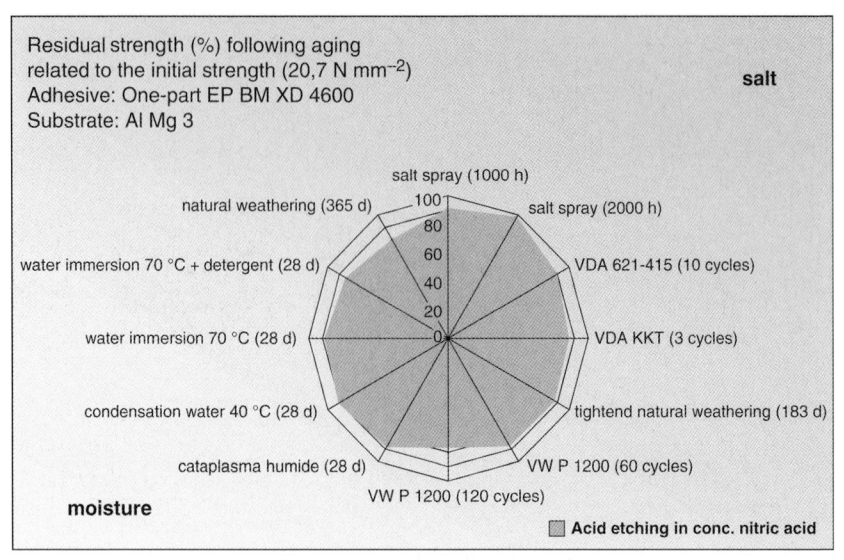

Figure 7.44 Aging characteristics after acid etching in concentrated nitric acid.

In order to assess satisfactorily the durability of epoxy-resin-bonded aluminum, at least in a primary manner, as wide a spectrum of aging simulations as possible is required that must be as close to reality as possible. For evaluating the aging behavior of bonded aluminum following different surface preparations, the aging simulations were suitably classified according to the major failure-inducing mechanism, such as exposure to humidity or corrosion, although in some aging simulations both mechanisms contribute to failure.

When exposed to corrosive media, bonded aluminum was significantly less durable following mere degreasing or grit-blasting with corundum compared to a chemical surface preparation (see Figures 7.36 to 7.38).

To further investigate this effect, so-called 'tensile shear specimens with thick adherents' (DIN 54 451) of a curable wrought aluminum alloy (EN-AW 2017) were subjected to a chemical or mechanical surface preparation and then bonded with a hot-setting, one-part epoxy resin adhesive. Once again, the standardized VDA alternating climate test 621-415 known from the automobile industry was used as aging simulation.

In service, the loads applied can considerably impair the durability of structural polymer–metal joints [16]. Hence, during the artificial aging simulation, some samples were subjected to a static load corresponding to 10% of the initial strength of the bonding, by means of special clamping devices. The shear deformation of the bondings was determined during the test by means of fine-strain extensometers supported close to the lap (resolution: 15 μm).

The τ-γ diagrams of samples aged under static load compared to non-aged reference samples and reference samples aged without load are shown in Figures 7.45 to 7.48. Likewise, in this test run there was a distinct loss in strength

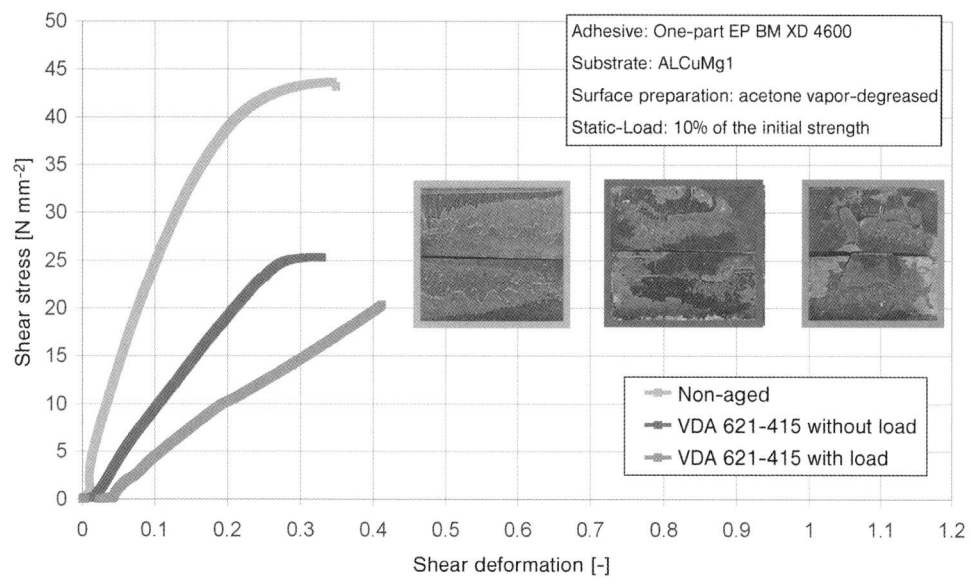

Figure 7.45 τ-γ diagram of acetone-degreased aluminum bondings.

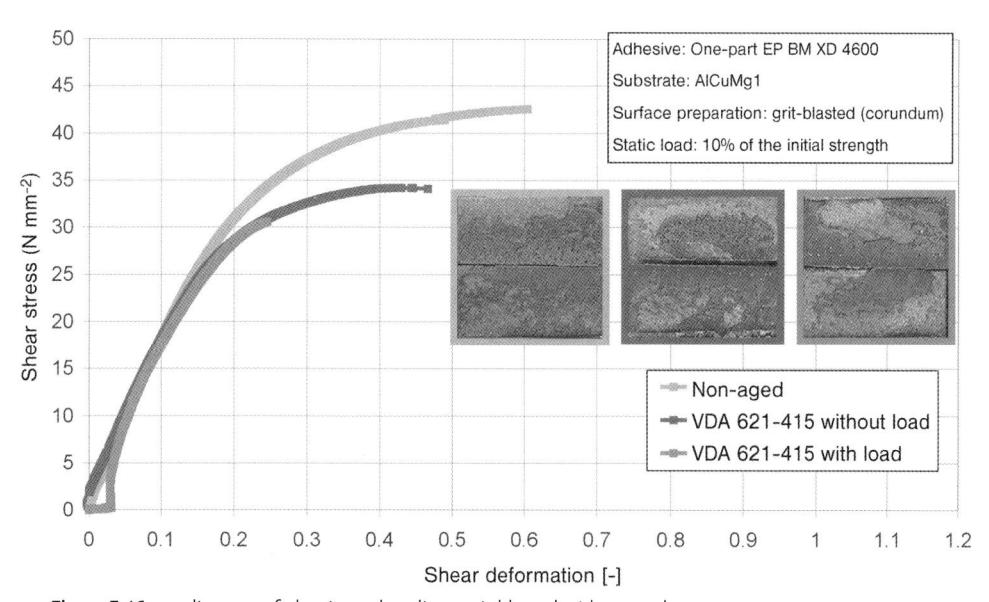

Figure 7.46 τ-γ diagram of aluminum bondings grit-blasted with corundum.

in bondings with degreased or grit-blasted (corundum) adherent surfaces following aging in corrosive media. When a static load was applied, this effect was further enhanced. It was noted that the shear modulus – that is, the quotient from shear stress and shear deformation – decreased in the case of vapor acetone degreasing.

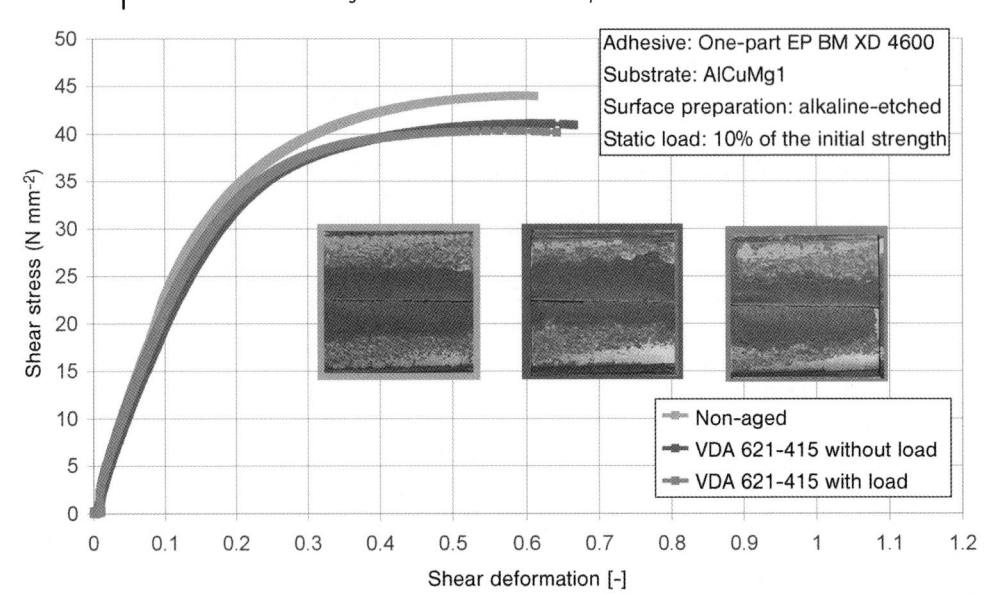

Figure 7.47 τ-γ diagram of aluminum bondings after alkaline etching.

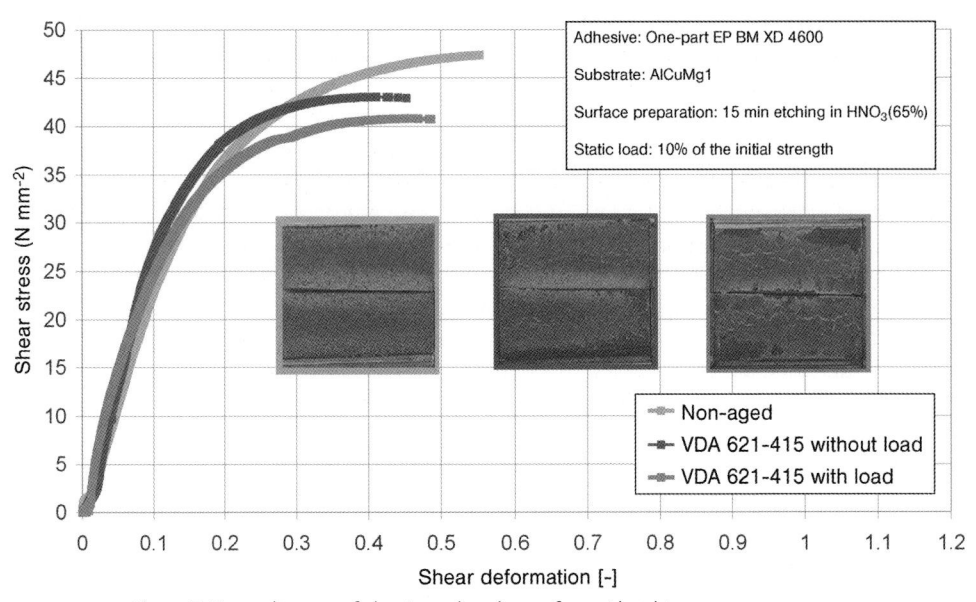

Figure 7.48 τ-γ diagram of aluminum bondings after acid etching.

Following 10 cycles of the VDA alternating climate test, a relatively minor loss in strength was detected when the adherent surfaces of bonded aluminum were prepared by alkaline or acid etching prior to bonding (Figures 7.47 and 7.48). The load additionally applied during the aging process did not induce any significant, technically relevant decrease in the shear modulus.

Furthermore, visual inspection of the fracture surfaces revealed that the failure pattern was in analogy to that of bonded, noncurable AlMg3 alloys where bond-line corrosion started from the unprotected edges and progressed towards the inner region of the bonding following acetone degreasing and grit-blasting (corundum) prior to bonding. Following alkaline or acid etching, there was cohesive failure (near the interface).

It is known that surface preparation can have a major influence on the electrochemical potential of metal surfaces. Mechanical surface preparations (e.g. blasting with white corundum) induce residual compressive stresses of around −70 MPa in the AlMg3 surfaces. By cold working, the metallurgical structure becomes susceptible to corrosion that usually starts at the edge and screw dislocations, respectively.

The depth profile of surfaces that were subjected to different surface preparations can be analyzed using surface analysis methods such as secondary neutral particle mass spectrometry (SNMS). In the case of vapor-degreased AlMg3 adherents, an accumulation of magnesium (an alloy component) is detected near the surface.

From the literature it is known that magnesium diffuses from the bulk along the grain boundaries and accumulates in the covering layer of aluminum, which can be explained by the metal's electrochemical nature. These heterogeneities can impair the resistance of the oxide layer to corrosion and consequently impair the durability of bonded joints [19, 20].

If, in contrast, the naturally grown covering layer is removed by alkaline etching with a sodium hydroxide solution, then no magnesium is detected by means of surface analysis following the generation of these fresh oxide layers.

Following acid etching, the samples have almost identical SNMS depth profiles. Concentrated nitric acid has the interesting property that it only slightly attacks aluminum oxide, despite its usual strong oxidizing effect; consequently, there is no removal of covering layers by etching in the conventional sense. In an acid environment, the magnesium (oxide) phases accumulated at the surface become unstable and are dissolved. In this case, too, the percentage of magnesium detected on the surfaces remains largely below that of the alloy.

At first, it might seem obvious to suspect the electrochemical properties of the adherent surfaces, which are changed by the surface preparations, as being the reason for the dissimilar aging behavior in corrosive media.

Based on this approach, the failure mechanisms detected on the fracture surfaces of the bondings are expected to correlate with electrochemical analyses performed with the adherents following different surface preparations. However, potentiodynamic current density–potential measurements of the aluminum surfaces discussed here do not conform to this hypothesis.

Only in the case of substrates etched with nitric acid is the breakdown potential shifted towards lower potentials. Likewise, only in this case does the passivity with regard to pitting represent an explanatory approach for the good durability of bonded joints exposed to corrosive media following alkaline etching performed prior to bonding.

It was noted that, in the case of etching with concentrated nitric acid, the bondings had a very good durability following aging in corrosive media, despite the rather

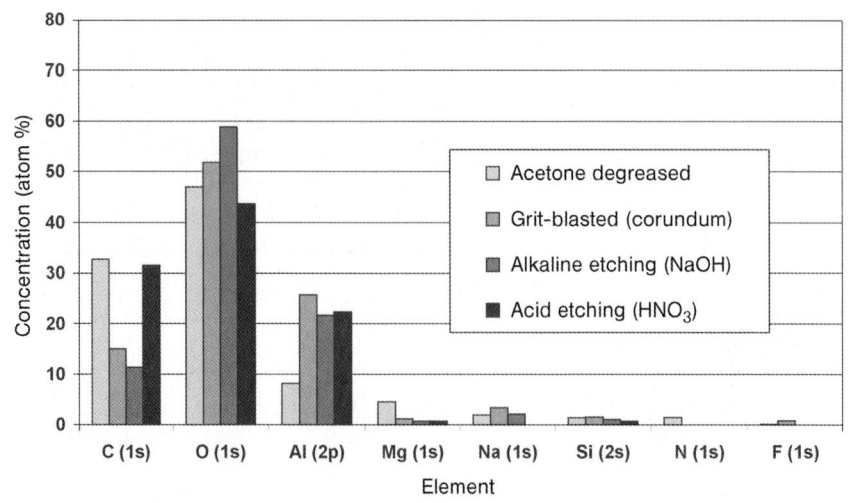

Figure 7.49 X-ray photoelectron spectroscopy (XPS) analyses of aluminum surfaces prepared with different methods.

moderate resistance of their surfaces to corrosion. This shows that the attempt to explain exclusively the aging phenomena of bonded joints by the electrochemical properties of the metal surfaces can sometimes be misleading because à priori this approach disregards the adhesion between polymer and substrate as a factor that significantly influences the aging process. The adhesional properties of a metal surface or its oxide layer are significantly determined by its chemical reactivity, its mechanical stability and, moreover, by its morphological characteristics. When the presence of carbon is taken as a hint on organic contamination, X-ray photoelectron spectroscopy (XPS) investigations show that both the abrasive effect of grit-blasting with corundum and chemical removal of covering layers by alkaline etching generate a 'clean' (i.e. freshly produced) and therefore reactive surface. In contrast to these surface preparation methods, neither acetone nor concentrated nitric acid can remove the organic contaminations of the aluminum covering layer (Figure 7.49).

As illustrated above, following mechanical surface preparation, aging takes place faster than in the case of samples which were etched with HNO_3 and had a good durability. Clearly, the cleaning rate alone, which is achieved by means of a specific surface preparation method, is not pertinent to the assessment and prognosis of the durability of epoxy-resin-bonded aluminum exposed to a corrosive environment.

In this context, a potential effect of the surface morphology generated by the surface preparation is subject to controversy. It has already been shown that the macroscopic dimension (i.e. the surface roughness) classified by several statistical engineering parameters, does not have any major influence. This has been demonstrated, amongst others, both on textured steel sheets (see Section 7.7.7) and on aluminum surfaces pretreated with different surface preparation methods [17]. There appears to be a more significant effect on a submicroscopic level in the dimension of the prepolymers applied. This effect has been extensively investigated on anodized aluminum surfaces [15].

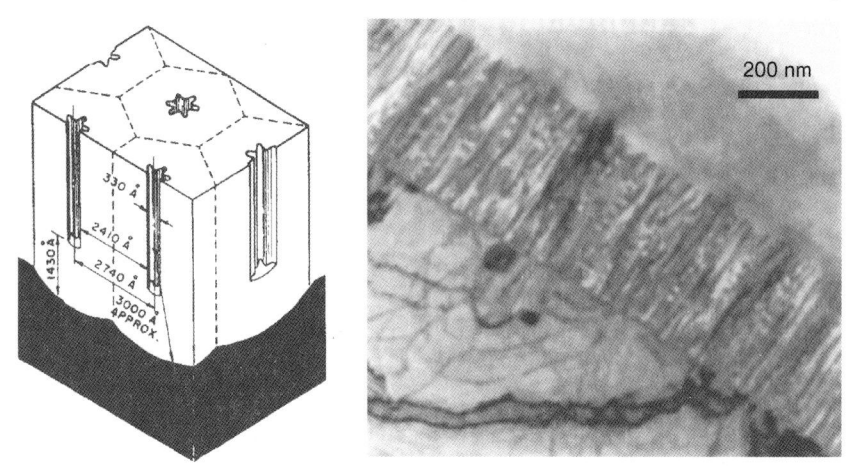

Figure 7.50 Model and real structure of an aluminum surface anodized with phosphoric acid [15].

Particularly in the aircraft industry, aluminum materials are subject to an excessively elaborate, multistage anodization procedure in order to optimally prepare their surfaces for bonding. The oxide coatings generated consist of a very thin, almost poreless, dielectric barrier layer, and a partly very fine-pored covering outer layer (Figure 7.50).

For example, with regards to bonding, it is beyond controversy that the anodization of aluminum surfaces with phosphoric acid generates the most favorable aging characteristics compared to other surface preparations. This is explained by a surface enlargement and improved electrochemical stability of the thick oxide layer generated.

Another explanatory approach is to assume that there is a micromechanical interlocking of the polymer after penetration of low-molecular-weight prepolymers into the structures of the porous oxide layer. This is quite conceivable when making a rough estimate of the dimensions in question. The diameter of an oxide layer pore generated by anodization ranges from 300 to 500 Å, according to the anodization procedure used. An epoxy resin molecule requires an average space of 138 Å^2, independent of the spatial arrangement of the molecule. The question remains as to how the low mechanical stability of the structures of the oxide coatings generated by anodization with phosphoric acid may contribute significantly to the strength of the adhesive joint system, operating on the premise that there is mechanical adhesion. Moreover, on the basis of this molecular approach, a clearly defined differentiation between chemical effects on the one hand and purely mechanical effects on the other hand seems to be a problem.

The effect of these different surface preparation methods on the morphology of the solid surface was investigated using transmission electron microscopy (TEM; Figure 7.51). By using ultrathin sections of acetone-degreased or grit-blasted (corundum) aluminum surfaces, a sharp transition with a defined interface can be detected between substrate and polymer, whereas the samples subjected to alkaline and acid

182 | *7 Adhesives and Adhesive Joints: Test Methods and Properties*

Acetone degreased

Grid-blasted (corundum)

Alkaline-etched (NaOH)

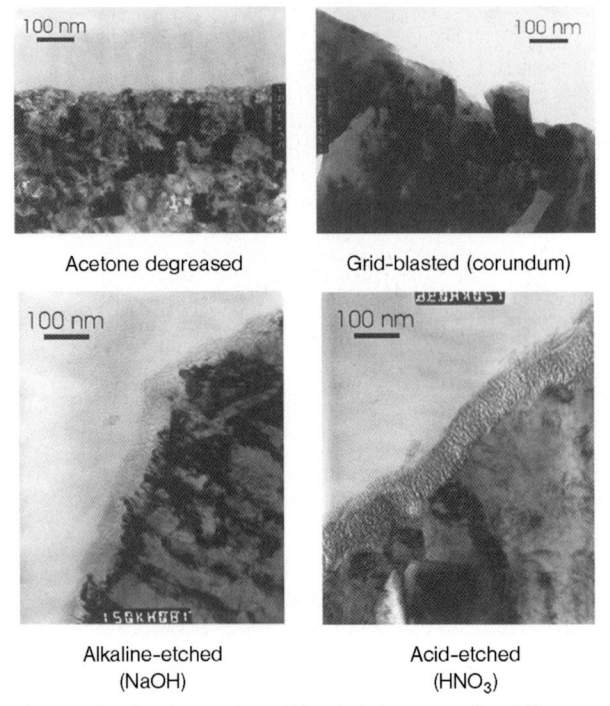

Acid-etched (HNO_3)

Figure 7.51 Ultrathin sections of bonded aluminum after different surface preparations.

etching had a nonuniform, thin, filigrane structure in this area which turned out to be aluminum oxide, as can be shown by oxide mapping (Figures 7.52 and 7.53).

Once again, the submicroscopic structures of the aluminum surface, together with their good aging characteristics in bonded joint systems exposed to corrosive environments, show that the chemical bonding and debonding reactions do not sufficiently explain the aging processes of metal–polymer joints. The displacement of

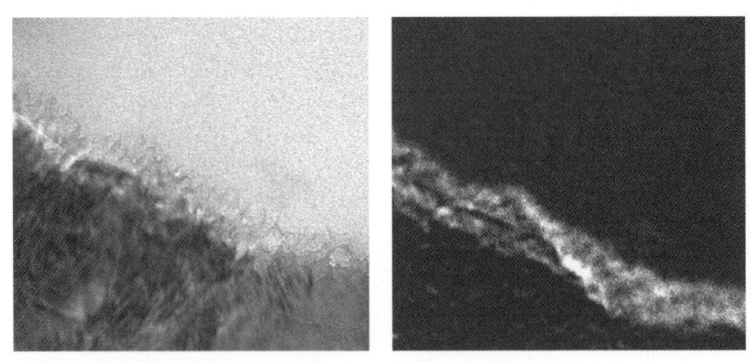

Figure 7.52 Oxygen mapping (right) on an ultrathin section (left) of bonded aluminum (EN-AW-5754, AlMg3) after alkaline etching (1 M NaOH).

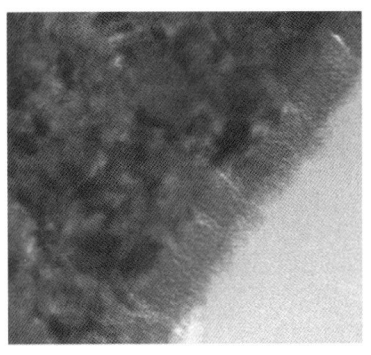

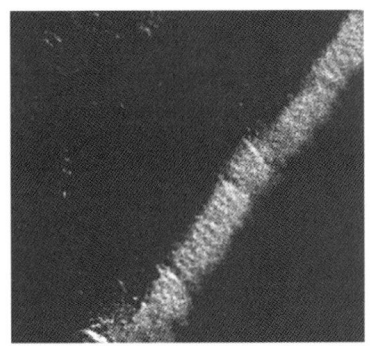

Figure 7.53 Oxygen mapping (right) on an ultrathin section (left) of bonded aluminum (EN-AW-5754, AlMg3) after acid etching (65% HNO_3 solution for 15 min).

the metal–polymer interface that moves into the nanomorphology of the surface generates an organic/inorganic interphase that is similar to a composite material. This seems to strongly inhibit the kinetics of the attacking failure mechanisms. Furthermore, a more homogeneous transition between solid surface and adhesive is created which may influence both the polymer network configuration and the fracture mechanics of the whole assembly in this area. Finally, it remains possible that the fine-structured oxides, which are found near the interface formed with the adhesive, modify the kinetics and the degree of crosslinking, respectively, inducing an increased molecular mobility and enabling the polymer to build up dynamic adhesion (as described in Chapter 3) which in turn may contribute to generate resistance to water.

Against the background of the problems associated with the aging of metal–polymer assemblies, it seems worthwhile – from an engineering point of view – to develop alternative methods to create such 'nanomorphologies' on the adherent surfaces and to take advantage of their obviously high effectiveness for the durability of bonded aluminum. The first attempts to replace elaborate etching and anodization procedures with a laser technology-based surface preparation of aluminum have shown great promise [21, 22].

To conclude this discussion it should be noted that the above-described results have been obtained with one-part, hot-setting epoxy resin adhesive-bonded joints. While similar qualitative results can be obtained with bondings using similar adhesives, they generally cannot be found with bondings using adhesives with different chemical formulations. This point is demonstrated in the next section, for bonded steel.

7.7.7
Adhesive Bonding of Steel

Long-term influences may completely change the strength characteristics of bondings. This is convincingly demonstrated by a comparison of the residual strength

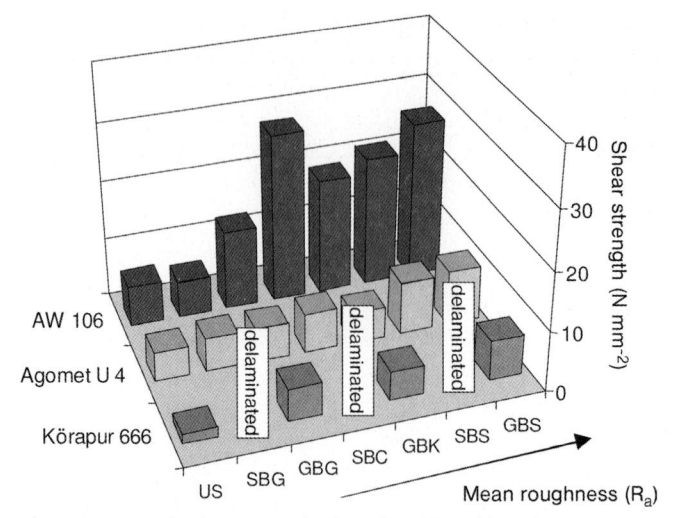

Figure 7.54 Tensile shear strength of aged (IT 80) mild steel bondings as a function of the surface preparation and the type of adhesive used.

values of mild steel bondings [9] following aging in 80 °C hot water (IT 80) for eight weeks (Figure 7.54) with the strength values obtained with non-aged bondings (see the discussion at the start of Section 7.7.1).

In the test run performed with the epoxy-resin adhesive AW 106, following an eight-week aging period in IT 80, the reference samples and the samples blasted with SBG and GBG failed, with an apparent adhesive failure pattern, at very low strength values (below 10 N mm^{-2}; Table 7.5), corresponding to 30–50% of the initial strength.

Table 7.5 Characteristics of fracture surfaces of aged (IT 80) mild steel bondings.

Adhesive	Characteristic	Treatment					
		US	SBG	GBG	SBC	GBK	SBS	GBS
AW 106 EP	Failure pattern	AF	AF	AF	SCF	SCF	SCF	SCF
	Related to non-aged ref. (%)	29	34	48	96	90	71	95
	Response, related to US	–	o	o	++	++	+	+
Agometacrylate	Failure pattern	AF	AF/CF	AF/CF	AF/CF	AF/CF	AF/CF	AF/CF
	Related to non-aged ref. (%)	25	29	26	32	26	44	43
	Response, related to US	–	o	o	o	o	+	+
Körapur PUR	Failure pattern	AF	delam.	AF/CF	delam.	AF/CF	delam.	CF
	Related to non-aged ref. (%)	12	–	41	–	36	–	50
	Response, related to US	–	–	+	–	+	–	+

AF = adhesive failure (see Section 7.2.6); CF = cohesive failure; delam. = delaminated; PUR = polyurethane resin.

Samples blasted with shot-blast ceramic (SBC), grit-blast corundum (GBK), shot-blast steel (SBS) and grit-blast steel (GBS) failed at significantly higher strength values (in the range of $20\,\text{N}\,\text{mm}^{-2}$), showing a cohesive failure pattern near the adhesive/substrate interface, corresponding to 90% of the initial strength in the non-aged condition, with the exception of samples blasted with SBS, which only achieved 70%. Surface preparation with SBG and grit-blast glass (GBG) did not significantly improve the durability in contrast to blasting with SBC, GBK and GBS, which clearly improved the durability compared to the reference samples. Within the standard deviation, there was virtually no loss in tensile shear strength and no modification of the failure pattern compared to the results obtained with non-aged samples.

When the acrylate adhesive was used, the reference samples failed at very low strength values (ca. $5\,\text{N}\,\text{mm}^{-2}$), corresponding to 25% of the initial strength. At visual inspection, adhesive failure patterns were found: the surface preparation with SBG, GBG, SBC and GBK resulted in near-identical strength values, but with different failure patterns. There was adhesive failure without bond-line corrosion at the edges of the bonding area, and cohesive failure in the core area. With these samples, around 30% of the initial strength of non-aged samples was obtained. Only when using SBS and GBS was there a slight increase in residual strength (to around 45%), indicating a tendency towards improved aging resistance. In this test run, the decrease in strength was caused by a diminishment of the bonding area due to aging. It should be noted that the size of the remaining bonding area with a cohesive failure pattern depends on the type of abrasive medium used. Compared to the reference sample, a slight improvement can be achieved with steel abrasives.

In the test run performed with a polyurethane adhesive system, the reference samples had a residual strength of $1.6\,\text{N}\,\text{mm}^{-2}$ following aging, corresponding to 10% of the shear strength in the non-aged condition, and an adhesive failure pattern was detected. It becomes obvious that the aging characteristics are significantly improved by a surface preparation using angular blasting media such as GBG, GBK and GBS, yielding residual strengths of around 40–50% of the initial value. On the other hand, all samples blasted with spherical blasting media such as SBG, SBC and SBS failed during aging by delamination of the bonding areas that became detached after an aging period of approximately six weeks. No further destructive testing was possible with these samples.

It is rather interesting that, with regard to the durability of steel bondings, the results obtained with the immersion test performed at 80 °C during eight weeks were similar to those obtained with a two-year outdoor weathering test performed in Kaiserslautern (Germany) without using any corrosive additives (Figure 7.55).

The adhesive used was a cold-setting epoxy resin-based model adhesive with an amine hardener that was similar to AW 106 and had similar failure patterns. However, the degrees of damage following short-term aging procedures are rarely similar to those following long-term climatic exposure, and they do not occur when exposure to corrosive media is involved which may impair the bondings in other ways. This applies especially to steels that are susceptible to rusting. Usually, though with some exceptions, these materials are not employed without a proper corrosion protection. Hence, these bondings are tested after subsequent application of a

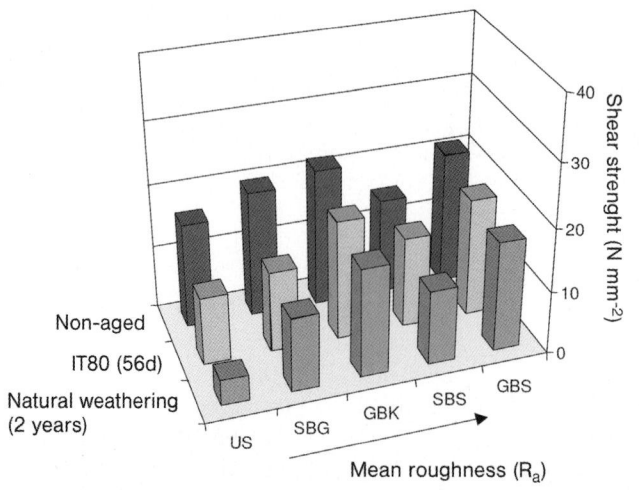

Figure 7.55 Comparison of results obtained after different aging procedures.

corrosion protection that is similar to that used on the planned component. With all test methods, bondings that are intended for use in automobiles, for example, are e-coated prior to long-term exposure.

Initially, it seems obvious to suspect the surface morphology in the micrometer range, generated by blasting with different media, as being the reason for the differences in durability. However, the following example, of results obtained with bonded textured body steel frequently used in automobile construction, shows this not to be the case (Figure 7.56).

Without discussing detail, may it suffice to say that all topographies were embossed on a steel coil using an embossing roller. The different textures had a similar chemical composition, as demonstrated by analysis, but the roughness and peak values of the sheets were quite different (Figure 7.57).

The adherents were bonded with hot-setting epoxy-resin-based Terokal 5026 adhesive and (as mentioned above) the samples were e-coated. The tensile shear strengths of all bondings in the non-aged and aged conditions, obtained following different aging procedures, are shown in Figure 7.58.

It becomes clear that the different topographies have virtually no influence on the strength and durability behavior of steel bondings and the failure patterns observed (Table 7.6). However, accelerated outdoor weathering, in combination with salt spray application once weekly, resulted in severe bond-line corrosion of the bonded areas despite e-coating, which usually does not occur in practice (e.g. with long-term use of bonded joints in a car). This is a typical example showing that accelerated long-term tests may also yield unrealistic results.

As expected, stainless steel bondings are less susceptible to corrosion. The results obtained with X5CrNi18-10 special stainless steel bondings with smooth surfaces, and with surfaces blasted in the steel mill (steels A and B) and intentionally contaminated with protective oil, are shown in Figure 7.59. Different hot-setting epoxy-resin-based adhesives were used, one of which was a formulation of epoxy

SBT texture

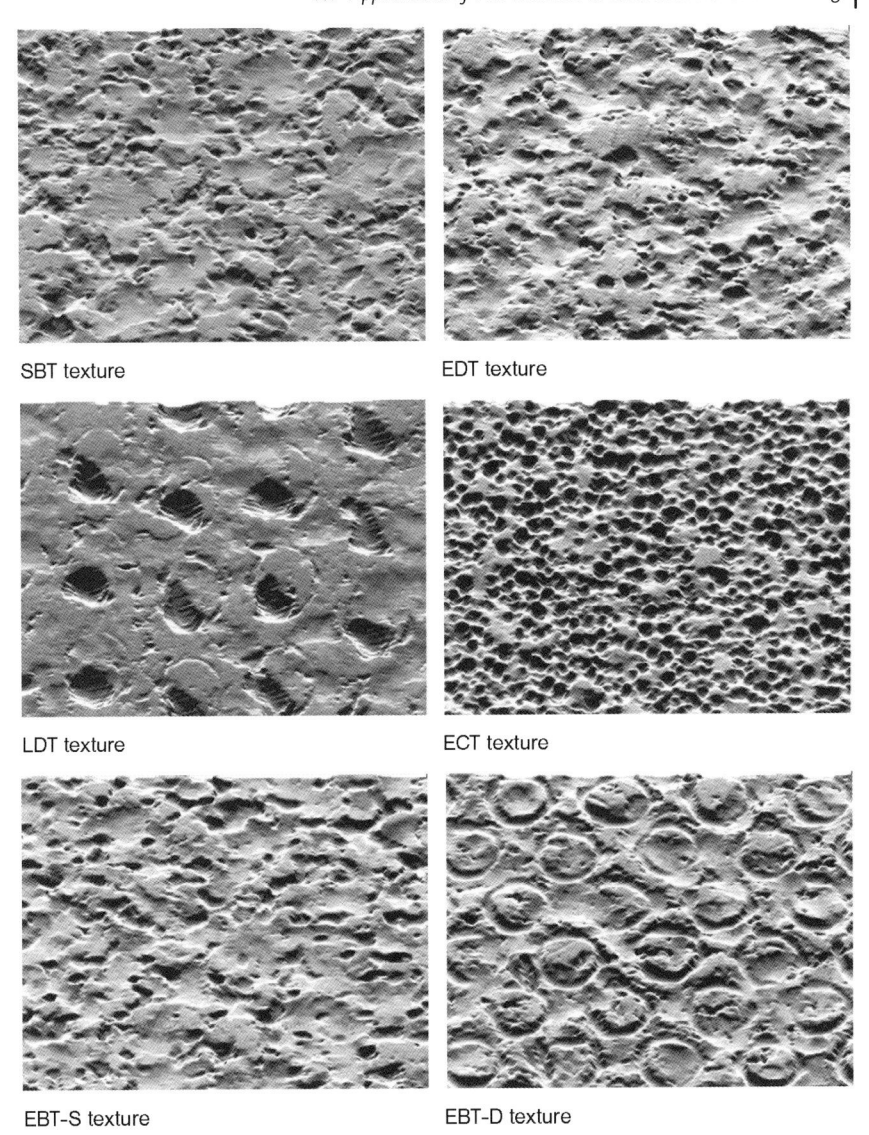

EDT texture

LDT texture

ECT texture

EBT-S texture

EBT-D texture

Figure 7.56 Scanning electron microscopy (SEM) images of different steel topographies.

resin and polyurethane. Again, without discussing detail, it was noted that the durability of special steel bonded with modern adhesives used in the automobile industry was quite satisfactory [14].

The above-mentioned investigation revealed that austenitic stainless steel, which formerly was considered to be difficult to bond, can be bonded perfectly with cold-setting adhesives to form a durable joint. Chemical or physical surface preparation only slightly improved the strength and durability, and in general could be dispensed with. With regards to the failure pattern, suffice it to say that with non-aged special

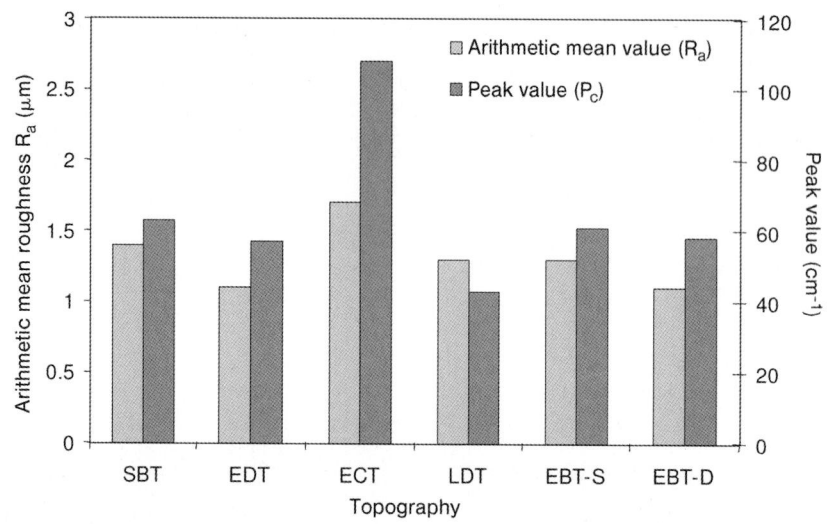

Figure 7.57 Arithmetic mean values of roughness (R_a) and peak values of different topographies.

steel bondings, a visual inspection frequently revealed adhesive failure or particular cohesive failure near the adhesive/substrate interface. However, in contrast to other adherent materials, this did not represent any major disadvantage.

7.7.8
Adhesive Bonding of Glass

It is difficult to bond glass durably unless special adhesion promoters are used (see Chapter and Section 5.10). When silane adhesion promoters are applied separately or

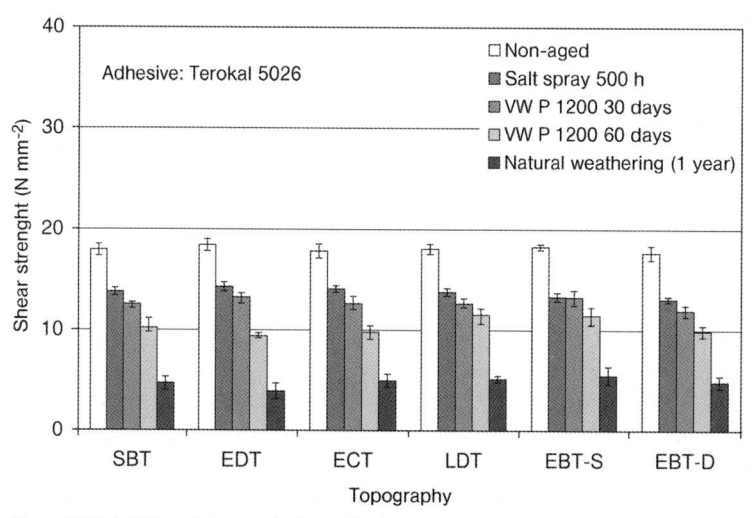

Figure 7.58 Initial and the residual tensile shear strengths of different topographies.

Table 7.6 Failure patterns of mild steel bondings with different surface textures.

Adhesive	Characteristic	Failure type	SBT	EDT	ECT	LDT	EBT-S	EBT-D
Terokal non-aged	Failure pattern (%)	SCF	100	100	100	100	100	100
		AF	–	–	–	–	–	–
	Related to SBT (%)		–	103	99	101	102	98
	Response			o	o	o	o	o
Terokal SST	Failure pattern (%)	SCF	85	85	85	85	85	85
		AF	15	15	15	15	15	15
		BC	–	–	–	–	–	–
	Related to non-aged ref. (%)		77	77	79	76	73	74
	Response related to SBT		–	o	o	o	o	o
Terokal P1200-30	Failure pattern (%)	SCF	65	65	55	50	55	55
		AF	30	30	40	45	40	35
		BC	5	5	5	5	5	10
	Related to non-aged ref. (%)		70	72	71	70	73	68
	Response related to SBT		–	o	o	o	o	o
Terokal P1200-60	Failure pattern (%)	SCF	35	40	35	50	50	30
		AF	50	55	50	40	40	55
		BC	15	15	15	10	10	15
	Related to non-aged ref. (%)		57	51	55	64	63	56
	Response related to SBT		–	o	o	o	o	o
Terokal OW 1 y	Failure pattern (%)	SCF	15	10	15	10	20	15
		AF	20	20	15	25	20	20
		BC	65	70	70	65	60	65
	Related to non-aged ref. (%)		26	21	28	28	30	27
	Response related to SBT		–	o	o	o	o	o

AF = adhesive failure (see Section 7.2.6); BC = bond-line corrosion; SCF = particular cohesive failure near the adhesive/substrate interface (see Section 7.2.6); OW 1 y = outdoor weathering for one year; P1200-30 = aging in accordance with VW standard 1200, 30 test cycles; P1200-60 = aging in accordance with VW standard 1200, 60 test cycles; SST = salt spray test for 1000 h; SBT, EDT, ECT, LDT, EBT-S, EBT-D = different textures (see Figure 7.56).

admixed to the adhesive, glass can be bonded perfectly, and with a good durability. The results obtained with float glass bonded with a two-part epoxy resin adhesive and an adhesion promoter are shown in Figure 7.60 [23].

Following exposure of the joints to outdoor weathering for five years the joints were found to be covered over with algae, especially in those areas where the adhesive had leaked. On visual inspection of the bonding, using a stereomicroscope, no damage such as cracks, discoloring or delaminating was detected. In fact, only a strip of adhesive that had leaked from the bond-line had delaminated, and its surface had become cracky. However, when a destructive testing was performed the glass broke in the lap area. The residual strength was 16.4 ± 3.7 N mm^{-2} – that is, approximately 70% of the initial strength determined before aging. These results are only obtained when the adhesive, which in principle can be cured at room temperature, is cured at 120 °C for 1 h, resulting in a better crosslinking and thus in an increase in the T_g-value.

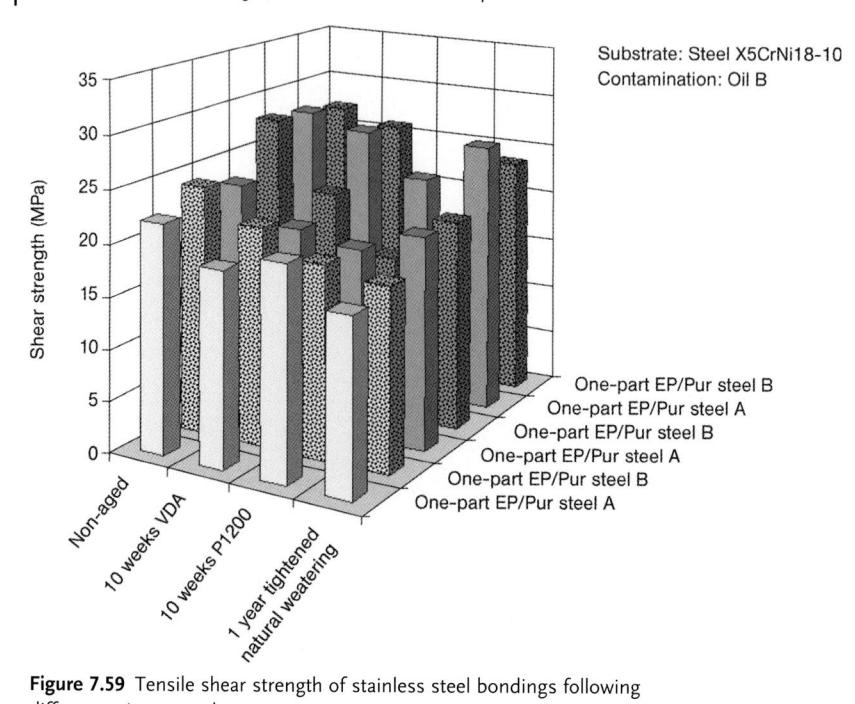

Figure 7.59 Tensile shear strength of stainless steel bondings following different aging procedures.

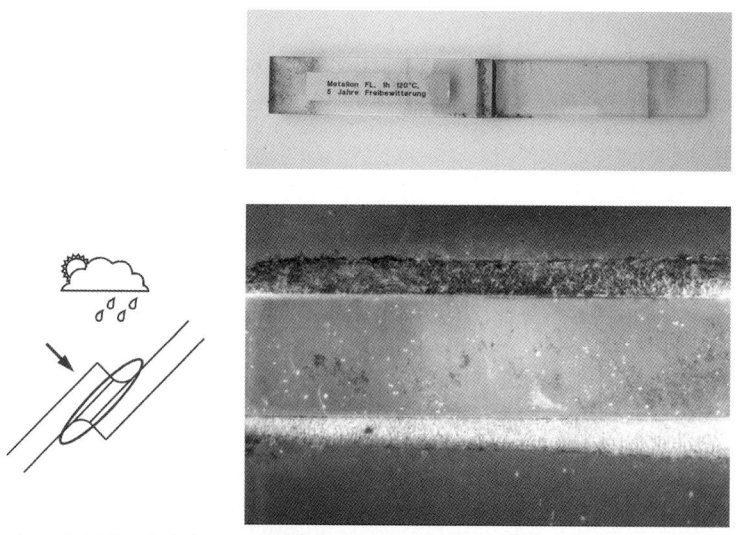

Figure 7.60 Bonded glass joint after five years in natural climate.
Right upper: shear specimen; right lower: bonded area.
Left: aging conditions.

With regards to floatglass, durable bonds can also be achieved with silicone adhesives (see Section 8.3) or polysulfides containing silane. Silane-crosslinking polyurethane as well as epoxy resins, can also be considered, although problems are sometimes encountered with imperfect crosslinking in larger bonding areas. Finally, it is possible to use acrylate PSAs. While silane primers considerably improve the durability, a general cleaning with water and thorough drying at 40 °C is an adequate surface preparation for glass.

7.7.9
Adhesive Bonding of Polymer Materials

Adhesive bonds with polymer materials generally achieve a good durability, provided that the adherent material contains polar groups or (as in the case of nonpolar polyethylene, polypropylene or similar polymers) polar groups have been generated by oxidation via chemical or physical surface preparation (see Section 6.2) that also improves the wettability. Media that diffuse into the bond may impair the strength. Based on common tables of resistance, the correct selection of the media to which the bonded joints are exposed prevents the media from irreversibly damaging the adherents and the adhesive. The changes in strength remain calculable because they can be determined in advance based on material characteristics. The adhesion zone, which can be critical in the case of inorganic materials (see above), is rather noncritical at long-term loading in the case of bonded plastic material; this is because, besides physical and chemical interactions, molecular diffusion between the adhesive and adherent apparently provides for a durable adhesion (see Chapter 3). However, this does not mean that all plastics can be bonded with all adhesives in the same way. Selection tests (strength tests without aging influence) make it possible to determine whether a plastic material can be bonded, or not. When the results are satisfactory, the durability can then generally be assumed also to be satisfactory, at least as far as modern industrial adhesives are concerned [24].

7.8
Experiences and Learning from Damage Cases

Failures range from minor flaws to catastrophes, and examples of failures can be found everywhere. Classifying an event as a failure implies that a deficiency has been identified, and that an improvement is sought. To prevent future failures an analysis is required, because one does not learn from the damage itself but rather from a proper failure analysis. Failure analysis almost always arrives at the conclusion that fundamental rules have not been observed. However, such as conclusion does not help to prevent future failures, because the complex interaction of relevant factors is only revealed by the current failure itself. Failure analysis is therefore a necessary and integral part of each developing technology, as it will only be accepted when failures can be explained and their recurrence prevented.

With regards to the discussion of failures occurring with adhesive bonding technology, a unique issue must be highlighted. Failures, which occur in welded, soldered, riveted and screwed joints are well known and well understood. The causes of these types of failure include crevice corrosion, heat-affected zones, stress concentrations, material incompatibilities and other factors. Based on a wide experience, failures can be explained rather quickly and conclusively, and no longer cause a designer to question the joining technique itself. With structural bonding, however, failures bring the technology itself under scrutiny. The reasons for this are twofold: (i) due to relatively little experience with these technologies there is a deficiency in, or a lack of, training in structural bonding; and (ii) there is an especially inadequate understanding of bonding as an interdisciplinary subject where rules must be applied. In practice, this situation becomes clear when, for example, following an adhesive failure, a designer requests a 'more suitable' adhesive. Here, simply trying to find an alternative adhesive misses the point – that the quality of the surface treatment, the construction, or other deficiencies within the process are more likely to be the cause. Yet normally, these points are not questioned. A lack of understanding becomes transparent by the quest for a 'better' adhesive, as this thinking ignores the main advantage of adhesive technology – which is to provide the capability of joining a wide range of different materials. The adhesive, in its low-viscosity phase, makes this possible since each time it adjusts itself to the energetically most advantageous condition according to the adherent surfaces available. Therefore, bonding is a universal and uncritical joining technique where the adhesive is the 'good-natured partner'. In adhesive technology, the majority of failures are caused by a breach of specific conditions during creation of the joint – quality defects of the adhesives themselves are hardly ever the cause of failure in bonded joints.

The following case studies of adhesive bonding failures illustrate the aspects of the complex interaction of different factors, not only when dealing with failures and their causes, but also in presenting the successful and highly demanding adhesive bonding applications which normally show no failures and which have been widely adopted because of their trouble-free service.

In the following sections we will demonstrate first, that failure analysis is essential in order to prevent failure recurrence. Such analysis can be performed by examining the failed parts yourself, or by learning from failures reported in literature. Second, the case studies will show that the failure of adhesively bonded joints can be explained conclusively. Despite its basic function for structural adhesive bonding, the phenomenon of adhesion is rarely in focus in the case of failure. Many years of failure analysis of bonded joints can be summarized with the statement that, in about 70–80% of the cases, damage was due to design deficiencies and other types of overloading. The saying that "...only adhesive bonding can save a bad design" is not completely unfounded, and it reflects the lack of sensitivity for other essential parameters. This is because, in only a few cases was the damage due to deficiencies identified in the bonding process, while in some cases it was due to an inadequate surface preparation, to problems with contamination, or to failure of the adhesive itself.

7.8.1
Delamination of Bonded Rotor Blade Pockets

Structural bonding of the rotor blades of helicopters is a particularly good example of the high performance of modern adhesive bonding technology. With a service life of several thousand flight hours, these components are exposed to the most severe dynamic and climatic load conditions. Adhesive bonding technology is necessary for the realization of modern rotor blades.

In the case of rotor blades of heavy transport helicopters, failures regularly occurred on a structural bonded joint between the upper outer cover sheet and the spar. The outer cover sheets delaminated during flight and induced strong vibrations in the rotor such that, in the majority of cases, an emergency landing was required. The rotor blades were put back into service following repair of the damage, until the next failure occurred. This type of failure led to repair costs of several million Euros, because on the rotor blades one blade pocket after another failed unpredictably and had to be repaired. Ultimately, this type of blade had to be replaced because of design problems. Yet, in only a few cases was nondestructive testing able to identify delamination in its early stages; consequently, a preventive repair was not possible (Figure 7.61).

A rotor blade consists of an 11 m-long aluminum spar with 25 bonded 'blade pocket' segments made from aluminum sheet. Together with the rotor blade pocket, the spar (leading edge of the blade) forms a wing profile. In the case of a failure the segmentation into rotor blade pockets limits the propagation of a delamination of the outer cover sheet of the rotor blade and maintains the overall function of the rotor.

On the upper and lower surfaces, the outer cover sheet of the blade pocket is bonded to the spar with a lap length of 40 mm each (Figure 7.62). Although, insufficient surface preparation of the aluminum adherents was suspected as the reason for the failure of bonded rotor blade pockets, microfractography revealed only cohesive failure of the adhesive.

Figure 7.61 Failure delamination in rotor blades.

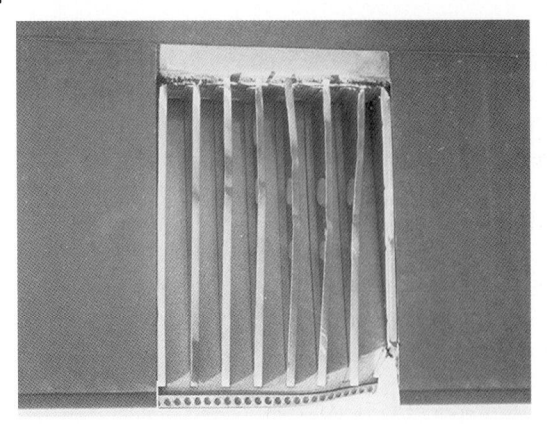

Figure 7.62 Delaminated rotor blade and starting crack in the bond-line between spar and outer cover sheet.

Adhesive interface failure did not occur in either of these cases, and consequently, inadequate surface preparation of the aluminum adherents could be ruled out as the cause of the damage. Indeed, the cause of the failure was determined to be a design deficiency. Due to the airflow around the rotor blade profile, there is a lower pressure above the blade than below (vertical lift), which bends up the trailing edge of the profile. Due to this aerodynamic loading of the rotor blade, the pocket-spar skin joint at the bottom side of the blade is under tensile stress, while the upper skin joint is in compression. Although both skins transfer these loads to the spar through the bond in shear, there is a difference between the upper and lower skins. With the existing construction of the rotor blade pocket, tensile stress of the lower bond implies a uniform stress distribution for the bond-line, whereas on the upper side the shear stress is combined with a peel load. Compression causes a peel force at the transition between the outer cover sheet and the spar bonding on the upper side of the blade. Additional peel forces generated by a torsion movement of the relatively narrow rotor blade pocket foster the failure. Crack initiation occurred at the left and right forward corners of the pocket segment bond and propagated by fatigue and cohesive failure of the adhesive, ultimately evolving into a rapidly progressing complete delamination and loss of the outer cover sheet under aerodynamic loading of the rotor blade. This failure was due to a fundamental design defect of the rotor blade. With regards to the layout of the bonded joint between the outer cover sheet and the spar, no measure was taken to prevent peel forces from developing between the adherents. As this design defect could not be repaired, and the design was not suitable for bonding, the rotor blades had to be removed from service and redesigned.

7.8.2
Total Separation of a Rotor Blade Profile

The next case study illustrates that segmentation into blade pockets is indeed very useful. Had it been used in the following case, the consequences of the failure might

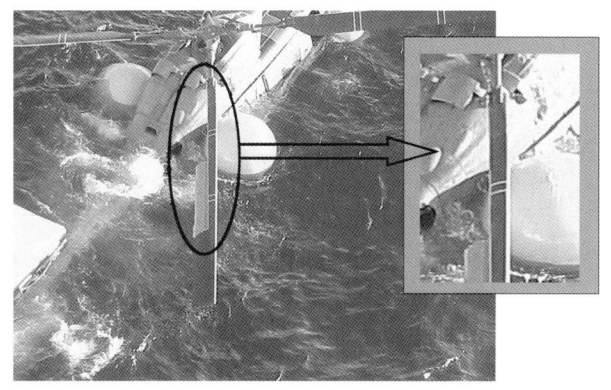

7.8 Experiences and Learning from Damage Cases | 195

Figure 7.63 Failure: total break-off of a rotor blade.

have been far less significant. During a flight over sea, severe damage occurred to the main rotor of a helicopter and an emergency landing on water was necessary; unfortunately, the helicopter could not be recovered (Figure 7.63). As the damaged rotor blade was not recovered, failure analysis had therefore to be performed by examining similar rotor blades with comparable flight histories. A photograph taken before the helicopter sank (Figure 7.64) showed a total loss of the rotor blade profile, at the root of the rotor blade, extending to one-third of the blade length, as well as a delamination of the upper outer cover sheet over another third of the length.

The failure pattern suggested that failure was initiated near the root of the blade and resulted in complete separation of the profile over the first third of the blade, subsequently followed by delamination of the upper skin over the central third. The exposure of the spar bond on the upper side of the spar was due to peel loading during separation of the upper cover sheet. Peel loading occurred when the profile was torn off upwards by lifting forces caused by the failure of the bond on the bottom side of the blade in the area of the root of the blade. The honeycomb core in the final third of

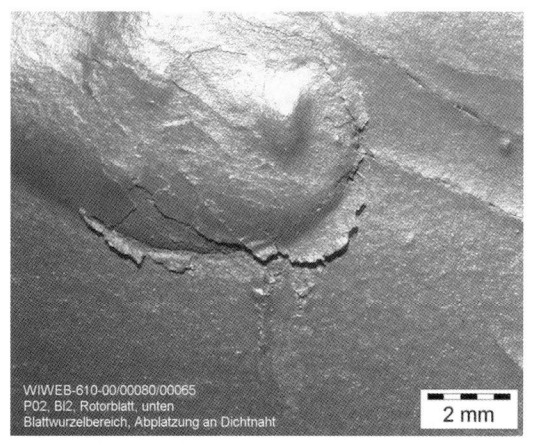

Figure 7.64 Cracks with capillary effects inducing accelerated aging of bond-lines in the rotor blade.

the length in the center remained intact, indicating that the profile still had sufficient stability under service conditions in the case of an intact outer cover sheet on the bottom side of the blade.

Destructive characterization of the same type of rotor blade revealed the incidence of cracks in the outer skins and along an elastic sealing compound material at the transition between the root of the blade and the pocket profile. Samples from these rotor blades were exposed to outdoor weathering tests and strength tests. It was shown that a failure of the bond is possible when there are cracks along the spar-pocket bond, and when seawater can penetrate via capillary cracks close to the bondline, thereby generating extreme aging conditions that are not covered by all normal tests. Hence, during routine inspections of the rotor blades the quality of the sealants must be checked thoroughly.

7.8.3
Bond-Line Corrosion in Wing Honeycombs

The following case illustrates the influence of a maritime climate and the classic pattern of bond-line corrosion (see Section 7.7.6). The aircraft concerned had been used for low-altitude maritime surveillance for about 40 years, and were therefore exposed to severe aerodynamic and corrosive conditions.

Both, the cabin and the wings are mostly made of bonded honeycomb structures, but when these aircraft were manufactured pickling, rather than anodization, was the typical surface preparation for aluminum adherents (AA 2024 T3) (see Section 8.2). The adhesive used was an epoxy resin adhesive for aircraft. Despite 40 years' of operation and loading, and the use of an inadequate surface preparation leading (from today's perspective) to poor resistance to corrosion and aging, the bonds were in a good condition. The only critical component was the heavily loaded center wing box, that carries and interconnects the engine, the cabin and the outboard wing (Figure 7.66).

As is typical for bond-line corrosion, adhesive failure started at the damaged areas of riveted joints and boreholes, and large-area delamination of the adhesive joint ensued. In such cases, delaminations are usually detected by means of ultrasound. The structure may then be opened, the corroded skin area removed and the area then

Figure 7.65 Failure: bond-line corrosion in wing honeycombs.

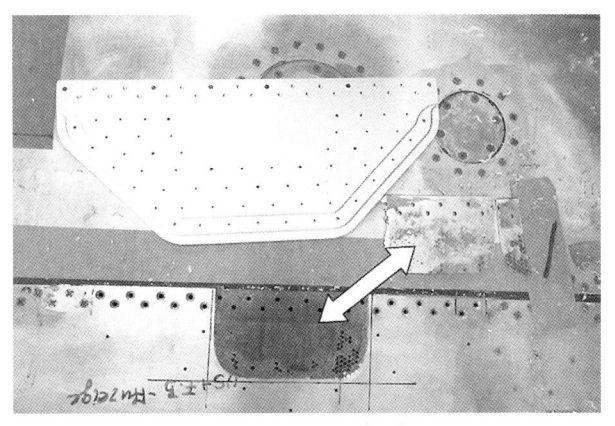

Figure 7.66 Bond-line corrosion in a sandwich.

re-bonded, with a second new outer skin sheet being additionally applied by bonding or riveting. As bond-line corrosion is a well-known risk, it is important to ensure a reliable sealing of the riveted joints and seams during maintenance and repair. Moisture and sea atmosphere must reliably be prevented from penetrating the bond and initiating bond-line corrosion and eventual failure.

The repair of corroded parts is standard for maintenance programs, and many years of experience have been acquired in this respect. Therefore, a sudden and striking increase in the corrosion damage of bonded joints attracted much attention. At first, the increase was assumed to be due to the age of the aircraft and suboptimal surface preparation. As a solution to the problem – and to prevent further damage – improved sealing measures were considered. When this was attempted, however, it was found that several years earlier, due to the introduction of environmental protection and occupational safety regulations, a change to chromate-free sealants was made such that anticorrosive chromated primers were no longer used. By doing this, a reliable corrosion protection system was eliminated, without taking into account its critical importance to the long-term performance of adhesive bonds and the structural integrity of the aircraft.

7.8.4
Delamination of the Outer Cover Sheet of Tail Units of Aircraft

Large-area delamination of the adhesive bonding between the aluminum outer cover sheet and honeycomb structures of the tail unit of aircraft resulted in several cases with a similar damage pattern (Figure 7.67). Due to these delaminations the outer cover sheets were even lost during flight, with both the bottom side and upper side of the cover sheets of the tail unit having been affected. The surfaces uncovered by delamination revealed a purely adhesive failure pattern and minor signs of aging by moisture penetration (Figure 7.68).

From the large-area failure of the bonding, it was deduced that the damage must have been caused by adhesive failure due to impaired adhesion as a consequence of

Figure 7.67 Delamination of outer cover sheets in the tail units of an aircraft.

contamination of the adherents. Surface analysis, however, did not provide any conclusive evidence for this assumption. It should be noted that surface analysis methods with sensitivity in the atomic monolayer range are preeminently well suited to making a statement on well-defined surfaces under laboratory conditions. Surfaces in an undefined aging condition, as they are encountered in practice, however, cannot be analyzed conclusively without further information.

In the present case, inaccurate temperature control with a too-low curing temperature during the autoclave process was found to be the cause, and led to several similar cases of damage. This was indicated by the following considerations. Impairment of adhesion by a contamination of the adherents was not likely. This could be deduced from the results of surface analysis, and from the fact that a 180 °C hot-curing foil adhesive was used. Indeed, higher temperature curing adhesive systems are expected to provide adequate contamination tolerance. A failure surface showing significant parts of cohesive failure would have been expected to occur. However, the large-area adhesive failure mode indicated that the foil adhesive must not have been able to wet the adherents properly at all, most likely due to a too-low

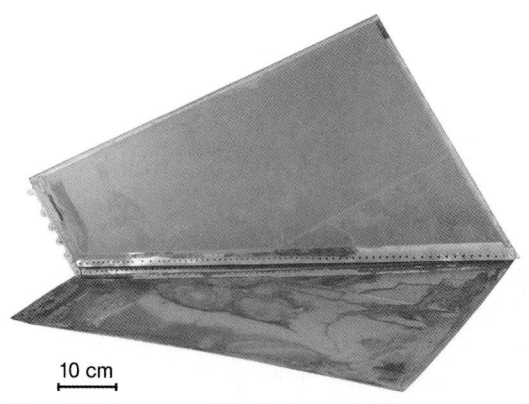

Figure 7.68 Large-area delamination under the outer cover sheet in the tail unit of an aircraft.

Figure 7.69 Heat-initiated delamination in a tail rotor blade.

curing temperature. Under the manufacturing conditions assumed, inadequate wetting due to an aged and precured adhesive could be excluded because hot-setting adhesives react very slowly. The occurrence of few, but identical, cases of damage was best explained by a defective autoclave process causing a too-low curing temperature and a resultant high viscosity of the adhesive film such that it could not properly wet the adherent surfaces.

7.8.5
Temperature-Induced Damage on a Tail Rotor Blade

Inaccurate temperature control during the autoclaving process may indeed be the cause for adhesive failure. This is demonstrated by the delamination of a tail rotor blade (Figure 7.69).

During inspection, a delaminated area was found in the bonding between honeycomb and outer cover sheet which was opened during a failure analysis in the laboratory. The distinct brown discoloration of the thermosetting adhesive, with large bubbles and separation from the honeycomb core cells, is shown in Figure 7.70.

In this case the adhesive must have been overheated for very short period of time while in the low-molecular-weight state – that is, during manufacture. An elevated temperature exposure after cure would have induced polymer decomposition instead of the observed large-pore formation in the cured thermosetting duroplastic resin. Furthermore, examination of the coatings on the external surface of the rotor blade did not reveal any signs of an external thermal damage due to operation.

Figure 7.70 Overheated adhesive film and adhesive separation from the core cells.

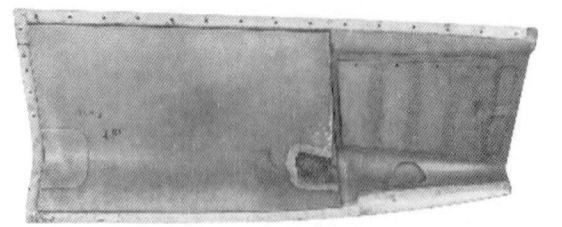

Figure 7.71 Delamination of a bonded glass-fiber patch.

7.8.6
Temperature-Induced Delamination of a Bonded Glass Fiber Patch

As illustrated in Figure 7.71, damage occurred in a honeycomb structure of a service hatch of an aircraft; here, a delaminated glass-fiber patch impaired the function of an engine control unit. The distance between the hatch and a heat exchanger nearby was less than 2 mm in some places, while in some areas they even came into contact with each other.

The honeycomb structure and the glass-fiber patch had been bonded with an epoxy resin system containing dicyandiamide as hardener. Dicyandiamide is identified, using IR spectroscopy, by two bands at 2190 and 2160 cm^{-1}, respectively. These bands result from a C≡N bond and can be used to detect the thermal load of the adhesive. Additionally, in the case of a high temperature caused by oxidation, a band structure is generated in the carbonyl stretch (C=O vibration) region (at around 1680 cm^{-1}). From the comparison between the results obtained with the defective hatch and those obtained with temperature-aged reference samples, two potential thermal histories were determined: long-term exposure of the hatch to more than 180 °C, or to 240 °C as a short-term exposure. Of these two possibilities, the long-term exposure was the more likely (Figure 7.72).

When assessing the possible thermal load, the thermal conductivity of the adherents and possible heat dissipation by the structure had to be considered. Compared to the glass-fiber patch, the thermal conductivity of aluminum was very high; hence, near the aluminum outer cover sheet the heat had dissipated into the

Figure 7.72 An enlarged view of the delaminated area.

structure, limiting the thermal load to which the adhesive was subjected where it contacted the aluminum. In contrast, where it contacted the glass-fiber patch, no heat dissipation had taken place. Consequently, the surface of the glass fibers was heated up and a considerably higher temperature reached. This thermal load at the glass-fiber patch eventually resulted in failure of the adhesive and delamination.

7.8.7
Incorrect Design of Bonded Filter Elements

Filter elements for hydraulic flight regulation systems (Figures 7.73 and 7.74) usually consist of a metal casing with a glued-in filter. The adhesive bond must be resistant to temperatures in excess of 150 °C. Failure of the bond and loosening of the filter from its seating allows the fluid and included debris to bypass the filter. In hydraulic systems, the release of contaminants represents a severe risk because it may cause the blockage of servo valves and total failure of the hydraulic control systems. When operating the filter it is important to note that there is a possibility of short-term pressure surges in the opposite flow direction.

Figure 7.73 Incorrect design of a bonded joint in filter elements.

In several filter elements used for aeronautical applications, the filter was found to be loosened due to the failure of an adhesive bond. Initially, only the filter elements of the defective type were available for failure examination. This version had been used generally for repairs owing to availability of stock, and such use would indeed have been continued for an extended period of time. In this version, the filter was bonded into a slot in a base plate, but as no adhesive residues were found in the slot an inappropriate surface preparation of the stainless steel base plate was assumed to be the cause of the failure (Figure 7.75).

The filter was adhesively bonded with a silica-filled, bis-phenol-A-based epoxy resin adhesive. The role of the filling was to reduce the coefficient of expansion and improve the behavior of the adhesive under thermal cycling.

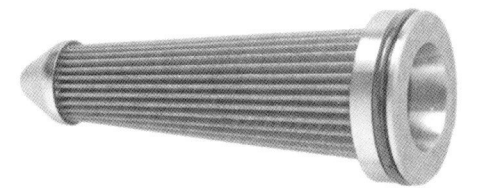

Figure 7.74 Normal flow direction, from left to right.

Figure 7.75 Base plate with slot for the filter bonding.

When the manufacturer was informed about the design problems of the bonded joint, a new version of the same filter was delivered. Although, visually the filter elements were of identical construction, the strength of the bonded joint was much greater. The new version of the filter elements could not be separated from the base plate without destruction. The bonding of the upper filter cap broke at a force of 786 N, whereas with the former filter elements the bond had detached from the slot of the base plate at a force of only 211 N. This improvement in bond strength of the new version was the result of a design modification, which made it more suitable for bonding under critical conditions. The design modification had added an undercut for mechanical locking as a fallback position. Only the undercut and the added shape support of the adhesively bonded joint design could prevent uncontrolled bond separation in the case of reverse-pressure surges. Under such load conditions, the problem would not have been effectively addressed solely by improving the surface preparation of the stainless steel base plate.

7.8.8
Deficiencies of the Peel Ply Treatment

One of the problems of adhesive bonding technology is the fact that bonded joints may fail unpredictably when the adherents are contaminated. Even pre-existing contaminations in the form of monolayer adsorbates on the adherent surface prevent a load-carrying intermolecular interaction between the adhesive and the load-bearing adherent. Contaminations induce adhesive failure and poor aging behavior of the bonded joints. With this in mind, it is easy to lose sight of the fact that an inappropriate surface preparation may cause the same symptoms, even in the absence of any contamination. An example of this is the 'peel ply' procedure, a method of surface preparation of fiber composite materials (Figure 7.76).

The purpose of the peel ply procedure is thought to generate a laminate surface with good bonding characteristics. For this purpose, during the manufacture of the laminate, a 'peel ply' fabric is incorporated into the surface and removed prior to bonding. The main purpose of the fabric is to absorb excess resin material during the curing of the laminate. Consequently, peel ply removal should also be suitable as a

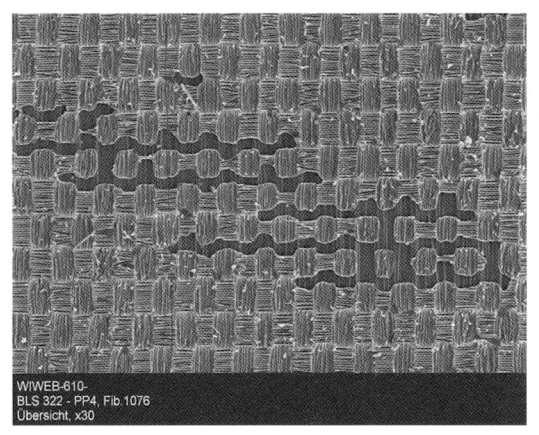

Figure 7.76 Insufficient peel ply treatment.

surface preparation for bonding, because removal of the fabric generates a fresh laminate surface that is activated by the fabric texture and the removal of excess resin. This surface usually has adequate bonding characteristics, without further treatment. The process result is good bond-strength and climate-resistant joints. When the resin is peeled off any superficial contaminations are removed, polymer chain scission is induced in the fractured polymer at resin bridges, and reactive groups are thought to be created on the surface.

Although this surface preparation has proven to be a suitable method for bonding purposes, localized nonsystematic adhesion problems frequently occurred with bonded laminates. This type of damage suggests an influence of the peel ply on the surfaces to be bonded. Surface analysis performed on the delaminated fracture surfaces revealed release agent contamination with considerable localized fluctuations in the degree of contamination. However, on the peel ply, there were no contaminants that would have caused the observed failure pattern, as could be demonstrated by comprehensive surface analysis performed on samples that were randomly selected from peel ply fabric.

The release agent used in the autoclave process was found to be the contaminant detected on the laminate surface after removal of the fabric. The degree of contamination of the laminate surface depends on the degree of resin absorption by the peel ply. In regions with a good resin absorption – that is, with a higher excess of resin in the laminate – a surface with little or no release agent results when the fabric and the resin adhering to it are peeled off, whereas contamination occurs in regions with incomplete resin absorption. In the latter case, peeling off the fabric does not represent an adequate surface preparation because the resin and contaminants are only partially removed (Figure 7.77).

Resin absorption by the peel ply – and consequently the quality and reproducibility of the surface preparation – are determined by several factors. These include the resin content of the composite-fiber fabric preimpregnated with the matrix resin, the processing parameters, and in particular the structure of the laminate (unidirectional,

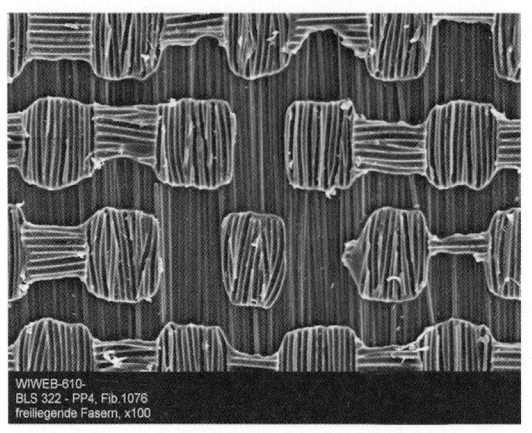

Figure 7.77 Scanning electron microscopy image of a carbon fiber-reinforced polymer (CFRP) surface following incorrect peel ply treatment.

multidirectional) – that is, by the packing density of the fibers. In unidirectional laminates, a larger amount of excess resin material is available due to the higher packing density of the fibers, which induces a decrease in the degree of contamination caused by the release agent. Unidirectional laminates are typically used in destructive testing (tensile test, double cantilever beam test) to investigate and to prove the suitability of surface preparation methods, whereas multidirectional laminates are relevant for practical applications where more matrix resin material is needed for the space between the diagonal fiber layers. This implies a lower excess resin, inducing a higher risk of local contaminations with the release agent. Furthermore, a shortage in resin material occurs with decreasing thickness of the laminate. This is especially critical as bonding is the preferred joining technique used for structures with thin walls.

However, the peel ply surface preparation must be regarded as critical for yet another, independent reason. Indeed, due to the deliberately poor adhesion between the resin material and fabric (otherwise, the peel ply could not be removed), the peel ply leaves a negative imprint of its structure on the prepared surface. When coming into contact with the peel ply, and during the curing process in the autoclave, the functional groups of the liquid matrix resin material 'see' only a nonbondable surface (see Section 3.2). In the autoclave, consequently, a similar structure with low surface energy is also generated as an interface image on the resin side. This is subsequently uncovered by the peel-off procedure, although unless further surface preparation is performed it does not fulfill the requirements of a surface with excellent, reliable bonding characteristics. The only part of the surface that has good bonding characteristics is the region on which fracture surfaces of the resin bridges have been generated during the removal of excess resin.

8
Adhesive Bonding Technology: Fields of Application

8.1
Introduction

Examples for the successful use of adhesive bonding technologies can be found in many fields of application. When choosing the best adhesive for a specific application, the limits of resistance of an adhesive joint to static, cyclic or impact load must be considered. The expectation in service life depends on the service conditions in terms of aging and detrimental environmental effects.

Since not only the strength capacity and the material properties of adhesives, but also their characteristic processing requirements, play an important role in evaluating the suitability for a certain application, adhesives have been classified according to their processing conditions and their chemical nature (see Chapters 4 and 5 and Figure 8.1). The following types of adhesive can be distinguished:

- physically setting adhesives
- reactive adhesives setting by mixing of two components, irradiation, moisture, absence of oxygen, heating, etc.

Due to a growing ecological awareness, the level of toxicity of different types of adhesive has gained significant importance. For example, in physically setting adhesives this has led to a reduction in the use of solvents (see Chapter 4).

In structural bonding, reactive adhesives provide the highest level of strength, temperature resistance and durability.

Once specific types of adhesives have been established, for example in the paper and packaging industry, or the woodworking or construction industries, they may be used for decades without any significant modification; this in turn leads to standardized, trade-specific adhesive product lines.

Today, the packaging and the building and construction industry represent the largest turnover in adhesives, while the strongest growth in adhesive demand is seen in automotive and commercial vehicle construction, in electronics, and in medical engineering in general.

In the following sections we will discuss the state of the art of adhesive bonding in different trades, on the basis of representative applications.

Adhesive Bonding: Materials, Applications and Technology
Walter Brockmann, Paul Ludwig Geiß, Jürgen Klingen, and Bernhard Schröder
Copyright © 2009 WILEY-VCH Verlag GmbH & Co. KGaA, Weinheim
ISBN: 978-3-527-31898-8

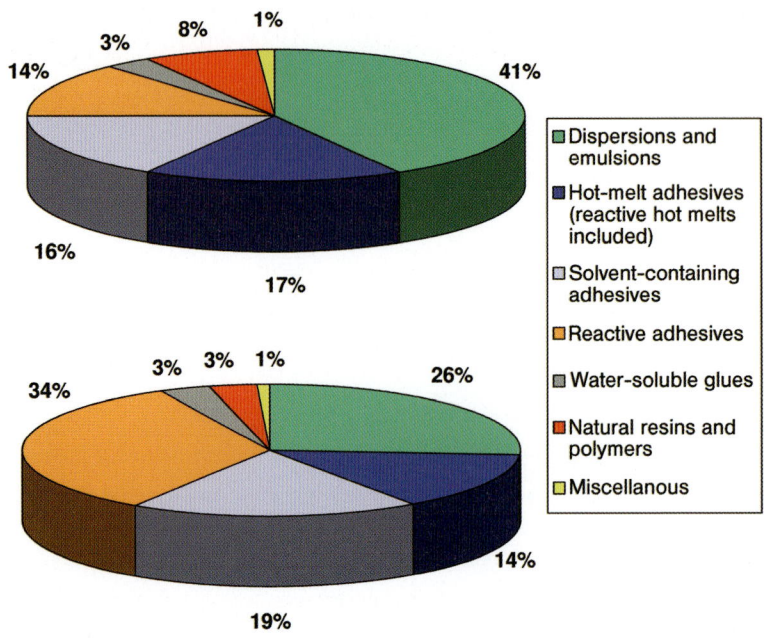

Figure 8.1 (a) Quantitative and (b) qualitative market shares of adhesive groups in Europe. (Illustration courtesy of Industrieverband Klebstoffe).

Legend:
- Dispersions and emulsions
- Hot-melt adhesives (reactive hot melts included)
- Solvent-containing adhesives
- Reactive adhesives
- Water-soluble glues
- Natural resins and polymers
- Miscellanous

8.2
Transportation

Perhaps the most diversified branch of adhesive use is in transportation, be it by water, by land, or by air. Private or public mobility, hobby activity or leisure time are relevant parameters for the classification of transportation systems according to their purpose.

In all fields of transportation, the higher the weight of a transportation system, the greater its energy consumption, wear and material cost. A railway car (ICE 1), for example, has an approximate weight of 54 t, and carries up 50 passengers which corresponds to a weight of approximately 5 t. One-tenth of the weight of the locomotive (weighing at least 150 t) must be added for each typical car in passenger traffic. A large commercial aircraft (e.g. Boeing 747) has a weight of approximately 220 t without any fuel and passengers, and transports 400 passengers at best (weight ca. 40 t). These examples illustrate the point that transportation systems have a poor weight-to-payload ratio which may be improved by new designs and construction methods, with adhesive joints being an important element.

Depending on the materials and engineering methods used, the adhesives must meet specific requirements. A military ship, for example, is especially stable and very heavy due to the materials used in its construction. A motor home, on the other hand, is lightweight, has a functional design and is low-priced, but is considerably less stable.

Important factors for the selection of an adhesive are the different fields of application, the quality assurance requirements (e.g. the service life of the materials used), accident protection or fire protection, and the manufacturing conditions, which can be further subdivided into highly automated quantity-lot production, small-batch production, and even customized production. It is also important to determine whether the adhesives are used for original equipment manufacture (OEM) or for subsequent repair. In general, there is a tendency towards shorter production times, reduced production costs and lower weight, which will further promote the use of adhesive materials.

The manufacture of transportation systems is subdivided into vehicle and aircraft manufacture, railway transport and marine; these areas of application are briefly reviewed, and examples provided, in the following sections.

8.2.1
Aircraft Manufacture

Aircraft manufacture is the number one field of application for high-performance bonding technology. However, with few exceptions, there has been a certain regression – or at least stagnation – in the application and further development of structural bonding. Problems which occurred during the 1970s contributed especially towards this trend. In metal aircraft manufacture, approximately 10–15% of the structural weight may be saved when the predominant riveting technology is consistently replaced by adhesive bonding; that is, up to 4 t may be saved in the case of a more important commercial aircraft. However, for weight reduction, interest has focused mainly on fiber-reinforced plastics, including systems reinforced with carbon fibers, that make it possible to save weight in approximately the same order of magnitude as adhesive bonding technology when used consistently and comprehensively. This method has certainly contributed to the optimization of aircraft performance, including greater safety for the passengers in the case of a fire – a fact which, in many cases, is not considered. However, the use of carbon-fiber-reinforced plastics has also resulted in an unexpected 'cost explosion' compared to the traditional structural metal design, with components such as spoilers or flaps made from fiber composites costing twice as much as their metal counterparts. Moreover, there is also some uncertainty about the standardization of repair processes worldwide. Hence, in the case of an isolated damage the components are replaced rather than repaired, further increasing the cost factor because these replacements are very expensive.

It should be pointed out, however, that the materials are not the *only* cost factor. In the majority of cases integral construction designs (i.e. monolithic structures) with very complex geometry are used, the manufacture of which is very expensive. The bonding technology that makes it possible to replace very large, complex structures with several simple components could lead to a considerable reduction in these costs. In fact, in some cases cost reduction has been achieved by returning to former metal construction designs, although this automatically involves adhesive bonding technology, possibly using magnesium and aluminum–lithium alloys. Although the latter approach may still be problematic to some extent, such components have been used recently, and the results of several series of experiments have indicated that, at least with regards to long-term bonding aspects, very few problems are expected with these materials.

Clearly, these aspects of weight reduction and cost savings will shape further developments in aircraft manufacture [1]. Indeed, today the situation is changing in dramatic manner and the saving of weight is now of great importance. Recently, ecological concerns have become more important than raw material aspects, so that fuel consumption will continue to be an important issue.

In all transportation systems, the most efficient way to save costs is to save weight. Although bonding *does* save weight, it is still quite expensive compared to riveting, which is basically an automated technology. Nevertheless, cost savings will most likely be achieved in the near future with the use of simplified surface preparation techniques and the automated application of paste-like, possibly cold-setting, adhesives. At present, two groups of adhesives are used:

- *Flyaway products* include impact-protection foils or paint replacement foils that are applied onto and remain within jet aircraft, and are subject to the most rigorous quality requirements (e.g. fire protection, mechanical loading). Examples include rotor blades or structures such as carbon-fiber Airbus tail-units, which are bonded with epoxy adhesives. The stringers and dampers of aircraft floors are also bonded, as are sandwich-structure engine fairings (Figures 8.2 and 8.3), aeration systems, fastening devices, insulations or decorative elements. An interesting structural

Figure 8.2 Bonded sandwich structure with aluminum cover sheets and aluminum honeycomb core in a fan reverser of a civil aircraft. (Photo courtesy of Lufthansa).

Figure 8.3 A bonded aluminum sandwich. The outer cover sheet (here separated) is perforated for noise absorption. (Photo courtesy of Lufthansa).

application here is the bonding of different metals, as adhesives may compensate for material tensions due to different coefficients of linear expansion, while their good insulation properties may prevent corrosion by electrochemical reactions.

- *Maintenance products* include masking tapes that are removed after use, and remain on the ground as they are not a permanent, integral part of the aircraft. The material requirements are less severe in such cases.

8.2.1.1 Historical Development

During the early 1940s, the first aluminum components were bonded by means of a hot-setting phenolic resin plastified with polyvinyl formal, called 'Redux' (see Chapter 2 and Section 5.4). High-strength and fatigue-resisting joints were obtained when aluminum alloys were pickled with an aqueous solution of sulfuric acid and sodium dichromate at 60 °C for 30 min before application of the adhesive. This so-called 'pickling process' was later used worldwide. Redux became history, and still today, it is sometimes used (almost unchanged) in aircraft construction.

Redux was a two-part adhesive consisting of a liquid phase of phenolic resin applied onto the metals to be bonded, which in turn were coated with a primer made from a similar type of material. Polyvinyl formal, which is solid (granular) at room temperature, was then sprinkled onto the metal adherents; any polyvinyl formal residues that did not adhere to the surface of the phenolic resin were then blown off using compressed air. By using this somewhat original procedure, the adhesive components were reliably and satisfactorily mixed and dosed. Following their application to both sides, the parts to be bonded were joined together, subjected to a pressure of up to approximately 8 bar ($8 \times 10^5 \, \text{N m}^{-2}$), and then heated for 30–45 min after a temperature of 170 °C had been reached. The granular polyvinyl formal (i.e. the thermoplastic component) melted, and the phenolic system cured in the form of a polycondensation. The water formed during the reaction was expelled from the bond-line due to the high pressure applied. About 75% of the adhesive layer was composed of polyvinyl formal and 25% of cured phenolic resin, in an inhomogeneous distribution. Such a distribution was found to be advantageous with regard to fracture mechanics, and a high polyvinyl content was required to obtain adequate peel strength of the joint, although the latter had a relatively low resistance to heat up to a temperature of approximately 60–70 °C [2].

The features of the bonded structures were so compelling that bonding technology was used in the 'Dove', an early commercial aircraft manufactured by De Havilland and, on a larger scale, in the first heavy commercial jet aircraft of the world, the 'Comet', which was produced by the same manufacturer [3]. Particularly in the fuselage of the Comet, both bonded stringers and doublers in the area of the windows were employed. Unfortunately, some of these four-engined aircraft suffered disastrous crashes due to structural failure. Originally, it had been assumed that the bonded joints might have been the reason for these catastrophes, but no evidence was found for this assumption. On the contrary, it is well known that the bonded joints used in the Comet, and later also in the 'Trident', did not present any problems and were in an absolutely perfect condition after a service life of 10–15 years.

In the twin-engine Turboprop Type Fokker 27 'Friendship', which entered service in 1953–1954, almost 70% of the total structure (nearly 550 components) was bonded with phenolic resin. This has been, to this day, the most consistent application of bonding technology in aircraft, both in the fuselage and in the airfoil structures. In total, 1000 of these aircraft were manufactured, with some operating for up to 30 years. The more up-to-date version of the Fokker F 50, with a similar configuration, is a reliable 'workhorse' for short-distance flights worldwide. The twin-engined jet airplane Fokker F 34 and its successor, the Fokker F 100, both of which were manufactured later, incorporated structures that were mostly bonded with phenolic resin and which proved satisfactory for a long time owing to the extremely good fatigue properties and excellent aerodynamic smoothness of the outer structures. Damage to the adhesive joints never occurred, owing largely to the fact that during construction of the Comet and later of the Fokker aircraft, anodization with chromic acid was used. This approach, following chromosulfuric acid surface preparation, made it possible to create relatively thick porous oxide structures on the aluminum which had good corrosion properties owing to their fully developed barrier layer (see Chapter 3 and Section 7.5). Originally, this anodization process was not introduced to optimize the bonding process, but rather to optimize the corrosion characteristics of the nonbonded area of the aluminum surface that had also been coated with a phenolic resin primer and cured separately before the adhesive was applied. It was assumed at that time that the water liberated during the setting of the primer, in the presence of phenolic resin molecules, contributed to a sealing effect of the aluminum oxides, thereby 'clamping' the primer within the rearranging aluminum oxide structures. Experience then taught that anodization did *not* affect the properties of the bonds, and consequently this manufacturing process was adhered to, without realizing that this combination of surface preparation, adherent and adhesive was near-perfect with regards to a resistant adhesion (see Chapter 3).

The bonding of high-strength aluminum alloys using phenolic resins was a worldwide success (which today can be explained scientifically), even in aircraft structures with long-term use. Little attention was paid to the problems of aging resistance of bonded metallic structures in humid conditions with possibly corrosive attacks, especially in aircraft manufacture. However, during the 1960s, in aircraft manufacture, the relatively brittle phenolic resins with low resistance to peel were gradually replaced by epoxy-based adhesives modified with nitrile rubber that also had a higher resistance to heat up to 80 °C. A formulation developed later required a curing temperature of only 120 °C instead of 170 °C, so that the heating process did not alter the fatigue characteristics of the aluminum alloy. Previously, when using phenolic resin adhesives this alteration was compensated by an appropriate dimensioning of the components. Adhesive joints bonded with epoxy resins were shown partly to have a higher strength and fatigue resistance than phenolic resins. The results of aging tests performed in humid and corrosive environments showed that anodizing had no better effect than merely pickling in chromosulfuric acid. Today, these tests are considered to have been completely inadequate, with little attention having been paid to the fracture surface analysis of the destroyed specimen following the aging processes. Nonetheless, anodizing was abandoned in some parts of the world in order to save costs. When long-term damage occurred in the form of a delamination near the interface, as well as corrosion penetrating the bond-line after a service life of about two years, anticorrosive primers based on epoxy and phenolic components were developed, since experience had taught us that phenolic resins were very efficient. Hence, strontium chromate was added as anticorrosive agent, especially for the case of damage. Although, especially in the United States and during manufacture of the Airbus, chromosulfuric acid was used for surface preparation, Fokker and the British aircraft manufacturers never abandoned the use of conventional adhesive systems with anodization [4, 5].

The decision to abandon anodization for adhesive systems based on a combination of epoxy/phenolic primers and epoxy nitrile adhesives proved to be wrong. As the author remembers, this decision was rather taken by intuition, since the interaction between phenolic and epoxy resins and aluminum oxides (which was an important prerequisite for resistant adhesion) was not yet well known, and hence neither was the importance of anodization recognized. Damage of these bonded joints occurred worldwide in the form of delaminations and bond-line corrosion after operation periods of two to five years in military aircraft (first during the War in Vietnam) and later also in civil aircraft (see Section 7.6). The delaminations reported by Boeing as early as the mid-1970s were found to be adhesion failure between the primers and the metal surfaces (Figure 8.4), partly in the form of corrosion of the adherents (which had rapidly invaded the bond-line from the poorly protected edges), and also in the form of delaminations between the primer and adherents occurring in front of the corrosion areas.

At that time, basic research into the long-term behavior of adhesion was started worldwide, with especially important contributions being made in Germany [6]. It was discovered that, in epoxy resin adhesive systems cured with dicyandiamide, a high alkalinity developed as soon as the system was attacked by water. This was due to

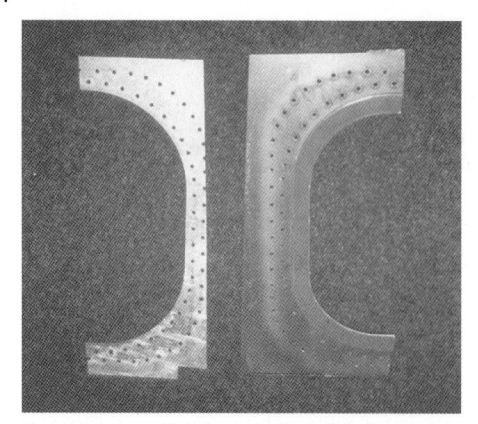

Figure 8.4 Part of a delaminated window doubler of a civil aircraft. The visible corrosion traces are caused by humidity invading after delamination.

the presence of rests of reactive dicyandiamide and reaction products of dicyandiamide and aluminum oxides that induced the liberation of amines. In the presence of the polymer, the aluminum oxides were destabilized by this alkalinity, such that an accelerated solvolysis was initiated that induced macroscopic damages on the surface. By nature, these damages occurred more rapidly on surfaces that were only pickled and where the oxide layers were only about 30 nm thick, than on anodized surfaces where the layer thickness was up to 2 µm. In the latter case, destruction of the bond by solvolysis would only occur after 40–50 years of operation, as may be estimated today. It is known that, at a pH greater than 7 (i.e. in the alkaline range), aluminum oxides are not stable with regards to solvolysis, but they are highly stable at acid pH (5.5–6.5). This explains the fact that no such damages occurred in phenolic resin bonds since, even in the cured condition, phenolic resins continue to remain slightly acid when water penetrates into the bond. The results are a stabilization of the oxides with regards to solvolysis. Phenolic resins may also form water-resistant chelate compounds with aluminum oxides (see Chapter 3). Whilst there is no doubt that chelate compounds may also be formed with phenolic epoxy resin primers, the remaining alkalinity of the overall adhesive layer may penetrate these primers in the case of an extremely humid environment and begin to destabilize the oxide layer.

As a consequence of these research investigations, anodization was reintroduced into aircraft manufacture worldwide, with Boeing, in the United States, applying phosphorus anodization, and the European aircraft manufacturers carrying out anodization with chromic acid. Today, with careful working procedures, high-performance adhesive qualities can be achieved when using either method, in combination with anticorrosive primers and epoxy resin adhesives, provided that the edges are protected both structurally and chemically against primary corrosion. In this way a structural service life of more than 30 years is easily obtained for commercial aircraft. At present, anodization with chromic acid is steadily being replaced by alternative anodization methods.

The development of bonding technology in aircraft manufacture was considerably impaired by the above-described problems, particularly with regards to metal bonding. However, this does not apply to the same extent to fiber composites, which will surely increasingly replace metal structures in large civil and military aircraft (as will be illustrated later), and which may be easily bonded.

8.2.1.2 Advantages of Bonding in Aircraft Manufacture

Bonding is a surface-to-surface and material joining technique that makes it possible to join any combination of similar or dissimilar materials. In aircraft manufacture, bonding offers the following benefits compared to riveting, joining by screws, soldering and welding:

- High aerodynamic surface quality of the bonded components owing to smooth, high-precision contours.

- Excellent properties with regards to stiffness, fatigue resistance and damage-tolerance owing to a flow of forces that is evenly distributed over the surface and is free of notches. This is combined with good damping properties of the adhesive layers in terms of vibrations and even ultrasonic fatigue that may be induced by the engines.

- A possibility to bond different materials without impairing the adherents via the influence of temperature, introduced stresses or damages of the surfaces.

- Easy construction of sandwich structures using lightweight core materials and outer sheets, and the possibility to manufacture multilayer structures in the form of metal laminates with good damping properties and high resistance against crack propagation through the adhesive layers.

- Gas-proof and liquid-tight properties, as well as electrochemical insulation of the adherents via the adhesive.

As mentioned above, in connection with the long-term resistance in humid environments, there are some disadvantages which must be considered when designing adhesively bonded joints. It is possible, however, to limit these disadvantages to a foreseeable extent. It is important to note (and will shortly be illustrated) that the very high-quality standards which must be applied during the production process are prerequisites for high-quality bonds. Furthermore, nondestructive testing of the strength characteristics of a joint is only possible to a certain extent. With today's common nondestructive testing methods, it is only possible to detect a lack of adhesive, a macroscopic delamination, and sometimes – especially with phenolic resins – the results of foaming that takes place during the course of curing if the pressure is not uniformly applied. Finally, with regards to long-term aspects, adhesive joints are characterized by some well-known degradation reactions which must be carefully investigated again and again with regards to their actual hazard potential. This must be done as soon as new adhesive systems are developed, although they do not represent any exclusion criterion with regards to high-strength structures intended for a long service life. Within the adherents, degradation mechanisms may also occur which represent a well-calculated risk, particularly with regards to long-term resistance.

Bonding technology is a *must*, especially in the manufacture of large metal aircraft structures. In the skin of a wide-bodied aircraft with a fuselage diameter of (commonly) 5.5 m and more, the fuselage shell must bear stresses of around 110–120 N mm^{-2}, which is easily calculated from a cabin pressure of 0.7–0.8 bar (0.7–0.8 × 10^5 N m^{-2}) that is needed to ensure passenger survival, and a common sheet wall thickness of 1.6 mm; *gust loadings* must also be added to this. The fatigue resistance of the aluminum alloys used, determined using polished round bar samples, is approximately 130 N mm^{-2} in terms of their fatigue limit during near-unlimited alternations of load; this does not leave much room for structural purposes, especially with regards to the weight. The same is true when considering that these types of stress only occur during flights, and that the number of flights will correspond to 60 000–70 000 loading periods (in terms of the outer cover of the fuselage and for the inner pressure at high altitude) for short-haul aircraft, and to a maximum of 40 000 periods for long-haul aircraft. Consequently, the aircraft are not sized in the area of their fatigue limits, but rather in the area of their fatigue strengths. It must also be taken into account that aircraft structures are extremely *notched* structures; apertures for cables, windows and doors and, in the case of a wide-body aircraft up to 3.5 million rivet holes, will weaken these metallic structures. In aircraft manufacture, bonding technology allows reduced stress concentrations to be achieved in the bond-lines. Furthermore, it is also possible to reinforce the common wall thickness of 1.6 mm by a doubling of the skin by using bonded sheets of 1.6 mm thickness or very often not more than 0.6 mm, in the areas of notches, windows, doors and longitudinal joints, as well as transverse joints. Bonding also has economic advantages when compared to the so-called 'integral construction' method, as the latter involves a reduction in structural surfaces to minimum thickness by means of chemical surface removal in areas where there is no need for a greater thickness. The service life may also be improved by the use of 'bonded doublers', most likely as a result of the good damping properties of adhesive layers (notably in the areas of longitudinal and transverse joints), although these joints are generally riveted. This situation can be demonstrated by the following example, where a typical longitudinal joint in a wide-bodied aircraft is created in the form of a simple lap joint that is then riveted. If, for the purpose of stiffening, doubler sheets are merely inserted, only one-fourth of the service life is achieved compared to a lap joint where doublers are bonded to both skin sheets to be joined. It must be noted that the riveting joint between both panels is not bonded. The fatigue resistance of the longitudinal and transverse joints, respectively, may again be increased by a factor of 2 if an adhesive layer is introduced into the joint, in addition to the rivets. This method was adopted long ago by Fokker, who used two-part epoxy resin adhesives.

Adhesive joints have another advantage: if cracks develop in the skin sheets, they will only run up to the adjacent stringers and frames, where they are then redirected by the adhesive layers and run back to the skin sheets (Figure 8.5). This effect is most likely due to the excellent damping properties of the adhesives. Although cracks

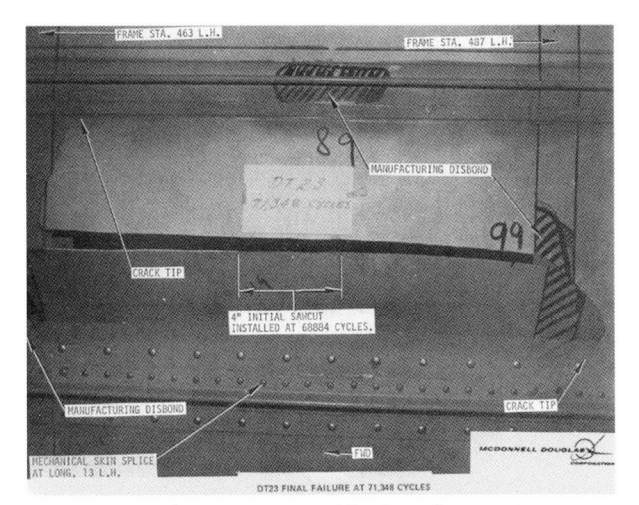

Figure 8.5 A crack in an experimental fuselage after 71 348 inner pressure cycles. The crack was initiated artificially after 68 884 cycles, and ran to the left in the bonded area of a stringer, and from there backwards. On the right side there was a delaminated stringer bond-line which could not stop the crack propagation [7].

cannot be avoided completely, in this way it is possible to control the risk with regards to the overall structure by specifically arranging the stringers and frames, thus preventing crack propagation. The thickness of the sheets may be reduced by using sandwich structures with lightweight cores and thin metallic outer sheets. This will save about 15% of the structural weight which, for an average wide-bodied aircraft, is approximately 35 t, as clearly demonstrated by investigations performed in the United States [7].

It is clear, however, that even today there are limits to the degree of weight reduction possible in metallic aircraft structures, and this point must be considered when building ever-larger, more economic aircrafts. The result has been the initiation of some revolutionary developments in aircraft manufacture. The most important factor here is the strength of the construction material, and in particular its *modulus of elasticity* in relation to its specific weight. High-strength aluminum alloys, for instance, have a tensile strength in a range of 350–400 N mm^{-2} and an elastic modulus of approximately 80 000 N mm^{-2}, the specific weight (density) of aluminum being 2.7 g cm^{-3}. Glass fiber-reinforced plastics, which have been used for many years, have similar values, their fatigue strength reaching more favorable values as compared to the initial strength, and their density being only around 2 g cm^{-3}. In recent years, considerable progress has been made when using carbon fibers (which today are available worldwide) for the production of fiber-reinforced plastics. This has resulted in an increase in the strength and elastic modulus by factors of 2, and a density of 2 g cm^{-3}. When building large aircraft structures, improving the elastic modulus leads to a considerable reduction in the problems of stable dimension and stiffness. Compared to aluminum structures, fiber-reinforced plastics also have a considerably lower risk of fire – an advantage which is often neglected. In time, therefore, fiber-reinforced plastics – and especially carbon fiber-reinforced plastics – should replace metallic structures far beyond the areas of application where they have been used previously in aircraft manufacture. Moreover, the high costs associated with fiber-reinforced components may be reduced by replacing complex, difficult-to-laminate monolithic structures with plates or profiles that are of a simple design and can be adhesively bonded. Bonding is the best joining technique for those inhomogeneous material structures that have a high notch sensitivity. For example, when using epoxy resin matrices (as is common practice today), bonded carbon fiber-reinforced materials have a good long-term resistance. It must be noted, however, that it is still necessary to provide a mechanically preformed surface preparation. For that purpose, fabrics are laminated onto the outer sheets of the areas to be bonded and then, after curing, peeled off before adhesive application; in this way a rough surface is obtained. Another possible approach would be to roughen the surface slightly, using mechanical means. The consequence of this is that there is no impairment of age resistance due to degradation of the oxide layers, as may occur on metal surfaces (see Section 7.5.4). It is very likely, therefore, that this approach will open up a new and very large area of applications for bonding technology in future aircraft manufacture. It is also likely that the hot-setting film adhesives which have been used to date will be replaced by two-part, cold-setting adhesive systems.

8.2.1.3 Adhesives Used in Aircraft Manufacture

As mentioned previously, in structural aircraft manufacture, phenolic resin adhesives are used which may today be applied in the form of prefabricated, supported or unsupported films. Furthermore, epoxy resin film systems supported with fabrics or mats are used which are processed on anticorrosive primers (also described previously). In order to obtain plastication, the epoxy resin matrices of these adhesives, which are cured at 120 °C, are modified by means of epoxy-terminated butadiene–acrylonitrile rubber components (CTBN) that segregate in the form of globular particles when the epoxy resin cures in the bond-line; in this way they act not only as a plasticizer but also as a crack stopper. For higher-temperature environments, other thermoplastic materials such as polysulfone resins also are used, whilst polyamide (which was used in former times) is still being used to produce adhesives for use in low-temperature environments. The epoxy resin component itself is produced from oxirane-terminated bis-phenol-A monomers and other oligomeric fractions; these are sometimes modified with regards to fire risk, smoke development in case of fire, and toxicity. Dicyandiamides continue to predominate the curing systems; at room temperature they have a crystalline form so that they are not able to react, and elevated temperatures are required for them to dissolve in the adhesive (albeit to a limited extent) and to have them react. Originally, a temperature of 170 °C was required for this purpose, but today specific accelerators in the form of amines make it possible to obtain the same results with a temperature of only 120 °C. Polyimide adhesives may also be used in very high-temperature environments – for example, close to the engines – as they resist temperatures of 250 °C and higher, at least for short periods of time.

With the exception of the longitudinal joints bonded by Fokker (see Section 8.2.1.1), cold-setting adhesives – for example, two-part epoxy resin systems which generally are cured at approximately 60 °C in order to improve their strength and resistance – have not yet been used in aircraft structures. However, they will increasingly be used in the future in combination with carbon fiber-reinforced plastics. Today, even cold-setting epoxy resin systems can be modified to obtain a globular rubber formation in the bond-line, which will result in an increased toughness. As a consequence, this type of adhesive may also be used in application areas where the temperatures are 80 °C or higher, thus offering an important development potential in the structural field.

8.2.1.4 Surface Preparation

At present, when manufacturing metallic aircraft the most widely used surface preparation method is a combination of cleaning, pickling and anodizing (as described above). Usually, the raw aluminum parts are precleaned with organic solvents, followed by alkali degreasing, pickling in diluted hydrofluoric acid in order to remove accumulations of copper on the surfaces and staining caused by the alkali pickling process, then pickling in chromosulfuric acid for 30 min, and finally anodizing in phosphoric acid or chromic acid. Unfortunately, this method is very labor-intensive, involving intermediate back-washings, and is highly problematic and expensive from both toxicological and ecological viewpoints. As the components to be bonded must undergo an overall treatment, the finally anodized surfaces must also be coated in the nonbonded areas, because the anodized surfaces have visually unappealing properties. Consequently, the surfaces are primed first and then painted in the nonbonded areas. This surface preparation method requires a very intensive control of the pickling and anodizing baths, with regards to their chemical composition and temperature and finally, during anodization, the current that is applied. This is the most important starting point for future optimization, since the chromic pretreatment methods will clearly need to be substituted by other methods on the next occasion. In general, however, aircraft manufacturers at present have special authorizations to continue using a pickling process followed by chromic acid anodizing, mainly because the modification of a permitted manufacturing program requires costly certification. Nonetheless, alternative methods of anodization have been intensely investigated. For example, in the United States a phosphoric acid anodizing process has been widely used in aircraft manufacture which is less critical with regards to environmental technology and toxicology, but which still requires an initial pickling in chromosulfuric acid. Other new anodizing methods currently under development in the United States include a boric-sulfuric acid anodizing process.

As mentioned above, the surface preparation of fiber-reinforced plastics is essentially mechanical in nature, and is therefore less critical with regards to ecological viewpoints. When the 'peel-ply' method is used – that is, when peeling off a fabric inlay – it is important to consider that the type of fabric used will influence the adhesive properties of the surface thereby created. In order to avoid such an uncontrolled influence the surfaces may also be ground, if necessary, using a mild process such as Scotch-Brite®.

8.2.1.5 Bonding Techniques

In modern aircraft manufacture, the bonding process starts with a chemical surface preparation of the components to be joined (see above). On completion of the final back-washing step, the surfaces are dried and then coated (as immediately as possible) with an anticorrosive primer, either by spray application or by using an electrostatic coating process. Before the assembly process, the primer is cured at 120 °C (see Section 5.10); this allows the surface conditions obtained to be preserved for several weeks until the bonding takes place. The components to be bonded are then joined using tools that may be very costly (such as clamping pins) to position the adhesive films, which have been pre-cut, between the adherents. As the epoxy resin adhesive films, which are able to cure without any application of pressure, would not adequately wet the primed surfaces, a foil that is sealed towards the border of the tool covers the joined and fixed components. The space between the foil and the tool is evacuated, and the whole tool is placed in a heated autoclave; the curing process is then initiated by the application of external pressure and heat. Clearly, this process is very labor-intensive and, indeed, in many other fields of industry such a system would not be used due to its associated high expense and time consumption. Yet, in aircraft manufacture the high-quality standards of bonding are met almost exclusively by this process. Although a reduction of the high manufacturing costs associated with this method represents an important future task, it must also be realized that riveting – a high-precision, largely automated method – is also very expensive.

8.2.1.6 Quality Control of Bonding in Aircraft Manufacture

In aircraft manufacture, the labor-intensive production process involves very intensive quality control measures. To start with, the personnel must be highly qualified, and every step performed to produce adhesive joints must be documented. Extensive measures are also taken during product incoming inspection, especially in the case of the primers and adhesives; after which every production step is, as far as possible, automatically controlled and monitored. Test specimens are produced which run along with the batches and assemblies during the manufacture process, and their properties are determined in destructive experiments at an early stage of production.

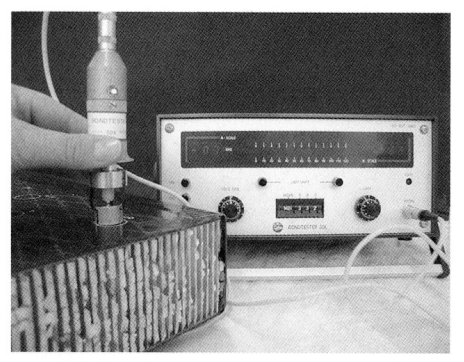

Figure 8.6 The Fokker Bond Tester with a CFR-aluminum core sandwich.

Finally, all of the finished bonds are tested by means of a nondestructive ultrasonic test method (see Section 7.6) based on resonance frequency (the Fokker Bond Tester; Figure 8.6), and the results documented. Any remaining test specimens are retained for subsequent control of the bonding quality obtained.

When damaged components need to be repaired, the same bonding processes are used. Although for technical reasons the costly chemical surface preparation processes are often replaced by other methods, such as grit-blasting with corundum. As the original structure integrity will not be completely obtained, the repaired components are closely monitored during operation.

8.2.1.7 Conclusions

Bonding technology as a high-performance joining technique for extremely long-life and heavy-duty structures was developed in aircraft manufacture, where it proved to be a success despite some critical periods during development. This technology clearly contributed to improvements in aircraft structures, and will continue to be essential in aircraft manufacture, notably when fiber-reinforced plastics are taken into consideration. Although, it seems unlikely that the costs of this labor-intensive production process will be greatly reduced, the development of bonding technology may in fact now occur in reverse sequence. Whereas, technology as a whole has learned from aircraft manufacture, it may now be expected that aircraft manufacture may draw lessons and gain experience from other fields of technology (e.g. car manufacture), in order to simplify and cheapen its processes (see Section 8.2.2). It is important, however, to take into consideration that the performance of aircraft differs greatly from that of other technical products. For mere reasons of amortization, a modern civil aircraft must achieve a service life of 20–30 years, most of which time it spends in the open air and often under difficult climatic conditions. During the flights, there are high mechanical stresses present, yet any adverse outdoor conditions must be considered noncritical. Within the aircraft fuselage, due mainly to exhalation by the passengers, considerable moisture accumulates and begins to freeze on the outer wall, later accumulating in the lower fuselage area as condensed water. In fact, when a wide-body aircraft lands in a tropical zone, between 1 and 1.5 t of condensed moisture is present on the lower fuselage area, containing highly corrosive components and requiring removal. While an aircraft is on the ground at airports, it is mostly unprotected with regards to corrosion and intense UV radiation. When compared to a car – which may have a maximum service life of 15 years, a total operating time of 4000 h, with a total distance covered of about 200 000 km – an aircraft would be expected to have a service life of 30 years, and to be actually flying for between 30 000 and 60 000 h. Within this operating time, the taxiing alone of an aircraft, especially when operated on short distances, would involve a distance of about 300 000 km far more than would be expected from a passenger car. Finally, it must be mentioned that, for a heavy airliner, a general check-up – with an inspection of all structural components – is scheduled after about 5000 operating hours. This means that absolute reliability of the joints must be ensured over a time period that largely exceeds the running time of a passenger car. Clearly, such reliability can only be achieved by means of a highly labor-intensive production.

Yet, in aircraft manufacture – when compared to automobile construction – a certain residual risk of the technology is possible without any loss of safety, because at all times every aircraft in operation is registered as to its whereabouts and operational period, worldwide, whereas cars will not be within easy reach of central organizations on a continuous basis. If it is necessary to perform certain constructive design optimizations on an aircraft – whether rework or repair – this may be done worldwide, utilizing a common practice. In this way it is clear that bonding technology has contributed considerably to reducing the predominant problem in aircraft manufacture, namely structural failure, which in earlier times would often require reworking to be carried out. Moreover, this approach has not only saved costs but also considerably improved passenger safety.

8.2.2
The Automotive Industry

Automobile construction is subdivided into the manufacture of passenger cars, trucks, buses and motor homes.

In the future, *weight reduction* will become increasingly important both in passenger car construction and industrial vehicle construction, although this may not be gathered from the development which has taken place in the past, in particular in passenger car construction. A reduction in fuel consumption through the further optimization of aerodynamics is essentially no longer possible and, in many cases, is not necessary because cars are mostly driven at speeds less than 80 km h^{-1} in town and in light traffic, and air drag is not important under such operating conditions. With regards to traffic development, driving conditions between 80 and 0 km^{-1} (traffic jam) will probably increase. Important reductions in fuel consumption have been achieved by optimization of the drive train, yet the vehicle weight was not reduced for a variety of (including fashion) reasons. With regards to foreseeable traffic scenarios and the resultant consumer behavior, these reasons will surely not persist, however.

While passenger cars continue to be built the way they are today, there will be very little room for weight reduction in the bodywork and chassis. Moreover, as vehicle history and development teaches us, the use of steel as the predominant material in self-supporting bodies, and of welding as the predominant joining technique, will surely continue. Although, lightweight metals or plastics may be used instead, from the point of costs the use of light metals (especially aluminum alloys) would require a drastic reduction in the number of components. This may be achieved by means of special body framings and simplification of the assembly process, such that a higher materials cost will be compensated by a lower production cost. Bonding alone, or in combination with riveting, is required when joining largely preshaped and pre-developed components. Although the recycling of aluminum materials is essential mainly for energy reasons, it is gratifying to realize that bonded aluminum constructions will continue to be recycled in years to come.

Today, *recycling* problems are the main reasons for not opting for the structural use of plastics. Although purely thermoplastic components – which are predominantly welded rather than bonded – may be recycled several times, fiber-reinforced plastics –

which are generally required to produce high-strength components – still encounter recycling problems that, as yet, have not been solved. There is, therefore, much reason to believe that steel in particular – being a 'good-natured' and low-priced, recyclable construction material – will continue to predominate passenger car construction, especially as the often-referred-to corrosion problems may be solved, or have already been solved. Again, the use of bonding may result in a weight reduction in the structural steelwork, because it offers more versatile design options compared to welding. During manufacture, the joints need not be accessible from both sides, and the minimum sheet thickness may be less than the 0.6 mm otherwise required by spot welding. A further reduction in weight may be obtained by using 'sandwich structures' for large-surface components, such as the roof or the underbody. In contrast to welding, bonding may simplify the production process, because the final assembly of a body – that is, fitting the underbody or placing the roof – is the final operation sequence, when all other processes such as surface finishing have been completed. Hence, the production conditions would be clearly improved.

For many years, fully bonded steel bodies have been tested to trial stage, without any problems. Compared to their welded competitors, their properties are surprisingly favorable, with the low strength values of adhesives complying with the requirements of hem flange bonding or the joining of glass and steel, because the surfaces being bonded are relatively large. The damping performance and fatigue strength of bonded steel bodies are considerably better than those of welded ones. Likewise, their deformation behavior – that is, the energy storage in case of imposed deformation, or accident safety – is extraordinarily positive (see Section 8.2.2.1).

Due to limitations in the production sequence, the processing conditions play an important role alongside those purely mechanical viewpoints when choosing the correct adhesive. For example, adhesives may show a good adherence to nontreated sheets, or at least be sufficiently resistant to 'washing away' in dip-painting baths.

Structural bonding must be performed without any special pretreatment on sheets coated with corrosion-protection oil, but this no longer presents any problems. Although bonding is produced 'against all the rules' here, the results have been successful, with even glued windscreens contributing considerably to the rigidity of the car body nowadays. Only one problem remains to be solved, namely the development of a simple, suitable technology for bonding in a repair shop in addition to bonding for manufacture. At present, very few cases exist where bonding can be successfully used for repair purposes, an example being the repair of a windscreen damaged by flying stones. If the windscreen cannot be repaired by means of adhesive setting by UV irradiation, then several alternative adhesives are available for gluing-in a replacement screen, with the final join retaining the original, factory, quality.

Repair technology will be developed further since it is expected that, in the near future, fully bonded vehicles made from steel sheets or perhaps even aluminum will enter the market. If bonding technology is consistently applied in bodywork construction, it is probable that structural weight reductions of 20% can be achieved. This means a saving of about 60 kg on the body-in-white of a medium class vehicle weighing about 300 kg.

Nerf bars, spoilers, hand straps, hooks and mirrors are all bonded by using pressure-sensitive adhesives (PSAs) in the form of high-performance foamed acrylic adhesive tapes, or thermally postcuring, pressure-sensitive self-adhesive structural tapes (e.g. mirrors bonded onto windscreens). Seals (e.g. in doors) are increasingly bonded by means of special foamed adhesive tapes owing to a noise reduction and simplification of the assembly process. In the production of seats, a variety of adhesive materials are employed, for example in the upholstery and seat covers.

Another interesting field of application are commercial vehicles, or vehicles used by public authorities such as taxis, police cars and municipal vehicles. Rather than being coated with a uniform paintwork, they are provided with a film adhesive applied on the original paintwork, which may be removed after the intended service life, increasing the resale value owing to the protection of the original paintwork.

Truck manufacture is an important area where adhesive bonding is also increasingly used. Here, many trailers are constructed from fully bonded sandwich structures that are self-supporting, torsion-resistant, and have excellent insulation properties as no fasteners are employed which may build up heat bridges.

Even safety-relevant components such as seat belts may be bonded to the sidewalls of the vehicle, as long as the adhesive force of the material used is higher than the stability of the sidewalls. As in the passenger compartments of cars, various bonded components are utilized in trucks, such as the floor coverings and seats. Bonding is used not only for joining but also for vibration damping and sealing; for these purposes, foamed polyurethane (PU) tapes and other high-build materials are employed.

Closed-top containers, as might be used in furniture vans, are manufactured by joining beams to sidewalls by means of high-build adhesives. Two-part epoxy resin adhesives are used for bonding planks, or foamed adhesive tapes for vibration damping. In the majority of cases, the optical results achieved with bonding are better than those possible with riveting. In addition, the outer wall of the truck can be used as an attractive advertising surface, by using film adhesives with large-area prints that are strongly visible even at twilight and in the dark, owing to their retroreflecting properties.

The manufacture of buses also relies on the bonding of strips, hinges, fabrics, seats and whole sidewalls, as well as various foils. The front screens, side windows and rear windows are glued to the body using moisture-setting PU adhesives. High-build adhesives may also be used here as sealing and vibration-damping materials.

Owing to the use of efficient, low-priced production techniques, a well-equipped motor home may cost the same as a 'superclass' vehicle with simple equipment, despite lower serial production quantities. As the cost factor is very important here, the manufacturers generally use glass fiber-reinforced plastics and other types of plastic that can be joined with adhesives in order to create the body shell. Other examples include the bonding of strips and damping/insulation material such as Styrofoam.

Adhesives are also increasingly used for the bonding of electronic components and identification labels or nameplates on the vehicle. By using high-performance adhesives, bonded joints can even be produced in motor cars that are exposed to important thermal and mechanical stresses.

8.2.2.1 Bonding in Passenger Car Bodywork Construction

In modern vehicle construction, high priority is given to weight reduction, improved long-term mechanical strength and corrosion resistance, increased comfort, crash resistance and higher body rigidity. Intelligent lightweight constructions can only be obtained by consistently using a material mix of steel, light metal and plastics which may be joined by conventional thermal and mechanical joining techniques, but only to a limited extent. It seems that bonding is a competitive joining technique with specific advantages for the above-described purposes.

The most important areas of application for bonding in the construction of vehicle bodyworks are direct glazing, hem-flange bonding and antiflutter bonding, the bonding of plastic components, and the structural bonding of bodies.

Direct Glazing Direct glazing constitutes the fitting of fixed glass panes with PU adhesives (Figure 8.7). In the majority of cases, one-part moisture-setting products are used which create an elastic joint between a painted body flange (finishing paint or cataphoretic dip coating 'e-coat') and the glass. The glass side is pretreated with an adhesion-promoting glass primer that provides for adequate light rejection. If a sufficiently low UV transmission is obtained by the ceramic screen process printing on the border of the pane, then transparent glass primers may also be used on occasion [8–10].

A conventional glazing system consists of a glass promoter (purpose: cleaning, adhesion promotion), the black glass primer (adhesion promotion, light rejection), the adhesive itself, and the paint primer (adhesion promoter for the paint side). The glass promoter contains silanes, and is especially efficient when applied as a thin coat. Therefore, it is common practice today that the promoter is wiped-off with a clean cloth after application (wipe-on/wipe-off process).

The one-part PU adhesive sets with moisture, taking the reactant (i.e. water) from the ambient air. The setting speed is determined by the diffusion of water vapor, with

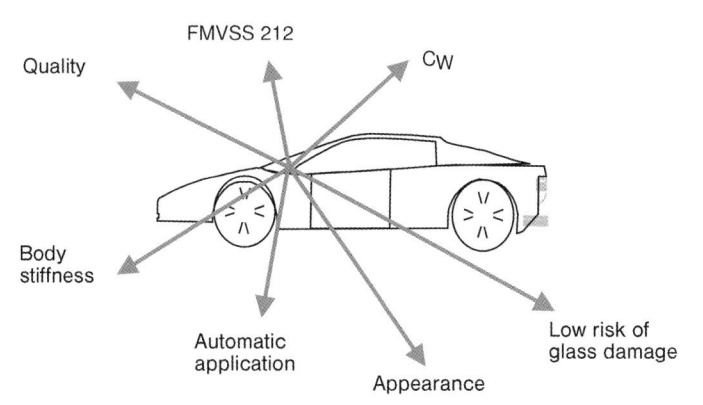

Figure 8.7 The different functionalities of direct glazing in cars. FMCSS 212 is a security standard (see text; C_W is the factor of air resistance).

low absolute ambient humidity delaying the setting process. Although about 20 years ago, metered water pastes were used in North America, so that the setting process was independent of the climate in the manufacturing plant, these have now virtually disappeared from the market because the dosage and mixing was labor-intensive.

Occasionally, the adhesive joints are created directly on the e-coat. In this case, after e-coat curing, the bonding area must be covered temporarily before any further painting process can take place. This additional stage is justified as e-coating offers a very well suited adhesion area because e-coats adhere well both to steel and to the glazing adhesive. Very little variety exists among e-coats in car manufacturing, although the various types and colors of paint require extensive adhesion testing. Furthermore, each additional paint coat represents another interface that may become a potential inherent weakness as to the adhesion forces and long-term resistance. The e-coat bonding area is covered by special adhesive tapes or polyvinylchloride (PVC) plastisol adhesives which must be easily detachable after curing of the finishing paint, without leaving any residues. When the tapes have been stripped off the area is left clean and perfectly suited for bonding with the glazing adhesive.

Direct glazing was first introduced in the United States during the early 1970s. Notably, in the event of a crash the pane will remain fixed in the flange, thereby complying with US safety standard FMVSS 212; this is in contrast to the rubber profiles which were previously used to fix the pane in the flange, and which tended to fail on impact. In cars fitted with airbags the bond must also be capable of absorbing the forces when the airbag impacts on the inner side of the windshield.

In addition to its safety functions, direct glazing offers the following benefits:

- the possibility of automatic application
- a small gap width between pane and flange for improved aesthetics and lower drag coefficient
- a distinct increase in torsion resistance through integration of the glass into the body structure
- highly reliable tightness
- virtually no glass cracking and breakage during manufacture

The tensile strength of the cured adhesive film is approximately 6–10 MPa, with a stretch at break of several hundred percent, and a physical shear modulus of around 3 MPa. However, it is common practice to measure the shear stress–shear deformation value at a shearing of 10%, using a tensile shear specimen with thick adherents according to DIN 54 451, with a higher adhesive thickness that is similar to conditions encountered in practice. By using this test method, lower values are obtained. The value issued is not a 'true' shear modulus but rather needs to be exactly standardized. Only those values obtained with the same thickness of adhesive coat, the same rate of deformation, and the same bonded area may be compared. For example, the effective rate of deformation of an adhesive coat of 6 mm will be only one-third of that of an adhesive coat of 2 mm at the same speed of the crossbeam of the tensile testing machine. However, above the glass transition temperature (T_g) all polymers basically have a viscoelastic behavior, and therefore the mechanical properties depend on the deformation rate. Hence, the two test results can no longer be compared, even if the bond-line thickness is taken into consideration when assessing the measurement of the shear stress–shear deformation value.

The T_g of the cured adhesive is lower than $-40\,°C$, so that the mechanical properties will hardly change within the whole range of the operating temperature.

Since the introduction of direct glazing, the basic design of the system has remained largely unchanged, and pane-fitting technologies that differ fundamentally from the firm (but elastic) bonding method are unlikely to be developed in the near future. Nonetheless, some processing details have been modified, not only improving the physico-mechanical properties but also facilitating the application of these systems at the assembly line:

- Hot application systems; after joining, the viscosity of the systems increases on cooling. This is partly enhanced by a crystallization of polymer components, and results in a good resistance to slipping off; consequently, auxiliary assembly tools may be dispensed with.
- The shear modulus is increased by a factor of 2 to 3; without any modification of the total weight, this further increases the torsion resistance of the body.
- Considerable improvements of the build-up of adhesion have been obtained with new paint systems that are more difficult to bond.
- More robust systems without any glass promoter or paint primer are used.
- Special two-part products accelerate the setting process and sometimes considerably reduce the influence of ambient moisture on curing.
- Glass primer products are already applied by the glass manufacturer and preserve their adhesion-promoting effect for months.
- Adhesives with low electrical conductivity prevent contact corrosion when joining metals that are less noble than carbon. With regards to strength and rheology, carbon black is used in adhesives as a highly efficient filler. Specific direct-current resistivity is used for assessment. This is useful, for example, when the insulating paint of an aluminum body is damaged during pane replacement. The complex impedance is also required to be sufficiently high, as radio antennas are often integrated into the panes.

For repair purposes, or for the recovery of end-of-life vehicles, the bonded panes are removed using vibrating cutting tools, or they are cut through with a high-strength steel wire. Adhesives that allow the joint to be easily debonded are currently under investigation, but the concepts developed to date have been largely unsatisfactory (see Section 8.16.5).

Safety at Work and Environmental Protection The noncured adhesive is mainly composed of a PU prepolymer, extremely low-volatile plasticizers and fillers as well as monomeric isocyanates in low quantities which react with the ambient moisture during the curing process. The isocyanate type used has an extremely low volatility (vapor pressure at 20 °C: 0.476×10^{-6} kPa). With correct handling, no traces have been detected in the ambient air even when using refined analysis methods.

In general, highly volatile solvents are not used for adhesives. These adhesives are safe to use provided that all mandatory regulations, ranging from storage and processing to disposal, are observed.

Glass Repairs The most important differences compared to OEM are as follows:

- Bonding onto the former adhesive surface, freshly cut during removal of the damaged pane. This surface is a good adhesion base for the repair adhesive. The preparation of the glass side is performed in a similar manner as during the OEM process.
- Priority is given to a speedy build-up of strength, influenced as little as possible by the ambient moisture, in order that the vehicle is returned to the road as soon as possible; therefore, two-part adhesives are frequently used.

Preference is given to repair systems that have a lower viscosity in order to allow hand application with a pistol. The functional characteristics, however, are required to correspond to those of the original quality (strength, modulus, electrical conductivity). The processing robustness is expected to be even higher, which allows the repair to be made at the roadside rather than in special repair shops.

Direct glazing has been the first high-volume and safety-relevant application in vehicle construction to prove the excellent performance of bonding as a joining technique, thus building confidence for further, high-demand fields of application. Adhesive systems that are comparable to direct-glazing products are used to bond sunroofs and roof modules. In particular, when joining dimensionally large adherents and material mixes (e.g. metal with plastics), the high flexibility and ductility of these systems represent important advantages since the different coefficients of linear expansion make joining difficult. In bus and truck construction, besides direct glazing, a similar application might be in the bonding of glass fiber-reinforced plastic (GRP) bus roofs to the metal structure. The relative movement of the large adherents, caused by temperature changes, must be absorbed permanently, and this requires a corresponding thickness of the adhesive layer. As a rule of thumb, the thickness should approximately correspond to the length difference expected to occur between the adherents. For example, in the case of a bus roof with a length of 10 m (GRP on steel) and a temperature of 80 °C at solar radiant exposure, the bond-line should be at least 5 mm thick. The operational temperature may differ by approximately 60 K from the room temperature prevailing during the bonding process and, in some cases, it may even increase further.

Hem-Flange Bonding and Antiflutter Bonding For the bonding of doors and bonnets (hoods), the build-up of adhesion must not be impaired by any oil located on the adherents (generally 3–4 g m^{-2}). If adhesives based on epoxy resins, butyl rubber and PVC plastisol are used, or combinations of PVC with rubber, or acrylate-plastisols, the oil is absorbed by the adhesive during the curing process in the e-coat curing stove.

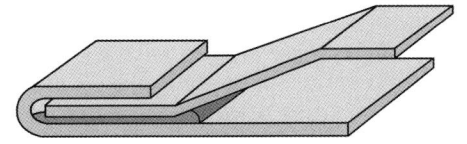

Figure 8.8 Hem-flange bonding.

If a high-modulus, high-strength adhesive is used for bonding the sheets to be joined in the hem area, the rigidity of the doors and the hoods is increased considerably (Figure 8.8). Protection against corrosion is improved by the sealing effect of the adhesive, with epoxy adhesives showing the best performance. The greatest protection against corrosion is achieved, however, by means of additional fine seam sealing.

When add-on components of the mostly complex structures are joined, and stiffening profiles for the hoods and roof are glued on, the bonding site is expected to damp vibrations while producing a stiffening effect, but without being outlined against the outer sheet. As the components pass through different cleaning and painting baths, and are heated to high temperatures (mostly up to 180 °C) in the e-coat curing stove [11] where they cure and cool down to ambient temperatures, very efficient adhesives are required for this purpose (e.g. PVC plastisols). Rubber adhesives are also being used increasingly for this purpose.

Large gap widths may be bridged by means of products which foam in the stove and prevent further outlining.

Bonding of Plastic Parts In lightweight vehicle construction, use is increasingly being made of plastic materials owing to their low material density on the one hand, and to the relatively easy fashioning of complex shapes on the other hand. Compared to metals, their strengths and moduli are considerably lower, with the exception of carbon fiber-reinforced plastics that are not taken into consideration here. Surface-to-surface joining by bonding is the first-choice joining technique rather than point-like joints with high local tensions.

For the joining of material combinations, bonding is also superior to other joining techniques in many cases, for example, when a plastic fender is joined to a metal structure, when a polycarbonate (PC) roof part is fixed or a PC side window, when a GRP bus roof is fixed, or when plastic rear gates including the glass pane are joined in combination.

No outlining is tolerated, either, in the case of visible vehicle parts that must be repainted, and the mechanical properties of the adhesive and the joint must not be impaired during the painting process, where temperatures of about 150 °C will be encountered.

As some of the plastic materials used (e.g. polypropylene) are nonpolar and have a low surface energy, adhesion promoters are often applied, or a physico-chemical preparation of the adherents is performed before the bonding process takes place (corona, low-pressure plasma, plasma surface treatment at atmospheric pressure, fluorination, pickling, etc.). Today, special acrylate adhesives are emerging onto the market that may build up adhesion on almost every type of plastic material by means of a graft reaction, while not requiring any labor-intensive surface preparation. Although no long-term results are yet available, this trend shows great promise.

The most frequently used adhesives, however, are polyurethanes, with preference given to two-part systems. Their strength builds up very rapidly, and independently of the ambient moisture. A wide variety of PU systems is available, ranging from soft, low-strength systems to high-strength, high-rigidity adhesives.

It must be noted that the T_g of the adhesives must not to be within the range of the intended operational temperature, because slight changes in temperature would otherwise considerably modify their modulus and strength, and the properties of the component would be largely temperature-dependent.

Structural Bonding of Bodies In the field of vehicle body construction, structural bonding is defined as the creation of a rigid long-life and force-transmitting joint of rigid high-strength adherents by means of an adhesive, even in areas which are exposed to the risk of a crash. 'Long-life' means during the entire service life of the vehicle, while 'rigidity' means moduli of elasticity in an order of magnitude of 10^3 MPa with regard to the adhesive, and 10^5 MPa with regard to the adherents.

These values make structural bonding considerably different from direct glazing and the majority of all other bonded joints created in vehicle construction. Structural bonding makes it possible to increase the rigidity of the components and the crash resistance, both of which are physico-mechanical properties. In general, the materials joined by means of bonding are metals, for example different types of steel, aluminum and magnesium.

One important aspect that helps us to understand the high performance of structural bonds is the T_g, or rather the range of T_g-values of the adhesive. The T_g should not be in the operational temperature range of the bonded body parts (i.e. $-30\,°C$ to $+80\,°C$). Rather, the T_g should be $>80\,°C$, thus ensuring both rigidity and crash resistance. As a consequence, the operating temperature should be $<T_g$. For the purpose of direct glazing, as described above, the glass temperature and the operating temperature must be chosen in reverse sequence. For direct glazing, the T_g must be below the operating temperature. With regards to polymer chemistry, adhesives are formulated so as to contain different components that induce a toughened rather than a glass-like brittle condition below the T_g. This is the main difference compared to rigid epoxy adhesives, which are not resistant to crash and therefore may not be used for structural bonding according to the above-mentioned definition.

High-strength adhesives suitable for different industrial applications have been known for a long time. However, for the structural bonding of bodies – which requires a high rigidity and high-energy storage capacity in the case of a crash – the brittleness of the adhesives in the case of an impact load, enhanced by lower temperatures, was opposed to these requirements. Although the properties of the adhesives were improved by developing so-called 'rubber- toughened epoxies', the scope of the improvement was not sufficiently important. The ultimate success was achieved by the development of a new method of toughness modification, involving an elastic component which is incorporated into the epoxy matrix like an island or by a reactive approach, and which makes it impact-resistant. Now, the deformability of the still rigid high-strength adhesive is high enough to prevent failure due to glass-like brittleness, in the case of stretching the adherent beyond the tensile yield point induced by the application of a load. It is important to ensure deformability of the adhesive even at lower temperatures, and the whole operating temperature range must be covered with regards to this safety-relevant issue. In the case of a crash, the stretching of the metal absorbs the most important part of the energy, while the contribution of the adhesive itself is only of secondary importance. On the other hand, however, the high-strength surface-to-surface adhesive joint is essential to allow the metal to store energy to the desired extent. Point-like joints (e.g. welding spots) will fail at less important deformations and local tensions, resulting in a lower energy storage. This principle helps to explain why steel and aluminum are – and will continue to be – the preferred materials for body structures exposed to the risk of a crash. *Magnesium* is less suitable for this purpose due to its brittleness, which is inherent to its hexagonal lattice structure. At the same time, this principle of action provides important indications for the structural design of the joint. The load-bearing capacity of the joint, resulting from the bonding area and the adhesive strength, is required to be at least slightly higher than the forces required to stretch the adherents. The cross-sections of the adherents, resulting from their thickness, are required to correspond to the areas of the bonded surfaces. If, for example, the bonding areas of thick adherents are too small, they may not stretch to the extent required for energy storage, as the joint will already have failed.

With regards to energy storage in the case of a crash, structural bonding is compatible with modern high-strength steel. With an increase in deformation speed, the strengths of both the adhesive and the steel adherent increase, albeit to different extents. High-strength steel, however, has softer dynamic deformation characteristics than 'normal-strength' steel; that is, with increasing impact speeds both types of steel will behave more and more alike. At a speed similar to real crash speed, their deformation characteristics are almost at the same level. In other words, high-strength steel has a high deformation stability in the case of an impact occurring at a lower speed. In this case, the adhesive is not exposed to any severe deformation, because the steel adherents are not stretched at all, or only to a very slight degree. In the case of higher impact speeds, however, high-strength steel will behave like softer steel. In any case, when joining high-strength steel adherents, the engineer must adjust their thickness to the relevant parameters, using thin-walled material that will save weight.

For structural joining purposes in vehicle construction, bonding competes with a variety of other joining techniques, including welding, riveting, clinching, joining by screws and soldering. In order to decide whether a technique is suited for a special purpose, it is important to know how it contributes to the fulfillment of present-day requirements in vehicle manufacture. These are mainly:

- suitability for multimaterial applications
- increased body stiffness
- increased operational strength
- increased energy storage in the case of a crash

- less corrosion
- increased comfort (acoustics)
- increased speed of application/processing
- functional cost savings

For nearly all requirements, bonding proves to be equal or superior to the other joining techniques, especially with its new generation of epoxy resins. The only disadvantage is that, immediately after assembling, there is no strength, as the strength will build up only during the course of the curing process. Consequently, hybrid bonding is increasingly being used, for example the combination of spot welding with bonding [12]. A comparison of joining techniques is provided in Table 8.1.

The number of welding spots may be considerably reduced, which will help reduce costs. In body construction, hybrid bonding is sometimes performed – that is, bonding in combination with other joining techniques (e.g. self-punching riveting) – especially when aluminum is used. The adhesive surface-to surface joint contributes to crash resistance and rigidity, while the aluminum saves weight. This technology has recently made it possible to produce a superclass vehicle with an aluminum body weighing 40% less than predecessor models, yet with an increase in rigidity of 60%.

In 1998, the new generation of high-strength adhesives with crash-resistance-improving properties was used for the first time in models with higher production quantities.

Calculation methods, especially for the behavior of adhesives in the case of a crash, are under intense scrutiny even today. Whilst the number of welding spots has already been reduced, this development is still at an early stage. In an experiment performed on a real body with 4000 welding spots, the torsion resistance was measured, followed by a hybrid bonding carried out on the same body (bonding plus spot welding). As a result of using bonded joints, the rigidity was increased by approximately 40%. When the experiment was repeated with a body in which the number of welding spots was reduced from 4000 to 2000, within the range of common tolerance the rigidity of the bonded spot-welded body was unchanged, showing that the potential of this method had not yet been reached.

Several years of experience, also with regards to dynamic operating strength, and gathered with millions of structurally bonded vehicles, have shown that there is a

Table 8.1 Comparison of joining techniques.[a]

Joining technique	Multimaterial design	Body stiffness	Crash resistance	Operational resistance	Corrosion resistance	Acoustics
Bonding	+++	+++	+++	+++	++	+
Spot-welding	–	0	0	0	–	–
Clinching	–	0	–	+	0	–
Riveting	0	0	–	+	–	–
Joining by screws	0	0	0	0	–	–
Laser welding	–	++	++	++	0	0

[a] Grading indicated as: +++, very good; ++, good; +, fair; 0, poor.

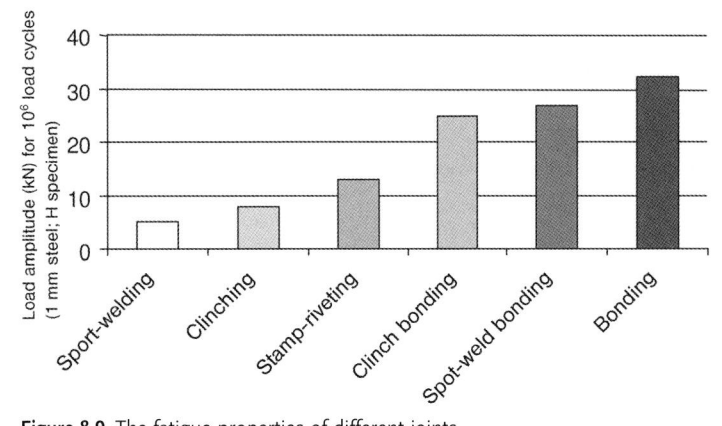

Figure 8.9 The fatigue properties of different joints.

considerable cost-saving potential here. For example, in a new vehicle, the rigidity of a bonded spot-welded body is about 20% higher than in its nonbonded counterpart, and this margin, which is important for both safety and comfort, will surely grow wider after several years of operation. The rigidity of a conventional body will decrease by around 20%, while that of a bonded version will decrease by only 5%; moreover, at that time the torsion resistance of the bonded structure will be 40% higher. Clearly, the absolute percentage changes obtained will depend on the model of the vehicle. These findings confirm the first indications provided by experiments investigating the dynamic fatigue limits of different joining techniques. The fatigue properties of different types of joints are shown in Figure 8.9.

In addition to the cost-saving potential, a reduction in the number of welding spots will further increase the joint's performance, since every spot avoided will reduce the thermally induced changes in the structure of the metal (this is also apparent in Figure 8.9). Another aspect to be taken into account is the enhancement of processing speed. Each spot requires around 3 s for both spot-welding and clinching, and almost twice that time for self-punching riveting. By comparison, the spray application of a suitable adhesive requires only 2 s per meter of bond-line. Even tacking by means of remote laser welding (which is still under discussion at present) requires 0.5 s per tacking point.

Bonding Process In body-in-white construction, the adhesive is applied onto the oiled sheets using robots. The application temperature is slightly elevated to around 50 °C, which induces a decrease in viscosity. Although on cooling the viscosity increases again. Adhesives with a high yield point are particularly well suited, because this improves rope stability and resistance to being washed away. At the same time, the viscosity must be as low as possible in order to facilitate 'pumpability'. Depending on the product, the adhesives can be applied by means of beads, swirl application or spray application.

In order to prevent pollution of the e-coat bath during subsequent cleaning and painting processes, it is advisable that pre-curing/pre-gelation be performed by means of heat application (e.g. induction heat), or preferably in a special body-in-white stove at 110–120 °C that may also be used for other products.

The final curing takes place in the e-coat stove at temperatures around 180 °C, with an adequate curing window between 150 °C and 200 °C, and curing times (e.g. 30 min) with similar tolerances. In future, crash-resistant cold-setting adhesives may replace crash-resistant hot-setting adhesives, at least partly, in order to dispense with the body-in-white stoves, or in order to join finished painted components to form the body.

These adhesives have a tensile loading shear stress of (sometimes considerably) more than 25 MPa, while the tensile strength of the adhesive film is even higher. The stretch at break is slightly above 10%. Formerly, rigid but glass-like brittle products had much lower values (of only few per cent).

Test Methods (see Section 7.2) Nearly all test methods generally employed for adhesive joints and adhesives are used here. As the adhesive joints must be high-performing with regards to the service life and the climatic conditions to which a vehicle is exposed, different accelerated aging tests are carried out in particular. Beyond the well-known test methods, a 'crash peel resistance test' is performed that takes into account the high rates of stress occurring in the case of a crash. For this purpose, the crash peel resistance is generally tested according to ISO 11 343. A wedge is driven at high speed (several $m\,s^{-1}$) into a suitable test specimen, and the peel forces related to the bonding width are evaluated. The testing is performed at temperatures that are relevant for vehicles and allow the crash peel resistance to be determined in $N\,mm^{-1}$, as well as a quantitative value for the energy consumed. The method is well suited for early identification of the different 'crash suitabilities' of adhesives in laboratory experiments (Figure 8.10; also see Section 7.2.5).

In addition, double hat boxes are shock-loaded in falling towers in order to determine the deformation behavior, similar to the conditions encountered in real components. Heights of fall of around 4–10 m are used to create a speed that imitates a crash situation; changing the weight may modify the energy of fall.

Quality Assurance The safety-relevant function of bonded joints exposed to the risk of a crash requires suitable measures of quality assurance. No nondestructive testing studies are carried out on a bonded body, and if a bonding defect is discovered it cannot be simply repaired; hence, *quality assurance* is more important than quality control. Visual inspection systems, robotic application and automatic dosage control ensure the application of the correct quantity of the correct adhesive, without interruption, at the correct location. According to the required performance, different structural adhesives will be used in a vehicle body, and these are commonly given different colors in order to identify them (the colors are of no importance with regards to their functions).

The viscosity of an adhesive is exponentially increased by an increase in the length of its polymer or prepolymer chains. The easy pumpability of an adhesive indicates that the storage life has not been exceeded, or that undue aging has not taken place as a consequence of elevated storage temperatures.

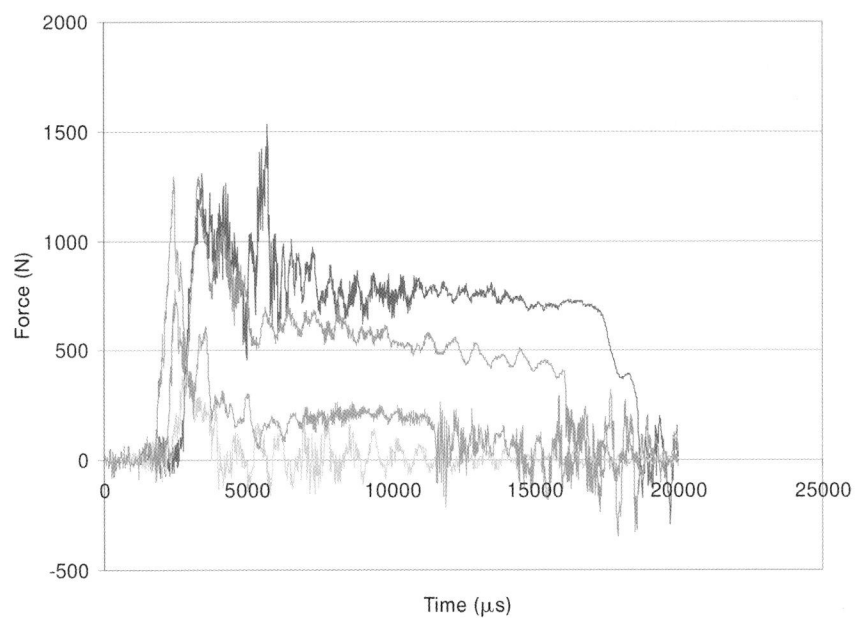

Figure 8.10 Crash peel diagram of different generations of adhesives.

The recommendation is to work with a bond-line thickness of approximately 0.2 mm, although the adhesives are rather tolerant in this respect. Most adhesives may be formulated to slightly foam in order to ensure proper filling of the gap. In order to take into account of the tolerances in gap width, however, it is common practice to work with a surplus of adhesive.

Environment Protection and Hygiene of the Work Place With the new generation of adhesives, the measures to be taken do not differ qualitatively from those already applied when working with epoxide adhesives for the purpose of non-crash-resistant stiffening bonding, which have already been used for a long time. The usual measures to be taken for the hygiene in the work place remain unchanged. It is advantageous to apply the adhesive by automatic methods only. When bonding is combined with thermal joining techniques (resistance spot-welding, possibly laser welding tacking in the future), the emission of volatile pyrolysates induced by the adhesive is considerably less important than that induced by the corrosion protection oils and forming oils. When the number of welding spots is further reduced, the total emission at the work place will be reduced even further.

At the end of its service life, the vehicle is shredded and the steel pellets are recycled in the electric furnace. Minor amounts of adhesive residues still adhering to the surfaces do not represent any problem.

Although some attempts have been made to develop a method for simple, controlled debonding, this has not yet been successful. Moreover, as for most of the time hybrid bonding techniques are used, a simple detachment of the welded, riveted or clinched body parts is barely possible.

Structural bonding is the driving force of modern lightweight body construction that will also reduce fuel consumption. During past decades, the typical weight of a passenger car of the same type has increased considerably; for a lower mid-size compact car produced at high production quantities the weight has increased from 750 kg (in 1980) to 1200 kg (in 2004). Experience teaches us that 1 kg of automobile structural adhesive has a weight reduction potential of around 20 kg, which in turn may reduce fuel consumption by 60 l for a total traveled distance of 150 000 km – equivalent to a decrease in carbon dioxide discharge of 150 kg.

Bodywork Repairs The high curing temperatures required by one-part epoxy adhesives can only be created in the e-coat stove, and not in the repair shop. Therefore, for repair purposes, new two-part epoxy adhesives are used that have an acceptable crash resistance but have an inferior performance compared to OEM systems. These are contained in suitable cartridges and mixed using static mixers. Curing takes place at room temperature and may be accelerated by using a repair paint oven, generally with temperatures of 60–80 °C. The parts to be bonded must be cleaned and free from oil before the adhesive is applied, as the epoxy systems are only capable of absorbing oil at higher temperatures. An alternative for repairing limited areas is to use a standard joining technique such as spot welding, with the welding spots set closer to each other.

8.2.3
Bonding in Rail Vehicle Manufacture

While for locomotives, bonding is used almost exclusively to fix protective foils, it also offers a wide field of applications in tramways, regional traffic or long-distance transport. The head-end of the German ICE, for example, is a bonded integral plastics construction. The windows are fitted by bonding, as are the sidewalls, ceilings, floorings and equipment such as lamps, hinges, strips and coverings. Anti-graffiti foils are applied by means of adhesive bonding to protect the vehicles against vandalism. The efficient and low-priced 're-design' concept for the modernization of old trains and tramways is also based upon bonding technology.

The profiles and joint elements are joined to form the body-in-white that is then precisely aligned. Only at this moment is the adhesive applied by injection via feeding ducts in the joint elements. The structure is not subject to any heat distortion, no straightening is required, and prefabricated components may be installed without any further adjustment (European Patent No. 0780279; see Figure 8.11).

Bonding is used not only in rail vehicles but also in rails. Inductive rail traffic safety systems are still based on insulating rail joints that interrupt the electrical conductivity of the welded rail system. These screwed fished joints, made from steel, are high-strain joints that may only be secured against loosening by means of adhesive bonding. The nonconductive adhesives interrupt conductivity together with another insulating element placed between the rail joints (Figure 8.12).

In rail joints that are exclusively joined by screws, important deformations occur at lower temperatures due to thermal contraction of the steel. Although the high-strain steel screws are not capable of preventing these deformations, if the fishes are joined

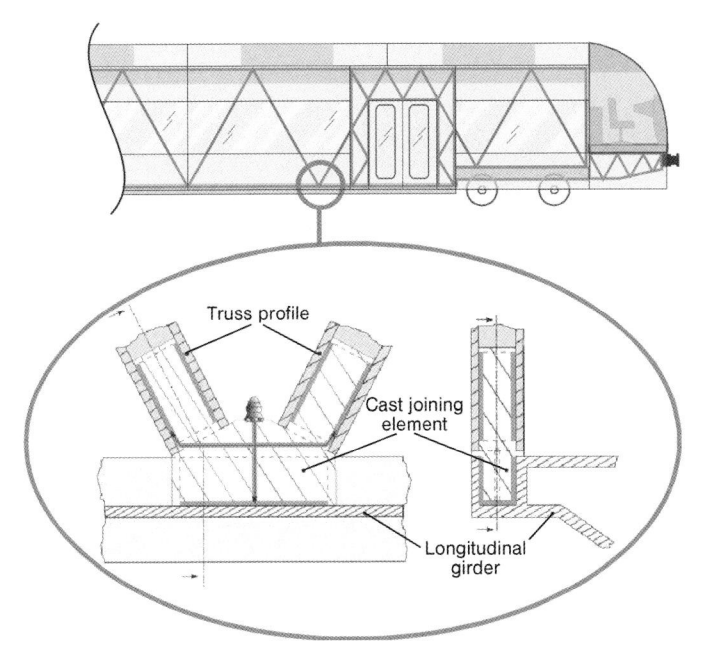

Figure 8.11 Injection-bonded framework in a railway car. (Illustration courtesy of Dorel Verlag).

to the rails by means of adhesive bonding in addition to the screws, these deformations do not occur.

For 'permanent ways', the rails are fixed to sleepers made from wood or concrete by means of so-called 'ribbed plates'. In conventional sleeper construction the rails are joined to the sleepers by means of screws (see Figure 8.12). However, for concrete grounds, underground railways and in tunnels, as well as in present-day fixed-bed installations, these screw joints may be replaced by adhesive bonds which also take on the damping function that formerly was provided by ballast beds. For this, the adhesive used must have a high elasticity and bond-line thickness.

Another example of the successful use of bonding technology is that of magnetic levitation vehicles. It is common practice in transformer and electric motor manufacture that the sheet laminates used as the core material in linear induction motors that form the monorail. These must be free from the eddy currents that occur at right-angles to the lamination direction. In the case of magnetic levitation vehicles, the sheets are made from special steel alloys and have a thickness of 0.5 mm. They are entirely bonded to each other, as no mechanical locking elements, such as screws, may be used for electromagnetic reasons.

8.2.4
Shipbuilding

The use of bonding in the shipbuilding industry extends to yacht, charter and military applications. In yacht construction, bonding has become an indispensable

Figure 8.12 An insulated rail joint with screws (right) and additionally adhesively bonded (left) at −28 °C.
(Photo courtesy of Henkel).

construction technique. As in the construction of motor homes, the cost factor is very important here, and consequently plastic composite materials are used in the majority of cases, to a point where entire ship structures may be manufactured by this method. Adhesives are also used for sealing purposes, with ship hulls, stringers and frames being bonded with two-part epoxy adhesives.

For charter applications, new construction techniques are increasingly being adopted, where in the past ship builders have conventionally regarded steel as being 'solid'. However, they are increasingly seeking alternatives in construction technology and, as a consequence, lightweight materials such as fiber-reinforced plastics are increasingly being used for the structures of very large, an example being in the construction of swimming pools.

One interesting example of the use of adhesives in charter applications is the building of a high-speed ferry that was 60 m long [13] and capable of transporting about 1000 passengers at a speed of 40 knots (ca. 75 km h^{-1}). This performance could only be achieved by using lightweight constructions, with the hull consisting of profile-reinforced integral aluminum plates with polycarbonate windows fitted by means of glass fiber-reinforced adhesives.

8.3
Building Industry

For millennia, bonding has been a predominant and successful joining technique in building and construction purposes. Examples include mortars and grouts, which were used in Mesopotamia for the bonding of bricks and tiles (see Chapter 2). Examples of so-called 'nonhydraulic' binders include gypsum, Sorel's cement, anhydrite and magnesia. Lime (and especially cement) are known as 'hydraulic binders', as they are capable of curing under water. Terms such as 'Roman cement', which relate to highly hydraulic limes rich in silicate, suggest that these materials were used in ancient Rome. Technologies to increase the strength of binders have included the addition of mineral agents such as volcanic ash originating from Pozzuoli, close to Mount Vesuvius. After the addition of further lime, this creates a hydraulic mortar (cement) during the burning process. Both, in Ancient Rome and during construction of the Palastaula (known as the Constantine basilica) in Treves (Trier, Germany), organic materials such as milk, casein, ox blood or urine were added to the mortars in order to improve their toughness, adhesive properties and freeze–thaw resistance (see Section 8.17.4.3).

8.3.1
Adhesives for the Bonding of Surfaces

For flooring, the laying of tiles and carpeting, a wide array of different types of physically and chemically setting adhesives is available [14, 15].

8.3.1.1 Adhesive Mortars for Tiling

Approximately 90% of the adhesives used for tiling are cement-based adhesive mortars. These consist of cement as the main binder, and additives such as sand fillers, plastics powders as adhesion promoters which also increase the flexibility, as well as special agents such as anti-foaming, wetting and thickening agents. When water is added to the cement mortar, it chemically sets by a 'hydraulic' reaction. Once cured, mortars are water-resistant.

New *fast-setting adhesive mortars* have significantly increased the working speed with such materials, from 24 h of setting time to as little as 2 h. Polymer-based tile adhesives are chemically setting products based on PU or epoxy resin bases, typically consisting of an A-component with fillers, and a B-component without any fillers. Depending on the desired field of application, PU or epoxy resin tile adhesives may be formulated to cover a large range of properties, from rigid to soft-elastic. Reactive resin products are preferably used for heavy-duty applications, such as commercial floorings exposed to a high level of mechanical stress, including forklift traffic.

Dispersion-type tile adhesives typically comprise styrene-acrylate dispersion binders, fillers such as chalk and silica sand, and additives such as thickeners and preservatives, defoaming and wetting agents. These ready-to-use adhesives cure physically by drying. As no mixing is required, tiling may take place under clean conditions, without any need for safe disposal. Dispersion adhesives are therefore particularly well suited to renovation purposes, for the installation of skirting boards, trimmings, etc., or for do-it-yourself work. The higher price of these types of product is balanced by the faster working speed.

8.3.1.2 Floor-Covering Adhesives
For flooring, different types of solvent-based contact adhesives, dispersions and even PSAs are applied.

Parquet is mainly bonded with three different types of adhesive, namely solvent-based synthetic resin adhesives, reactive resin adhesives and dispersion adhesives.

In spite of ongoing efforts to reduce the emission of volatile organic compounds (VOCs), solvent-containing synthetic resin adhesives still dominate the global market in flooring because they are inexpensive and have good application properties. They comprise a mixture of solvents (methyl acetate, acetone, ethanol, methyl alcohol), dissolved synthetic resin (polyvinyl acetate), and mineral fillers (e.g. chalk). The overall VOC content typically ranges between 20 and 25%.

Solvent-containing synthetic resin parquet adhesives do not contain water, in order to avoid any swelling of the parquet woods. Parquet-floor layers also benefit from this type of adhesive as they are easy to use. The most environmentally friendly way to bond parquet floors is to use waterborne dispersion adhesives, consisting of polyvinyl acetate plastics dispersion as binder, and chalk as the mineral filler.

Another example of surface bonding adhesives in building is that of wallpaper adhesives. These water-based adhesives have as their main components methylcellulose and thickening agents such as lignocellulose, modified starch and mineral fillers [15].

8.3.2
Adhesive Anchors in the Building and Construction Industry

An *adhesive anchor* is a reinforcing bar or threaded rod inserted into a drilled hole in hardened concrete with a structural adhesive acting as a bonding agent between the concrete and the steel. Typically, the hole diameter is only about 10–25% larger than the diameter of the reinforcing bar or threaded rod. Structural adhesives for this type of anchor are available prepackaged in glass capsules, in dual-cartridge injection systems, or as two component systems requiring user proportioning. The typical types of adhesive anchor systems and types of adhesive are shown in Figure 8.13.

A *grouted anchor* may be a headed bolt, threaded rod with a nut at the embedded end, or a deformed reinforcing bar with or without end anchorage installed in a preformed or drilled hole with a Portland cement and sand grout or a commercially available premixed grout [16]. Typically, the hole size for grouted anchors is about twice the diameter of the anchor. The details of grouted anchors are not described in this chapter.

8.3.2.1 Adhesive Anchor Systems
The most common version of the *capsule anchor* consists of a cylindrical glass capsule containing a polymer resin, an accelerator and a mineral aggregate. The capsule is

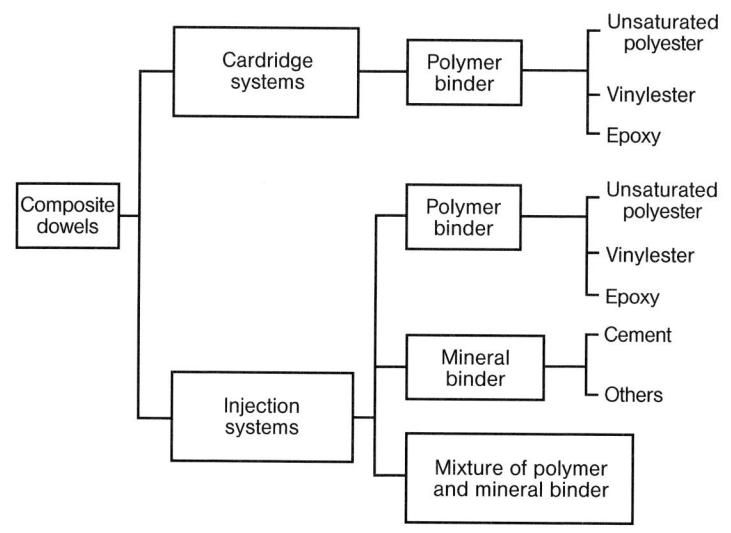

Figure 8.13 Classification of composite dowels. (According to Comité Euro-International du Béton CEB, 1994).

inserted into a drilled hole; deeper embedments are typically achieved by stacking multiple capsules in the hole. Setting of the anchor is accomplished by direct boring through the capsule with a threaded anchor rod (usually equipped with a chiseled end) chucked directly into a rotary drill. Straight reinforcing bars may be installed in the same way. The drilling and hammering action of the drill mixes the contents of the capsule with the fractured fragments of the capsule to form a relatively fast-setting polymer/glass matrix (Figure 8.14) [17]. As well as stiffening the polymer matrix and reducing shrinkage, the fractured glass and aggregate components serve to improve bond by scouring the sides of the hole during installation.

As an alternative to glass capsules, foil capsules have recently been developed; these are better suited for use on construction sites as they are more robust. Because of their flexibility they adapt themselves to the hole geometry and can easily be installed overhead. In injection anchor systems, plastic cartridges containing

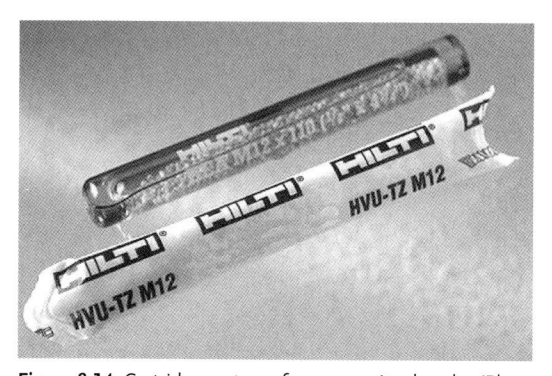

Figure 8.14 Cartridge systems for composite dowels. (Photo courtesy of Hilti).

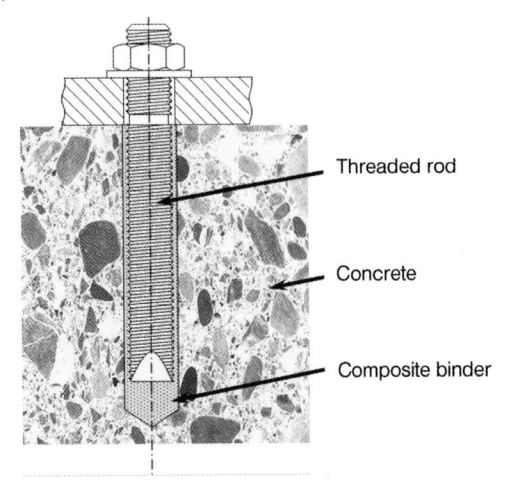

Figure 8.15 A composite dowel.

premeasured amounts of resin and hardener allow controlled mixing of polymer components. The components are typically mixed through a special mixing nozzle as they are dispensed, or are completely mixed within the cartridge immediately before injection. Typically, the catalyzed resin is injected into the hole first and the anchor rod (straight or deformed bar or threaded rod) is pushed into the hole and rotated slightly to promote complete contact between rod and adhesive (Figure 8.15). Care must be taken to prevent the formation of air bubbles in the adhesive during insertion of the rod. Other systems utilize a plastic pouch to contain the polymer components, which are mixed by manual kneading of the pouch. Immediately after mixing a small incision is made in the pouch, and the resin is poured into the hole.

Polymer components may also be purchased in bulk and mixed either manually or with a power mixer in a bucket and used immediately, or they may be pumped through a mixer and then injected into the hole. Both epoxies and polyester resins are available in bulk packaging. Care must be taken to assure correct mix design and adequate mixing of the resin components on site, and to protect personnel from exposure to fumes and direct contact with the polymer materials.

8.3.2.2 Adhesives

Epoxy adhesives in anchoring have proven to be durable, to have a long shelf life, and to undergo little shrinkage during curing. By national standards, the epoxy resin is designated as component 'A' and the curing agent as 'B'.

Polyester adhesives for this type of application usually consist of a polyester resin and a catalyst, typically dibenzoyl peroxide. Because of their chemical nature, polyester adhesives usually have faster exothermic reactions and curing times than do epoxy adhesives. However, the limitations of polyester adhesives can include a short shelf life, a tendency to degrade under exposure to ultraviolet light, and a tendency to selfpolymerize (without the addition of a catalyst) at elevated temperatures.

A *vinyl ester adhesive* is a thermosetting plastic consisting of a vinylester resin and a catalyst, typically dibenzoyl peroxide. Vinylester adhesives usually have exothermic reactions and curing times, which are faster than those of epoxy adhesives, but slower than those of polyester adhesives.

Hybrid systems consisting of organic and inorganic bonding agents have recently been developed. The polymerization reaction of the resin component ensures good bonding and a rapid curing injection system with good handling characteristics. The cementitious reaction especially improves mechanical properties in the cured state at higher temperatures. The combined action of the two components results in negligible material shrinkage.

8.3.3
Building Reinforcement by Means of Bonded Surface Reinforcement

Reinforcement by bonding of laminates made from steel, fiber-reinforced polymers (FRP) and especially carbon fiber-reinforced polymers (CFRP) are used for strengthening concrete structures. Due to its excellent material properties, such as high fatigue resistance, high stiffness and high tensile strength, CFRP is used to influence the flexural stiffness and ductility, as well as to strengthen and repair aging concrete structures. In addition to the traditional repair and strengthening methods, such as replacement or reinforcement via additional structural members, this practice has become part of the applied rehabilitation methods in civil engineering in many countries.

The result has been the publication of numerous guidelines and codes, which allow the designer to apply this modern strengthening technique to concrete structures [18]. The bonding of reinforcement is mainly used in the following cases:

- a need to improve the load capacity of concrete structures due to errors in the original design process, accidental damage or extension of service life
- a need to increase load capacity due to adaptation of use

Steel plates for reinforcement are pretreated by grit blasting and coating with protective and anticorrosive primers. Due to their low weight, the advantage of CFRP reinforcement straps is that they can be bonded without using devices temporarily to support (mechanically) the strap until setting of the adhesives begins.

The use of permanent mechanical devices to prestress the adhesively bonded reinforcement strap at both sides is an additional method to further increase the static load capacity of a concrete structure.

8.3.4
Structural Bonding in Structural Steel and Facade Engineering

Steel structures in the building and construction industry are currently those with the greatest degree of prefabrication. The established joining technologies in structural steelwork find limitations in the joining of high-strength steel grades and lightweight constructions in terms of consistent load distribution, destruction of the basic material by boreholes and the durability of welded joints under dynamic load conditions. At present, bonding (as an alternative to welding and screwing) finds only a limited range of application for structural joining in steel constructions.

The first attempts to introduce bonding technology into structural steel engineering date back to 1953, when two pipeline bridges were built using a hybrid bonding technique (high-strength friction-grip HSFG bolts + bonding) across the Lippe-Ems-Kanal, a canal in Marl/Germany, and in the bonding of corrugated steel to supporting girders in roofing applications.

In cladding and glazing, adhesive bonding and sealing today meets an increasing number of successful and impressive applications. Due to the low specific weight (ca. $2.8\,\text{g cm}^3$, close to that of aluminum) and brittleness of glass, adhesively bonded glass structures show a superior mechanical performance compared to mechanical fastening of glass by means of bolting and clamping.

The European Technical Approval Guideline ETAG 002 for Structural Sealant Glazing Systems (SSGS) regulates the use of building elements made from adhesively bonded glass panels as facades and roofs, or parts thereof, with glazing at any angle between vertical and 7° above horizontal. The guideline covers the general requirements for system assessment and the specific requirements for mechanically supported and unsupported systems, where the structural seal adhesion surfaces are glass (either uncoated or with an inorganic coating) and anodized aluminum or stainless steel. This guideline specifically addresses the design, manufacture and aging properties of silicone adhesives.

Recently, PSAs have been used successfully in glazing applications to bond glass panels to supporting metal framework. The viscoelastic adhesive fixture avoids stress concentrations in crack-sensitive lightweight glass panels, and also compensates for the difference in thermal expansion between glass and other materials such as an aluminum framework (Figure 8.16).

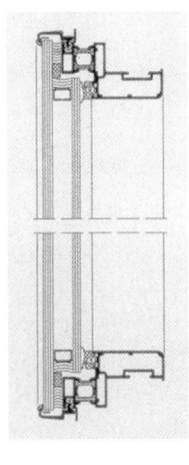

Figure 8.16 An example of structural glazing.

8.4
Wooden Constructions

8.4.1
Introduction

For various reasons, wood belongs to one of the most ancient and popular materials of mankind. With the exception of deserts, it is available everywhere and in abundance. Unlike metal or stone, wood is easily obtained and processed by means of machining technology. The life cycle analysis of wood and wood-based materials is excellent, since it is a regrowing raw product. Another important aspect of wood is that it is associated with positive feelings; people have a sense of well-being if their living and working environment is entirely or partly made from wood. The reason for this is that wood feels 'warm' owing to its excellent insulating characteristics. By absorbing and emitting moisture, it regulates the interior climate and is therefore a precious material in terms of building biology. Its acoustic performance – that is, the propagation and shielding of sound waves – also differs from that of other materials. This property is not only taken advantage of when building musical instruments, but also very often when furnishing concert halls, since wood creates excellent sound effects.

From the technological point of view, wood is a complex material. Due to its pronounced fibrous structure, most of the physical characteristics – particularly the mechanical ones – are highly anisotropic. For example, depending on the species, the tensile strength perpendicular to the direction of the fibers (perpendicular tensile strength) amounts to only 10–15% of the parallel tensile strength. The latter, however, is quite considerable. Beech, for example, has a parallel tensile strength of 120 N mm^{-2} at a green density (12% moisture content) of 0.7 kg dm^{-3}. By comparison, mild steel has values ranging between 310 and 630 N mm^{-2} at a density of 7.8 kg dm^{-3} [19]. Consequently, all things being equal, lightweight constructions are possible when using wood instead, provided that the designer arranges for the majority of the important forces to produce only tensile or flexural stresses.

But wood is not just *wood*; it rather reflects the natural variety of the species from which it is obtained. Depending not only on the species but also on the site of growth and its climatic history, there is a wide array of wood densities, strengths, colors and grain patterns, and this is why construction engineers and designers are so much attracted by this material. At a glance, wood appears to be a 'good-natured' substrate for bonding purposes. In the drying process, the capillary water evaporates, creating a porous structure and enabling the adhesive film to build up mechanical interlocking in the wood structure. The chemical structures of the main components of the wood matrix (i.e. cellulose and lignin), with their hydroxyl groups and other polar functional groups, offer various possibilities for the formation of hydrogen bridge bonds and even covalent chemical bonds in the adhesive layer. This is the reason why the most ancient wooden objects and artifacts of the history of mankind ever found by archaeologists already also bear witness to the bonding of wood. These assemblies were nonstructural – that is, they were nonloaded assemblies – and the glues used to create them originated from Nature, being obtained from plants and animal organs. The production of glutine glues (i.e. glues made from skin or bones) required an advanced craftsmanship (see Section 5.9), and it is no wonder that works of art and articles for daily use produced with these glues were considered extremely precious. As a consequence, they were found, for example, among the grave goods of the pharaohs of Ancient Egypt. Moreover, even the glue itself was considered precious; among the treasures of the tomb of Tutankhamun was found a glue tablet that had accompanied the pharaoh on his last journey, together with gold and gems. Pliny the Elder (23–79 AD) was the first to scientifically and systematically describe gluing in his 37-book work *Naturalis Historia* (see Chapter 2). He noted that specific species of hardwood were more difficult to bond that softer types of wood – a fact that still applies to modern adhesives. During the first half of the 20th century, wood and wood adhesives were the key materials for the cutting-edge technology of the time. Today, designations such as 'Propellerleim' (a glue once developed for the assembly of wood propellers), are reminiscent of these applications. Even jet aircraft such as the British De Havilland 'Vampire' (Figure 8.17) were made from wood. Immediately after the Second World War there was a flood of new testing methods and standards aimed at increasing the safety and reliability of wood constructions. Yet, it is in the nature of standards to be conservative and to hinder the market entry of new systems. This applies in particular to standards that prescribe the use of specific materials, and in this case the new materials had to fight for acceptance via lengthy special approvals. Unfortunately, all of this proved detrimental to the advancement of technology.

In the modern context of wood bonding, it is useful to consider two dimensions. On the one hand, there is a difference between 'structural' and 'nonstructural' bonding, and on the other hand between 'industrial' and 'handicraft' bonding. Handicraft production is characterized by small series sizes, most often even one part. Generally, when fabricating unique pieces or small batches, process optimization in terms of type approval or load calculations is too labor-intensive or uneconomic, and therefore, structural (i.e. load-bearing) bonding is rather an exception here. For loaded joints,

Figure 8.17 The De Havilland FB.6 'Vampire' fighter bomber. This aircraft was mainly produced with wood; it was used until 1990 by the air force of Switzerland as a training aircraft.

such as shelves in furniture, mechanical fastening techniques (e.g. dowels, screws or nails) are used, the advantage being that special fixing or positioning tools are not needed to hold such joints in place. The assembly of structural elements at the construction site is also a handicraft situation, since on most occasions on the construction site there is no possibility of applying pressure to a bonded joint. Therefore, frictional fastening techniques are preferred for fitting or roofing purposes. In industrial production, however, there are no such limitations, as large numbers of pieces justify investment in complex facilities, and the development of refined processes, predominantly motivated by aspects of economy and sustained quality. Economy is an important factor because high investment costs need to amortize, while sustained quality is required to prevent the multiplication of flaws in series production that may have disastrous consequences in terms of liability and damages [20–22].

8.4.2
The Structure of Wood

8.4.2.1 The Molecular Structure of Wood

The dry weight of wood is composed of 40–50% cellulose, 20–40% hemicellulose, 20–30% lignin and 2–7% extraneous materials including resins, fats, peptides, tanning substances and coloring matters. Depending on the species of wood, the site and the lumbering time, freshly cut wood will contains between 50 and 150% of water in relation to its dry weight (oven-dry weight). Wood contains 'free' water which is stored in the cell cavities and spaces, and 'bound' water which is directly associated with the cellulose fibers. Cellulose molecules have long and unbranched chains and are composed of repetitive cellobiose units. From the chemical viewpoint, cellobiose is a disaccharide consisting of two glucose units. *Cellulose* itself belongs to the polysaccharides (sugar) class, characterized by a high content of hydroxyl groups. These polar functionalities not only provide for the formation of hydrogen bridge bonds, but also for the production of covalent bonds with reactive adhesives. The cellulose molecules are arranged into fibrils, which are organized in parallel to make up the walls of the cells, which in turn are aligned along the vertical direction of the stem. *Lignin* is a mixture of polyphenolic molecules. At the end of cellular growth, these substances are integrated into the cellulose cell walls where they act as a binder and induce lignification, during which process the density and mechanical strain to which the wood may be exposed are substantially increased. *Hemicellulose* also consists of sugar molecules, without being structurally related to cellulose. The constructive polymers are composed of D-xylose which is a pentose. The molecules have considerably shorter chains than those of cellulose; furthermore, the chains have branches of arabinose and other sugars. Hemicellulose is easier to decompose than cellulose, and acts as a wall-building and reserve substance. Water is integrated into the fiber structures via hydrogen bridge bonds as 'bound' water, and then stored in the cell cavities and spaces where it can move freely as 'free' water. This has a decisive effect on the dimensions of wood-based materials. Up to two-thirds of healthy green wood consists of water, and this is known as the 'maximum swelling condition'. During the drying process, the free water evaporates first, leading to a fiber-saturation range of 23–35% of wood moisture. By means of a further natural or technical drying processes, an equilibrium moisture content of 6–15% is obtained, within which range the wooden work piece will later adapt to the ambient conditions.

8.4.2.2 The Macroscopic Structure of Wood

For material and bonding technologies, an important factor here is the fact that, in a stem section, from the center of the stem to the outside layers, the features of the wood change. In many species of wood, at the center of the stem there is the 'heart' where the storage cells (pith rays) are dead. Storage cells are located in the *sapwood*, which is mostly lighter in color and wrapped by the *cambium layer* (growth region) and *bark*. Each year, a new growth ring (annual ring) is formed, within which there are two portions: (i) *early wood*, which is more open-textured and formed early during the vegetation period; and (ii) *late wood*, which has narrow-volume cells that are formed in the autumn. The function of early wood is to conduct water and mineral salts, while the function of late wood is mechanical stabilization of the stem.

8.4.3
Consequences

8.4.3.1 Strength, Swelling and Shrinking

The radial symmetry of the stem and the arrangement of the cellulose fibrils in longitudinal direction have an important effect on the mechanical and adhesive properties. It has already been mentioned that the parallel and perpendicular tensile strengths differ considerably. The important swelling and shrinking behavior, which depends on the equilibrium moisture content, must be carefully taken into account when wood is joined to wood, or wood to other materials. This behavior shows different patterns depending on the directions within the stem. As a rule of thumb, swelling occurs at a ratio of 1: 10: 20 along the direction of the fibers, radially to the fiber (in the direction of the pith rays) and tangentially to the fiber (in the direction of the annual rings) (Figure 8.18). Within sawn wood, changes in the ambient

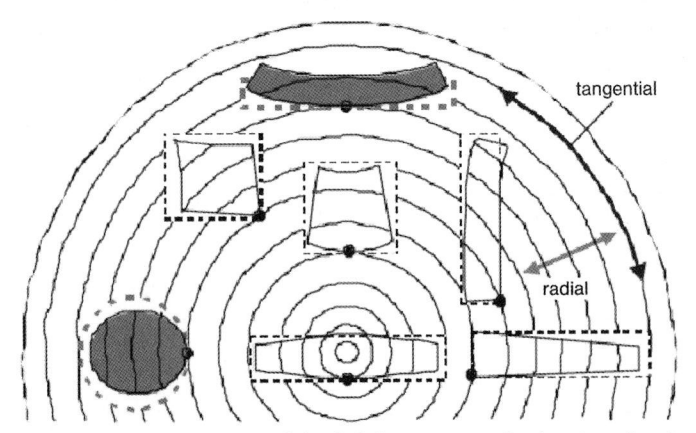

Figure 8.18 The dependence of shrink deformations on the direction of cutting the butt.

conditions may therefore induce considerable shrink deformations, while very high shear stresses may occur in the interphase between wood and materials which do not swell in the presence of moisture (metals, plastics, glass, stoneware).

8.4.4
Adhesive–Wood Interphase

It is well known that the adhesion of an adhesive at the interfaces of the substrates depends on physico-chemical interactions and mechanical interlocking [23]. *A priori*, in the case of porous surfaces, there is a larger area per joint section, and cavities enable the 'tentacles' of the adhesive to penetrate into the substrate. Machining techniques such as sawing, planing, filing, grinding, brushing or slicing all involve the processing of wood. The so-called 'front face' or 'front end' generally has larger pores as the cell cavities and interspaces are aligned along the direction of the fibers. Although as-sawn surfaces are less even than planed or ground surfaces, in the latter case 'fragmental layers' made from cell material may be detrimental to the quality of the bond, whereas during planing or slicing the cell walls are cut neatly and cleanly. Another important factor with regards to the behavior of the adhesive at the interphase is the density of the wood, and this depends on the species. Within any one species the density of early wood will differ from that of the late wood, as illustrated in Figure 8.19.

8.4.5
Adhesive Systems

A wide array of adhesive raw materials and formulations is available for the purpose of different applications, such as required performance of the adhesive joint, and the chemical and physical properties of the wood-based materials (as described above).

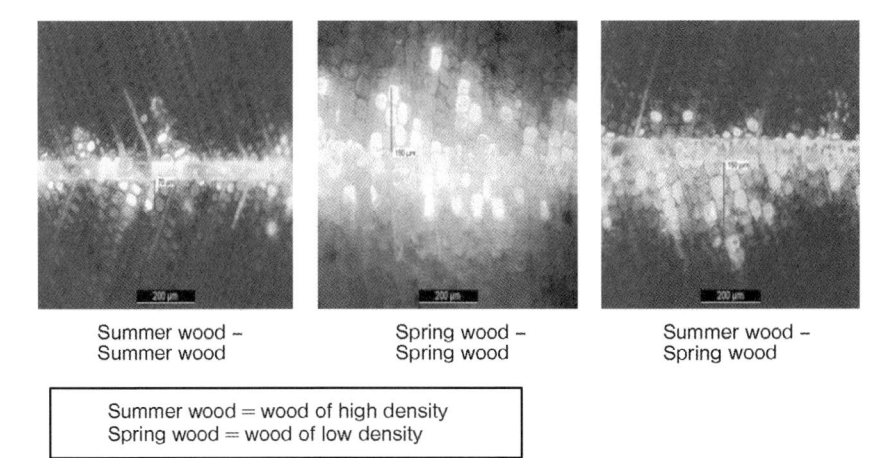

Summer wood – Summer wood   Spring wood – Spring wood   Summer wood – Spring wood

Summer wood = wood of high density
Spring wood = wood of low density

Figure 8.19 The dependence of penetration behavior of an adhesive (bright regions) on the wood density.

All of these are required to fulfill the following conditions, however, whenever wood is bonded:

- The adhesive must be compatible with water or water vapor, both when applied and during continuous operation since, in practice, wood always has a residual moisture content of 6–15%; oven-dry wood is only significant for scientific purposes.
- The dimensions of wood change depending on the ambient humidity and climate; this results in the occurrence of high shear stresses at the joint interphases that must either be compensated by the elasticity of the adhesive, or supported by sufficiently high adhesion and cohesion forces.
- Wood adhesives are required to have certain gap-filling characteristics, since even when planed or ground, the wood surfaces are never absolutely flat. Furthermore, wood is generally bonded under pressure that is applied by different means, ranging from clamps or cramps in the case of handicraft applications, to fully automated stack or forming presses in the case of industrial applications.

8.4.6
Adhesives Based on Renewable Resources

Historically, and up to the early 20th century, aqueous natural glues were used for bonding wood (see Section 5.9). For structural purposes, animal glues were preferred to plant glues as they generally produced higher strengths in the bond-line. Animal glues are composed of proteins that are responsible for both adherence (adhesion) and strength (cohesion). They are applied in a hot or warm condition and, owing to jellification during cooling, a high early strength is achieved. The glue film is produced by the evaporation of water, resulting in an elastic bond-line that swells or shrinks with changing ambient moisture, similar to wood.

When synthetic adhesives were developed in the mid-20th century, the use of natural glues for the purpose of bonding wood grew economically insignificant. Although within the context of ecological aspects, science and research are again taking interest in natural products, there is no general economical trend, with the exception of niche applications such as the restoration of antique objects. The following problems which occur with natural glues have not yet been solved satisfactorily:

- The manufacturing costs of natural binders are clearly higher than those of synthetic adhesives.
- The work-up of animal and plant raw materials gives rise to concern with regards to aspects of hygiene and ecology.
- Natural glues must be improved by the addition of synthetic agents in order to obtain the technical performance required.
- Products originating from different sources may differ considerably in quality.
- Existing facilities for the bonding of wood-based materials must be modified for the application of natural glues.

- As with all other agricultural products, fluctuations in the price and supply of natural adhesives may be relatively important, making planning difficult.

8.4.7
Synthetic Wood Adhesives

8.4.7.1 Physically Setting Systems

The 'workhorse' of the wood processing industry, apart from the manufacture of engineered wood products made from strands and fibers, is *white glue*, an adhesive based on polyvinyl acetate dispersions, either in the form of homopolymers or copolymers in combination with acrylates or functional monomers (e.g. malein acid derivatives). The adhesive film is physically produced by the evaporation of water. This process requires a minimum temperature at which the film is formed; this is known as the MFFT (minimum film-forming temperature), and typically ranges between 2 and 15 °C. If the drying process of the bond-line takes place at a temperature below the MFFT, a chalky or crumbly layer without any efficient strength will be formed rather than a compact adhesive film. Generally, the setting of adhesives is accelerated by heat (i.e. by using heated presses). However, this method does not work efficiently in the case of a greater thickness of the wood layer, because wood is a good heat insulator and hinders heat transfer from the pressing jaws to the adhesive bond-line. Owing to the high dipole moment of water, radiofrequency fields may be used instead. These are absorbed by the bond-line to a greater extent than by the surrounding wood, and are able to penetrate even thicker components.

8.4.7.2 Chemically Setting Adhesives

In chemically setting systems, a polymerization takes place in the bond-line during compression, followed by crosslinking of the polymer molecules. This results in an increase in the average molecular weight of the basic polymers, inducing in turn a considerable increase in the strength values compared to the noncrosslinked matrix. Furthermore, the crosslinking results in transfer from a thermoplastic to a thermosetting behavior, which is prerequisite for joints exposed to a consistent load. Under these circumstances, thermoplastics would be subject to creep, and this would result in failure of the bond-line in the course of time. Reactive systems adhere over a wide spectrum of materials and can be formulated under the aspect of very different mechanical properties (e.g. modulus of elasticity). Therefore, selection of the correct proper adhesive requires consideration of the type of substrates to be bonded on the one hand, and the future technical performance of the bond-line on the other hand.

The most extensive class of chemically setting adhesives is the group of *formaldehyde condensation resins*. The reaction partners of formaldehyde are ureas, phenols or melamines, as well as all possible combinations of these. The reaction partners are used in the form of two-part systems in aqueous solution. For industrial purposes, the components are mixed directly in front of the glue spreader, whereas for handicraft purposes powder premixes are used which are mixed with water to create a reactive product that can be applied to the substrates. An intermediate form of highly crosslinking reactive resin and thermoplastic aqueous systems is that of *postcuring dispersion adhesives*. Here, shortly before application of the thermoplastic emulsion, a bifunctional isocyanate component is added which reacts with the hydroxyl groups of the polymers and protective colloids. With this adhesive system, the water resistance and thermal endurance achieved generally correspond to durability class D4 of DIN-EN 204. Suitable dispersions with a high content of crosslinking agent are also called *emulsion polymer isocyanate* (EPI) adhesives. These have almost the same properties as thermosetting resins, and so may be used for specific structural purposes.

One-part moisture-curing polyurethane adhesives have recently gained importance for wood-to-wood bonding, especially in civil engineering when wood is used as a building material [24]. Compared to conventional formaldehyde condensation resins, this class of adhesives offers the following benefits:

- lower quantities need to be applied
- almost loss-less application
- no need for a mixing process, as it can be used as received from the manufacturer; no need for a glue room, and no waste-water
- reaction times may be chosen from short to extremely short
- unobtrusive bond-line, with an aging behavior similar to wood
- approval for indoor and outdoor use as well as for effective lengths
- ecologically harmless

When bonding wood to other materials, all known reactive systems may be used. Selection of the adhesive system depends on the required performance profile of the intended application. A high modulus may be required, for example, when fixing carbon fiber-reinforced lamellae to structural beams within their tensile loading area by means of two-part epoxy adhesives. Yet just the opposite would be needed when bonding parquet to mineral grounds by means of reactive systems based on silane or polyurethane. Owing to the toughened behavior of these joints, impact sounds are dampened, and important tensile and shrinking stresses which are induced by climatic changes in the wood parquet are absorbed.

For mechanical high-speed manufacture processes, hot melts are preferred because their adhesive strength builds up via solidification or crystallization; this is an advantage compared to other physically or chemically setting systems. Conventional hot melts are based on thermoplastic polymers that are adapted to the desired application by means of tackifiers, rheology modifiers and stabilizers. Polymers used for the bonding of wood are mostly EVAs (copolymers of ethylene and vinyl acetate), polyamides and atactic poly-alpha olefins (APAOs). Resins made from natural raw materials (colophony, terpenes, etc.) or of synthetic origin (hydrocarbon resins, acrylate resins) are used as tackifiers. These hot melts are – and will remain to be – thermoplastic, which limits their potential thermal endurance. They are also subject to retardation and may therefore not be used for structural purposes.

The benefits of conventional hot melts and thermosetting adhesives, however, can be combined by means of *reactive hot melts*, which post-cure via a moisture effect following application. In PU resin hot melts, the basic polymers carry reactive isocyanate groups which, after the thermoplastic hot melt-bonding process, undergo a crosslinking reaction initiated by the moisture contained in the substrates. The thermoplastic joint is strong enough to allow destacking and processing of the work pieces. The increased strength values induced by crosslinking of the isocyanates build up over the course of hours and days during subsequent storage. Hot melts containing reactive silane groups show a similar behavior but have a different adhesive profile.

8.4.7.3 Adhesion Promoters (Primers)

Today, in all fields of application of bonding technology, customers are demanding a universal adhesive and, moreover, an adhesive that develops infinite adhesive and cohesive forces on nonprepared surfaces without being sensitive to environmental or aging effects. According to current knowledge, there is no such 'all-rounder' in the offing, at least not for the purpose of refined professional bonding purposes. The adhesion between material surfaces in general, and wood in particular, may not be sufficiently explained by a model that is based on the existence of a sharp interface between the substrate and the adhesive. As shown in Figure 8.19, there is rather an *interphase* than an interface – a zone of finite dimension – in which the substrate and the adhesive amalgamate to form a new material. The adhesive failure which is often referred to corresponds in fact to a failure within this interphase. Another impediment to the bonding of wood is the fact that some wood species are rich in components that rediffuse to the surface shortly after a mechanical surface preparation. These components include low-molecular-weight resins and oils as well as aqueous solutions of sugar or peptides, and they may weaken the joint by acting as a release agent or by softening the wood structure. Several authors have therefore recommended increasing the durability of bonded wood joints by using *adhesion promoters*. It has been demonstrated for example that, following appropriate treatment with primers, the relevant delamination values of glulams obtained in the water test were clearly below the maximum prescribed by standards, even if difficult woods were bonded and if different adhesive systems were used [25]. The dilemma of primer treatment is that an additional manufacture step is required; hence, it will only be adopted for bonding purposes where quality is more important than quantity.

8.4.8
Wood Composites

Currently a wide array of wood-to-wood or wood to other materials composites is available, varying almost indefinitely in material and geometry. The only limits are set by the imagination of the engineers and designers, and any attempt of classification will be largely in the eye of the beholder. However, for the purpose of the present discussion, the classification according to Dunky and Niemz [19] was chosen (Table 8.2).

Compared to boards and beams cut directly from the trunk, solid wood materials and veneering materials have decisive advantages. For example, layered veneer

Table 8.2 Classification of wood materials according to Dunky and Niemz [19].

Solid wood	Veneering	Chip wood	Fiber	Composite
• Solid wood board • Gluelam • Two-way joist • Laminated wood • Compound board elements • Prefabricated elements	• Laminated veneer lumber • Plywood • Parallel strand lumber/parallam	• Particle board • Oriented strand board (OSB) • Laminated strand lumber • Waferboard • Extruded particle board • Scrimber • Special boards	• Medium-density fiber board (MDF) • Porous fiber board (SB) • Hard fiber board (HB)	• Blockboard • Blockplywood • Layer engineered parquet • Doors • Windows

sheets or wood lamellae may be joined in a stacked array in order to optimize the properties of the wood:

- *Drying behavior*: Owing to their limited thickness, lamellae may be adjusted to a specific uniform moisture content that is in a state of balance with that of the ambient air, even in the case of high-speed industrial drying processes.

- *Stability of form*: In thin boards, dense and large-volume cell structures are distributed more homogeneously than in thick elements. Swelling and shrinking therefore induce less important warpage. The layers located above and below intensify this effect.

- *Wood quality*: Defective spots in the wood, resulting from prunings, resin ducts, or growth irregularities, are easily discovered on thin boards. They may be cut off, and the board may be joined together on the front side by means of wedge finger-joints, making it possible to produce highly elastic components (e.g. parallel bars).

- *Free dimensions*: Using wedge finger-joints on the front sides, lamellae of (in theory) indefinite length may be produced. By using breaking finger-joints, these in turn may be joined to produce laminated lumber.

- *Forming*: Wood may be plastically deformed to a certain extent by means of thermal moisture treatment. This is easier to handle with thin boards or veneer material than with solid wood boards. Bent elements may therefore be built from laminated lumber bent and glued in the forming press.

- *Decorative elements*: Lamellae made from woods of different shades, joined in a stacked array, may be used to create visually appealing components for furniture, interior finish of vehicles and ships, for ship decks or musical instruments.

In combination with other materials, wood composites offer an almost unlimited field of applications. Among the most varied possibilities, some examples are presented in the following section.

8.4.9
The Supreme Discipline: Laminated Wood Constructions
(Civil Engineering Using Wood as Building Material)

8.4.9.1 Development

The groundbreaking technology of laminated wood constructions, invented by the carpenter Otto Karl Friedrich Hetzer (1846–1911) from Weimar in Germany, is the supreme discipline of wood bonding. Until the 1950s, Hetzer's name was synonymous with beam constructions made from layers of laminated wood boards, today called 'glu(e)lam'. Several types of adhesive have been used for this technology; Hetzer himself invented – and was granted a patent for – formulations based on natural raw material, especially casein. The adhesive joints obtained were of good quality, but they had the disadvantage of not being resistant to water, so that they could not be used for outdoor applications. In 1926, I.G. Farben (today BASF) introduced synthetic resins based on urea/formaldehyde, called 'Kaurit glue', which overcame this disadvantage. Due to their brittleness, only perfectly fitting joints could be obtained with this glue, but the addition of different fillers such as industrial flour eliminated this shortcoming, without impairing the adhesive force.

In general, the strength of the adhesive joint is required to be higher than the perpendicular tensile strength and the shear strength. In practice, this becomes apparent in fracture tests when the wood 'tears out' (cohesional failure in the wood adherents), where the resultant fracture surfaces are seen to have a high fiber content. This strength in a normal climate is considered to be a necessary, but not sufficient, property of an adhesive joint in structural wood elements. It is well known that the strength of bonded wood joints decreases under the influence of moisture and temperature changes. Therefore, standards have been developed for a classification of structural and nonstructural adhesively bonded wood according to the climatic conditions and fields of applications.

Urea-based adhesives already covered a wide range of applications as they were appropriate for elements not exposed to outdoor weathering or special climatic conditions. However, only synthetic resins based on resorcin and melamine are absolutely resistant to water and weathering. From the early 1990s, a new adhesive technology was introduced with the development of one-part polyurethane adhesives, the advantages of which have already been described. One problem which made their market entry difficult, however, was the fact that the existing – and very extensive – technical standards for wood and wood materials were tailored to conventional gluing procedures, and the new technology was only established through special approval (Radovic) [24]. Today, this still applies to some geographical regions – for example, in Japan, where modern polyurethane adhesives have not yet been generally approved for the structural gluing of wood as the relevant Japanese Industry Standard (JIS) does not include this class of adhesives. Moreover, there has been very little advancement in any striving for the standard to be modified.

8.4.9.2 Laminated Wood Constructions Today

Today, wood constructions are becoming increasingly popular worldwide. However, they are no longer limited to rustic, single-family houses and bridges or austere factory buildings, but rather cover all modern designs. Certainly, ecological reasons (wood is a CO_2 neutral, regrowing raw material) play an important role here. The fact that wood constructions may be prefabricated industrially is also important as vital processing steps may be transferred from the construction site to the factory. This is turn changes a handicraft operation into an industrial prefabrication technology based on processes that are similar to a series production, including a refined working schedule, a high degree of automation, precisely working, electronically controlled processing machines and refined, closed quality assurance systems. A series production also involves advanced cost control systems that increase the competitiveness of wood constructions compared to other construction methods.

When the prefabricated structures are finally assembled on the construction site, these operations are performed in a very quick and efficient way. Erection times of few days are the rule, not the exception, reducing to a minimum the expenses for assembly aids such as cranes and scaffolding. In many climatic zones, another advantage worth considering is the fact that it is possible to wait for dry weather before final assembly on the construction site; this not only allows the building times to be shortened but also prevents water damage in new buildings.

Another cost-saving factor is the fact that scheduling is made very easy, with the different teams of craftsmen being scheduled tightly so that they do not interfere with each other, as before. Wood constructions also offer a new dimension of design options to the architect; for example, when taking the specific properties of wood and

Detailed view of the vertical girder Longeron made from gluelam, 42 m effective height

Unfinished state of the load-bearing structure with helical struts for drive-on ramp; without the lamellar covering

Figure 8.20 Palais de l'Equilibre, Schweizerische Landesausstellung (Swiss national exhibition) 2002.

the possibilities offered by glued wood constructions into consideration for his/her structural analysis, the architect can erect buildings of filigrane elegance and beauty. A good example is the exhibition pavilion of the Swiss Confederation built for the national exhibition in 2002 (Figure 8.20). Here, the main hall was suspended on a girder with an effective height of 42 m and covered with an outside facade made from wood lamellae. The space between the interior construction and the girder was used for a helical ramp from which the different floors were accessible by wheelchair.

8.4.10
Prefabricated Elements

In modern wood construction, not only girders are prefabricated by means of laminated wood technology, but also various other elements including stairs, windows, doors, whole attic storeys and walls. The motivation for using prefabrication technology has been described in the above section. Figure 8.21 shows the fabrication of a stair string where the laminated veneer lumber is formed around a gauge. The adhesive joint is then cured by means of a vacuum press under atmospheric pressure. The stair is assembled later as a cantilever stair, that is, it is fixed only at the top and at the bottom.

Often, prefabricated elements are composed not only of wood, but specific properties may be conferred to a material by including metals, plastics or fibers. For example, the introduction of an aluminum plate in an external door prevents the door from warping due to weathering. Furthermore, the aluminum plate represents an absolute barrier for different moisture conditions prevailing both inside and

Figure 8.21 Prefabrication of a staircase element (stair string) which is formed by a vacuum press (foil bag) around a gauge.

outside the building. A third function of the metal plate is to improve the behavior of the door element in the case of fire. The insulation properties with regards to heat and sound play an important role in the case of wall elements and edges of windows. In contrast to mechanical joining techniques, using for example nails or screws, adhesive joints do not build up any heat or sound bridges. Sandwich elements consisting of wood bonded to specific insulators allow the highest energy and sound protection standards to be met. It is even possible to construct a building without any heating system, even in a central European climate.

Engineered Wood Products Made from Strands, Fibers and Composites Engineered wood products are fabricated by pressing strands, laminated strands, fibers or short-board elements into panel elements, mostly using synthetic resins based on formaldehyde condensation products or polyurethanes or, more seldom, mineral binders. For this technology, woodcuts that may not be used as building material, or rapidly growing plantation wood species, are employed. The classification of these materials with regards to their load-bearing capacity has been standardized. For example, Dunky and Niemz [19] have provided an extensive discussion of the technology and the influencing factors, as well as an overview of the standards concerning wood and engineered wood products, including their testing.

Behavior of Wood in the Case of Fire The everyday experience that wood will burn may lead to an erroneous assessment of the safety of buildings made from wood. On the contrary, even if they do burn, solid and thick layers burn slowly and in a foreseeable manner. In contrast to steel girders – the strength behavior of which is not foreseeable in the case of fire (called 'catastrophic failure') – burning wood constructions do not bend without early warning. The thermal degradation of wood starts at around 150 °C, inducing the expulsion of flammable gases while simultaneously producing charcoal. Starting at this temperature, but critically between 300 and 400 °C, spontaneous ignition of these gases then occurs. The burning reaction, in being exothermic, causes a thermal feedback with regard to the surface to be produced, and this results in a sustaining flame. However, if the temperature rises further (i.e. above 500 °C), the ratio of charcoal production to gas expulsion shifts in favor of charcoal, which has a considerably lower thermal conductivity than wood; this in turn results in a limitation of the thermal feed-back mechanism. The burning process reaches a quasi-equilibrium that is disturbed as soon as further joints open, resulting in an enhanced burn-up speed. Consequently, the behavior of adhesive joints in the case of fire is of utmost importance for the suitability of gluelam as a structural material. Tests performed on beams of gluelam in a fire resistance test according to ISO 834-1975 have shown that, under the influence of extreme heat, adhesive joints do not fail earlier than the wood adherents they hold together (Figure 8.22). Indeed, even next to the fire zone there was no delamination of the wood layers. The shear strength of the adhesive joints within the noncarbonized part of the beam still corresponds to that of the intact parts [21, 23].

Figure 8.22 Gluelam beams before and after a fire test [20, 22].

8.4.11
Prospects of Wood Constructions

Within the scope of the COST project, two working groups of the European Cooperation in Science and Technology studied 'Wood Adhesion' and 'Glued Wood Products' [20, 22]. These State of the Art Reports not only present an extensive collection of material and references, but also provide recommendations for potential future research and development activities. As the saying goes, 'the proof of the pudding is the eating' and, as a final consequence it will be the end consumer who will decide whether products made from wood will be highly sought after, or not. And it is for the architects, designers, engineers and adhesive professionals to translate new ideas into highly sought-after, wood-based products.

8.5
Paper and Packaging Industry

8.5.1
Introduction

Paper, board and cardboard are important materials for many daily life products, with their frequent use in the manufacture of packaging materials and graphic products, for example. Packaging materials are essential for the handling, safe distribution, warehousing and storage of the magnitude and multitude of modern goods.

Marketing aspects are an increasingly important factor for consideration in the design of packaging materials, as in today's shopping centers and supermarkets all goods must be 'self-sellers'. Packaging materials can give goods a commercially alluring appearance [26], and this is an important aspect because up to 70% of all shopping is based on 'impulse buying'. Today, on average consumers barely allows themselves 3 s to evaluate a product. When packaging materials are used for foods, aspects regarding food law also play an important role.

In the packaging sector, in addition to paper a wide range of other materials is used to confer predefined characteristics to the packaging material. Adhesives make it possible to implement these requirements and to meet the demands resulting from the processing of the materials. This applies increasingly to the print-finishing sector, where new printing stocks and print methods, as well as ever-increasing production speeds, have resulted in the development of new adhesives that offer of a wide variety of graphic products.

Industrial adhesives are primarily required to join materials in a fast and safe way, and to provide for trouble-free, inexpensive production on existing manufacturing lines (using automatic machinery as far as possible). Furthermore, they must resist the conditions to which the finished product will be exposed during its later use. Price is another important aspect that must be considered when selecting an adhesive, while ecological aspects are also playing an increasingly important role in the selection of an adhesive system. Today, more and more care is taken that adhesives can be processed in such a way that as little waste and wastewater as possible accrues from their production and use. The impact of adhesives on the reuse (e.g. returnable products) or recycling of bonded products is also a growing matter of interest [27].

8.5.2
Manufacture of Packaging Materials

Adhesives have always been a decisive factor in the manufacture of packaging materials made from paper, board or cardboard, or even from plastics or other materials. At the start of the 20th century, paper – a natural product – was bonded with adhesives based on natural raw materials such as proteins, starch or cellulose (see Section 5.9). Today, a wide variety of different materials is used in the packaging industry, and the many requirements can be met only by synthetic adhesives.

8.5.2.1 Corrugated Board and Paperboard Containers

On a quantity basis, the most important paper product bonded with adhesives is *corrugated board* (Figure 8.23), which is a very important material for the production of shipping containers.

The development of the corrugated board industry began in 1874 when a patent was issued in England describing the bonding of a flat linerboard to a fluted corrugated sheet to generate a material with excellent cushioning effect that was preeminently well suited as a packaging material. The first adhesives used for the bonding of corrugated board were conventional paper glues, such as a mixture of cereal flours and water. Due to the large amount of adhesive needed for this application, its commodity price has in the past played an important role, besides its technical properties. In 1939, the Stein Hall glue was invented; this was an adhesive system based on starch which made it possible to considerably increase the efficiency of corrugated board production. Stein Hall corrugating adhesives are a mixture of fully gelatinized starch, available as a colloidal solution, with native,

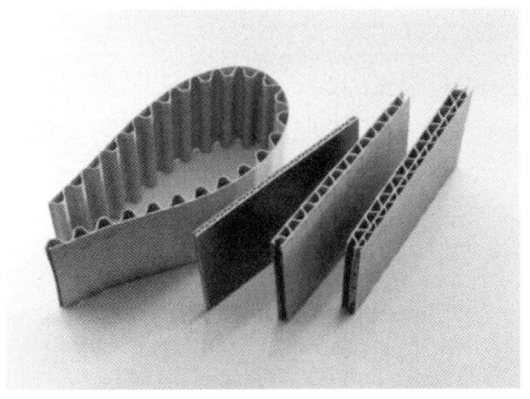

Figure 8.23 Corrugated board.

ungelatinized starch dispersed in water. This combination makes it possible to obtain high-solids adhesive formulas, without increasing the viscosity to such an extent that the adhesive can no longer be processed on the roller application system. Unfortunately, in this state the adhesive does not have a satisfactory adhesive property for bonding. Following its application, during compression of the corrugated board, the adhesive layer is heated for a short period to such an extent that the dispersed native starch gelatinizes. This results in a steeply rising viscosity which allows the freshly bonded corrugated board to resist the forces occurring in the further production process. With starch-based adhesives, excellent technical properties can be combined with a low price, and the production of corrugated board has therefore become the most important field of application of natural polymers in adhesives (see Section 5.9). Currently, corrugating adhesives are undergoing continued development to cope with ever-advancing manufacture technologies. The trend towards faster production speeds requires the constant adaptation of technological parameters of adhesives such as viscosity, gelatinization temperature and initial 'tack' [28].

8.5.2.2 Lamination

Lamination is an important field of application in the further processing of corrugated board and carton paper. For this purpose, plasticized polyvinyl acetate homopolymer-based dispersion adhesives are normally used. Special flatness and properties which provide for a high punching speed are important requirements for lamination. As this application is relatively simple, the price of the adhesives used is crucial, and the addition of fillers (e.g. chalk) which are admixed with the dispersions offers an opportunity to reduce raw material cost. The adhesives are applied by means of rollers.

8.5.2.3 Folder Gluer

Corrugated board is used for outer package or transport package purposes after having been transformed into containers by means of so-called 'folder gluers' which glue folded sheets of corrugated board to convert them into boxes using plasticized polyvinyl acetate homopolymer based dispersion adhesives (side-seam gluer). On modern machine tools, the adhesive is applied by means of electromagnetically controlled nozzles; depending on the thickness of the board up to three nozzles may be necessary.

8.5.2.4 Folding Cartons

In the retail package sector, folding cartons are the most popular product. These are manufactured by bonding folded paperboard using plasticized polyvinyl acetate homopolymer based or ethylene vinyl acetate (EVA) copolymer-based dispersion adhesives (side-seam bonding), thereby converting flat sheets into boxes. Very often, the package material must be sales-promotional, and so the folding cartons need to be available in a wide variety of designs and with finished surfaces. From coating and lamination to premium printing and varnishing, all types of paper finishing modes are applied. These surface properties are crucial when selecting the correct adhesives for side-seam bonding. Historically, dispersion adhesives were applied by means of discs, but the newer machinery relies increasingly on the use of nozzles with electromagnetically controlled valves (Figure 8.24).

8.5.2.5 Flexible Packages

Plastic materials, particularly plastic films that are converted into flexible packages, are becoming increasingly popular in the packaging sector.

Flexible packages are manufactured from monofilm or, more often, from multilayer film made from different plastic films or from combinations of plastic films, metal foils or paper. The underlying principle is the combination of materials with selected properties to achieve a performance synergy effect, including temperature stability, barrier properties, mechanical strength and alluring appearance (Figure 8.25).

A multitude of materials is used for packaging composites, ranging from polyethylene (PE), polypropylene (PP) and paper to polyvinyl chloride (PVC), polyvinylidene fluoride, cellulose film, polyester, polyamide and aluminum. The plastic films are

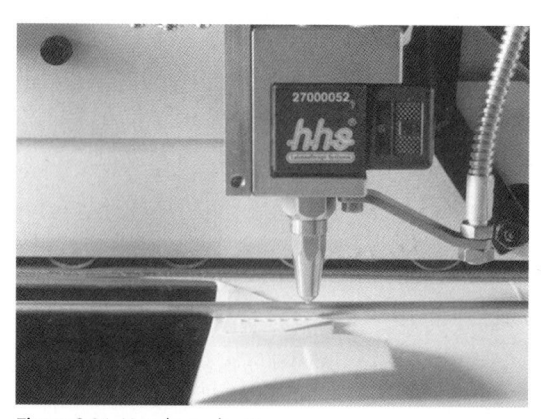

Figure 8.24 Nozzle application.

8.5 Paper and Packaging Industry

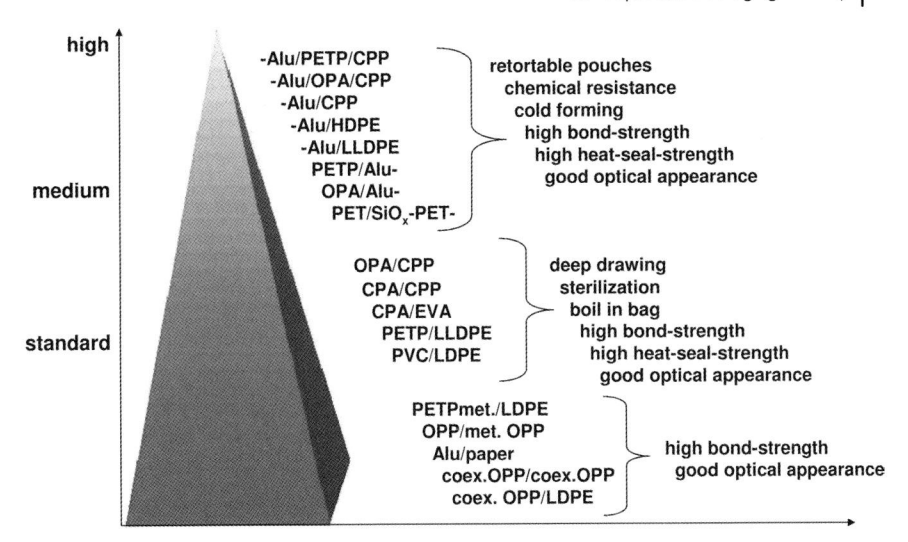

```
high │
     │           -Alu/PETP/CPP          retortable pouches
     │           -Alu/OPA/CPP           chemical resistance
     │           -Alu/CPP               cold forming
     │           -Alu/HDPE              high bond-strength
     │           -Alu/LLDPE             high heat-seal-strength
medium           PETP/Alu-              good optical appearance
     │           OPA/Alu-
     │           PET/SiOx-PET-

               OPA/CPP                 deep drawing
               CPA/CPP                 sterilization
               CPA/EVA                 boil in bag
standard       PETP/LLDPE              high bond-strength
               PVC/LDPE                high heat-seal-strength
                                       good optical appearance

               PETPmet./LDPE
               OPP/met. OPP
               Alu/paper               high bond-strength
               coex.OPP/coex.OPP       good optical appearance
               coex. OPP/LDPE
```

Figure 8.25 Laminating composition.

frequently vapor-coated with aluminum or SiOx, or coated with polyvinylidene chloride (PVDC) or acrylic to enhance the barrier or light protection properties, respectively.

A great number of special laminating adhesives have been developed for different applications [29]. One important group here are the polyurethane adhesives (PU resin or PU laminating adhesives), which are based on aromatic or aliphatic isocyanates, respectively (see Section 5.6). Although PU resin laminating adhesives containing solvents have been used in large quantities, the production of composite materials using solvent-containing PU resins is very energy-intensive due to the drying process, the large amounts of raw materials required for the adhesives, and the amount of solvent added. These systems are suboptimal in terms of their sustainability, even if the production system is coupled to a solvent recovery system. In order to relieve this problem, high-solid and solvent-free PU resin laminating adhesives have been developed, starting with one-part systems which set under the effect of moisture. As NCO-terminated polyether or polyester prepolymers, they depend on moisture either to be available either in the air or to be added during lamination to cure into a polyurea/polyurethane in a crosslinking process. At present, these systems are used only for paper composite material, such as oriented polypropylene laminated with paper. To avoid the limitations set by solvent-free one-part systems crosslinking under the effect of moisture, two-part systems have been developed which are usually based on polyester or polyether with a polyisocyanate hardener. Both parts are liquid at room temperature and can be processed at low temperatures, preferably between 25 °C and 45 °C. Recent years have witnessed the development of systems with processing temperatures between 40° and 70 °C. This new generation of solvent-free laminating adhesives combines the advantage of a high solid system (initial adhesion or 'tack') and that of a solvent-free liquid phase (easy processability). The prepolymers are synthesized in such a way that a fast curing can take place in

Curing method	Step 1	Step 2	Step 3	Step 4
Electron beam curing	Coating	Laminating	Electron beam	Fast curing
UV radicalic	Coating	Laminating	UV light	Fast curing
UV cationic	Coating	UV light	Laminating	Curing in 24 h

Figure 8.26 Curing methods.

several stages, starting with the desired and complete reaction of the monomers (Figure 8.26).

For sterilizable aluminum/plastic film composite materials, solvent-free laminating adhesives based on aliphatic isocyanates have been developed. These so-called 'aliphatic laminating adhesives' are processed at 70 °C with conventional lamination machinery, dosing and mixing devices, respectively. A significant advantage here is the lack of aromatic isocyanates that plays an important role with regards to food-law regulations. One feature of all the above-mentioned PU resin laminating adhesives that is the prerequisite for their suitability for laminating purposes is that, within the composite material, thermal and mechanical resistance only fully develops when they are completely cured. The curing period may be one week or longer, depending on the type of adhesive used, the film/foil combination, and the curing temperature. Increasing the temperature shortens the curing period. In either case, the time needed to produce composite materials (including printing, cutting, etc.) is of the utmost importance, as the capital tied up by long curing periods may become a significant expense, especially for companies with a high competitive pressure. It is not surprising, therefore, that there is a strong desire for a laminating adhesive that cures so to speak 'on command'. One possibility would be to use radiation-curing systems, which have the advantage of achieving a rapid initial adhesion ('green strength'). When comparing the different laminating adhesives, the 'green strength' of solvent-free systems is clearly considerably lower than that of systems containing solvents. This situation can be explained by a significantly lower molecular weight range and distribution, respectively, and is dictated by the low viscosity required to apply the adhesive without further dilution with solvents. Laminating adhesives which set under the effect of irradiation can be applied without solvents and at low viscosity, and provide for both a high 'green strength' and a very rapid attainment of ultimate strength following irradiation. The different crosslinking mechanisms of radical and cationic polymerization allow different processing steps to be performed with regards to irradiation and lamination (Figure 8.26).

Another possible way to avoid solvents is to use aqueous laminating adhesives; these systems are based on one-part or two-part polyurethane or acrylate dispersions, respectively. They can be processed on existing lamination machinery, and require only a minor investment for a retrofit. Aqueous systems also stand out for their high 'green strength' and fast curing; neither do they leave any solvent residues in the composite material. One disadvantage, however, is the large amount of energy needed for the drying process. Unfortunately, waterborne adhesives generally do not provide the end-use performance level of solvent-containing or solvent-free systems.

8.5.2.6 Bags, Pouches and Sacks

Many articles for daily use are packed in bags, pouches and sacks. For this, special adhesives have been developed for manifold applications (e.g. side-seam bonding, bottom pasting, partial lamination), ranging from simple paper bags that are still bonded with starch-based adhesives to elaborate pouch designs bonded with dispersion adhesives or hot melts.

Depending on the adhesion requirements, the dispersion adhesives used are based on plasticized polyvinyl acetate homopolymers, EVA copolymers or acrylates. Resin dispersions can be admixed to improve adhesion. For nonabsorbent materials such as polypropylene films, hot-melt adhesives are used that are based on thermoplastic rubbers (e.g. styrene–butadiene–styrene or styrene–isoprene–styrene block copolymers) mixed with resins and oils, depending on the application. *Bags and pouches*, ranging from simple cornet bags to elaborate block-bottom and cross-bottom bags, can be manufactured from a wide variety of materials and material combinations. Occasionally, these materials are imprinted, and this must be taken into consideration in the bonding process. As bags and pouches are produced on high-speed machines with a performance of 150 to 200 bags or pouches per minute, the adhesives must also provide for a clean machine run.

In the manufacture of *paper bags*, starch-based adhesives are still widely used for side-seam bonding and even for bottom bonding. For the gluing-in of valves and for the bonding of internal layers made from PE, dispersion adhesives with a large adhesion spectrum are used; these may be EVA copolymers, acrylates or natural rubber, mixed with resin dispersions.

In the manufacture of pouches and sacks, one special challenge is that of the handles, particularly when using recycling paper, bogus paper or soda/kraft paper. Occasionally, dispersion adhesives bleed through, and although this might not be visually discernible, it may hinder a hassle-free production process. A possible solution here is to use coarsely dispersed dispersion adhesives containing minor amounts of low-molecular-weight substances, or starch-containing dispersion adhesives (provided that they do not dry too slowly). The bonding of the handles may give rise to further problems. Very fast-curing dispersion adhesives or, alternatively, hot-melt adhesives, can be used to avoid displacement of the handles during the production process on machines running at ever-increasing speeds.

Polyethylene sacs are bonded with reactive two-part polyurethane adhesives, with the adhesive being applied by means of rollers for the bottom pasting. The valves used to fill the sacs are also bonded with reactive, two-part polyurethane adhesives or, alternatively, with hot melts based on synthetic thermoplastic rubbers mixed with resins and oils.

8.5.3
The Setting-Up and Closing of Trays and Boxes

Adhesives are used widely not only in the manufacture but also in the filling of packaging material. A pivotal requirement here is the secure closing of packages, especially as a tamper-evident closure.

8.5.3.1 Setting-Up a Tray
Hot-melt adhesives with extremely short setting times are generally used for the setting-up of trays (Figure 8.27). The majority of hot melts employed for this purpose are EVA copolymer-based systems, although new polyolefin polymer-based adhesives are becoming increasingly popular owing to their interesting technical properties, which make them very suitable for this type of application.

8.5.3.2 Carton Closing
Bonding is the most important method of closing packages made from paper, cardboard or paperboard. Depending on the automation level of the packaging line, the design of the machinery and the demands placed on the packages during transport and storage, one of the following systems is used:

- hot melts
- PSA packaging tapes
- water-activated adhesive strips

On occasion, dispersion adhesives are also used, particularly with the two-shot method (i.e. in combination with hot-melt adhesives). This combination is used, for example, when a rapid setting process is required by the production machinery (hot melts), and a good thermal and chemical resistance, for example to essential oils, is also needed for safe and durable bonding (dispersion adhesives).

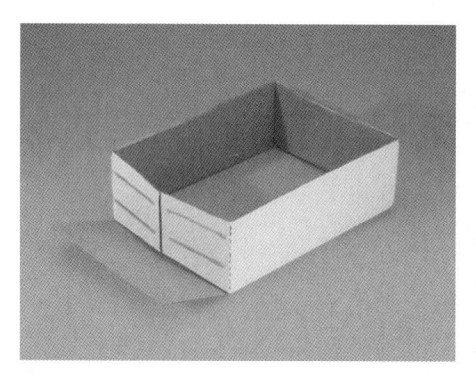

Figure 8.27 A tray.

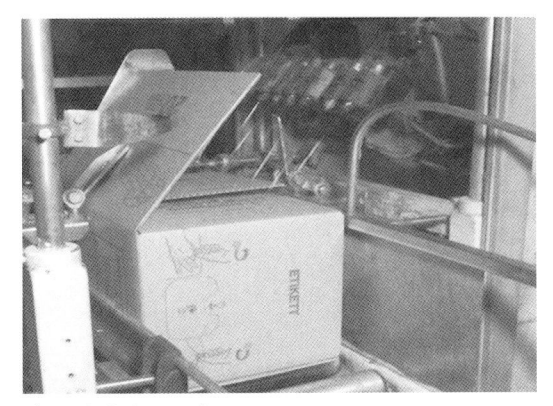

Figure 8.28 Carton closure.

Hot Melts The hot-melt adhesives used in the packaging sector are thermoplastic water-insoluble polymer mixtures (see also Section 5.2). For carton closure purposes (Figure 8.28), mainly EVA copolymer-based products are used, but the new polyolefin-based adhesives are also becoming popular for carton closing as well as for tray set-up.

For special applications (e.g. hot filling), copolyamide- or copolyester-based hot melts, with substantially higher melting ranges, are chosen. The adhesives can be tailored to meet the needs of a desired application by mixing them with resins (hydrocarbon resins or natural resin derivatives), waxes or other polymers. The hot melts are solid at room temperature, are applied in the molten state (Figure 8.29), and set within the subsecond time range.

The adhesive films formed when carton flaps are joined have a thickness of approximately 0.2–0.8 mm. It is possible to obtain high speeds of machinery by using hot melts, but such applications require the correct equipment with an appropriate initial investment expense. This disadvantage is outweighed, however, by a good running performance on the machinery, the result being production lines that run more efficiently. In recent years the development of low-temperature hot melts has

Figure 8.29 Hot-melt application.

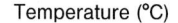

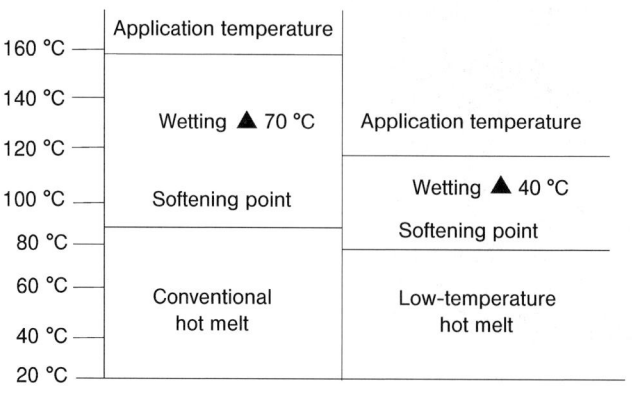

Figure 8.30 Application temperatures for conventional and low-temperature hot melts.

become increasingly popular [30]. Compared to conventional products, which were processed at operating temperatures between 160 °C and 180 °C, low-temperature hot melts can be processed at substantially lower temperatures ranging between 120 °C and 140 °C (Figure 8.30).

Owing to lower application temperatures, these hot melts are subjected to considerably lower thermal stress, which results in an improved viscosity and color stability. Because the application equipment is also exposed to lower temperatures, it has a longer service life and requires less maintenance and energy input.

PSA Packaging Tapes These tapes are used for the closing of cartons (Figure 8.31) and have a width of a few centimeters (preferably 30–100 mm). They are coated with a PSA, which means that at room temperature they adhere to a surface upon application of pressure (see Section 5.1). Biaxially oriented polypropylene film (BOPP) is the most frequently employed base material in the packaging sector (ca. 80%). PSA tapes with PVC as the base material are only rarely used, and in some exceptional cases paper is accepted as the base material. Today, the following three systems are in common use for the manufacture of PSA tapes: solvent-based PSAs (ca. 45%); waterborne dispersion systems (ca. 25%); and hot melts (ca. 30%), which are becoming more popular. The adhesive film thickness on the base material is generally 16–25 µm.

Water-Activated Adhesives Water-activated adhesive tapes or strips are commonly made from high-strength (50–150 g m^{-2}) kraft paper coated with a water-activated (usually nonthermoplastic) adhesive (e.g. animal or plant glue). The strips are sometimes reinforced with filament yarn to increase their tensile strength. Before application of the tape, the gumming adhesive is reactivated with water that penetrates into the carton, resulting in the setting of the adhesive.

Reclosure Systems for Liquid Packaging Boards Consumer-friendly closure systems are an integral part of up-to-date liquid-packaging boards (also called 'fitments'), as

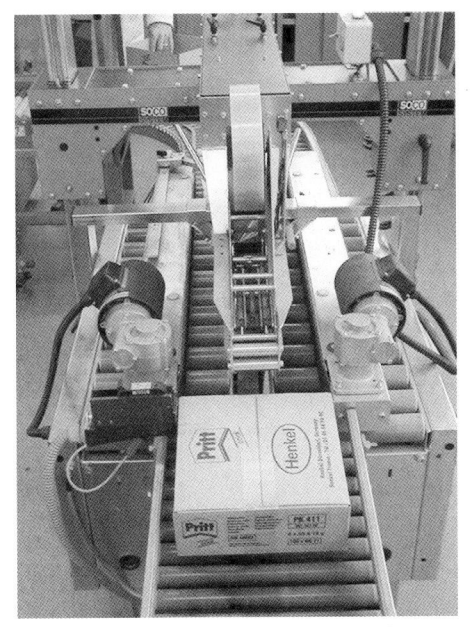

Figure 8.31 Closing of cartons by means of PSA tapes.

their convenience adds customer-appeal to many packages. These devices are expected to allow easy and clean opening (without auxiliary aids such as scissors or knives), pouring and hygienic reclosure; evidence of tampering is often also required. There are two reclosure options: (i) screw caps that provide for a reliable reclosure and horizontal storage; and (ii) snap closure systems. These are usually made from polypropylene and are safely and economically applied by means of hot melts. As liquid packaging boards usually have a polyolefin cover, only special hot melts can be used for this purpose. Polyolefin copolymer-based hot melts, for example, have a sufficiently large adhesion spectrum that enables them to bond reliably to the relatively nonpolar polyolefins. Adhesives are applied to the spouts by means of rollers or special nozzles, and have an adhesive film thickness of up to 0.5 mm [31].

As many beverages are stored in a refrigerator, the bonds must resist low temperatures. Usually, bonding stability at storage temperatures between 3 °C and 50 °C must be guaranteed.

8.5.4
Pallet Securing

Cargo is commonly transported in palletized unit loads, with individual packages stacked on each other on a pallet base. In order to secure the load to the pallet during transport, one solution is to use plastic wraps; however, because this creates much packaging waste, such wraps are being increasingly replaced by palletizing adhesives (Figure 8.32) [32].

Figure 8.32 The securing of pallets.

The sticky pallet-securing products are usually called 'adhesives' and are even made from similar raw materials. However, in fact they basically differ from packaging adhesives on several points, and are better designated as 'antislipping products'. When a carton, for example, is closed by means of bonding with an adhesive, owing to the strength of the adhesive bond, its material will be destroyed and fibers torn away when the bond is opened. This is due to an intimate contact between the adhesive and the surface of the substrate (adhesion), the adhesive itself having a higher inner strength (cohesion) than the substrate. A pallet-securing product, however, is expected to hold the load units firmly together during transport and storage, yet allow an easy separation of the packages, without damaging their surface. 'Antislipping products' must therefore provide either for adhesion failure between the 'glue' and the surface of the package, or cohesion failure within the adhesive film when packages are de-palletized. These 'antislipping products' can be either waterborne or hot melts.

Waterborne 'antislipping products' are usually two-phase systems (dispersions) where the sticky substances (the inner phase) are finely dispersed in water (the outer phase). They are processed at room temperature, their antislipping effect being generated during the drying process. With waterbornes, the cartons and boxes to be palletized must be brought together while the 'antislipping product' is still in aqueous condition ('still wet'). It is of the utmost importance not to exceed the open time of the antislipping product. Waterborne products can be applied either manually (brush, roller, handheld gun) or by means of fully automated nozzles (spray or bead application). Manual application requires a relatively low investment, but dosing is difficult and the pallet-securing quality is not consistent. With automatic equipment, the application process and quantity can be tailored to the individual package, providing for consistent pallet securing quality.

Hot-melt antislipping products are 100% solids that are heated to temperatures ranging between 100 and 180 °C in order to transfer them into the liquid state; this then allows them to be conveyed and processed. Setting takes place by cooling via heat transfer to the substrate and to the environment. As palletizing is generally a slow operation, the package to be fastened is not placed into the 'wet' (liquid) hot melt, but onto a more or less set, solid product. This results in a surface bond with little or no penetration into fiber, and hence little or no fiber tear when separated.

8.5.5
Labeling

Today's produce is unimaginable without labels, which are used to inform, decorate and aid logistics. In modern shopping centers and supermarkets, labels help to provide an alluring appearance to the products that must be self-sellers. Labels play a particularly important role in the beverage, cosmetics and cleaning products industry sectors as they bring products 'to life' – without labels, beverages would be anonymous bottles or cans. To meet the increasing demands of the market, different development trends have occurred with regards to container materials, labeling methods and the adhesives used for this purpose. Due to a wide variety of requirements, various very specialized labeling and adhesives solutions are now available commercially [33].

8.5.5.1 Pressure-Sensitive Labels

Although, currently the European label market (Figure 8.33) is dominated by precoated pressure-sensitive labels, this market share of about 50% is expected to grow further.

Self-adhesive labels are usually applied as tags, they adhere well to the majority of materials, and they can be made from either papers or plastic films. Precoated PSA labels are widely used in the cosmetics and cleaning products industry, but in the beverage industry their use is limited to products of high value or to special promotions, as they are rather expensive. Another factor which has limited the use of self-adhesive labels in the past has been the performance of the dispensers, although today dispensers can perform at 60 000 containers per hour. Precoated PSA labels are manufactured by coating adhesive onto ribbon-shaped base materials such as paper, plastic films or metal foils with an adhesive layer thickness of 18 to 50 μm. The adhesives applied for this purpose are based on different technologies. Besides the natural rubber-based PSAs (which usually are available as solvent-containing

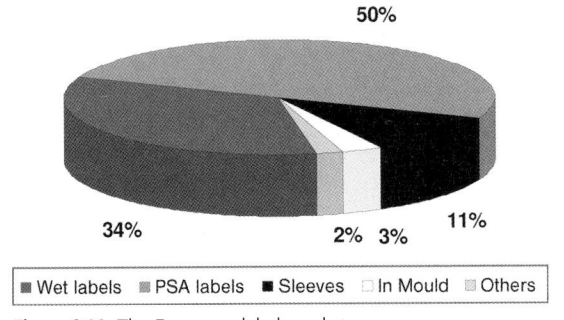

Figure 8.33 The European label market.

systems) and thermoplastic rubber or acrylate-based PSAs, dispersion adhesives are becoming increasingly popular. Acrylate dispersions in particular have remarkable technical properties; for example, they are a transparent labeling material, and precoated PSAs are increasingly being used for so-called 'no-label look' labels. They provide for an especially alluring container design since, at first glance, these transparent film labels (which are only partially printed) appear as though they have been printed directly onto the containers. The adhesives used for this purpose are totally transparent and not visually discernible at all.

In addition to the conventional water-resistant PSA labels, plastic film-based label materials are available that are suitable for reusable containers. In this case, lye-soluble PSA coatings are used which lose their adhesion capacity when exposed to the conditions of a bottle cleaning machine and are removed together with the plastic base material.

8.5.5.2 Inline Labeling

As with precoated labels, adhesives used for so-called 'inline' labeling must be selected correctly to ensure high-quality labeling. For inline labeling, either waterborne adhesives or hot melts are applied during the labeling process – that is, at the same time as the labels. Both adhesive systems can be combined for special applications.

Waterborne Labeling Adhesives Labeling with waterborne adhesives has a long tradition. The first adhesives used for labeling were waterborne products based on natural polymers such as dextrin, starch or casein (one of the most ancient raw materials for adhesives [34]). Even today, casein-based adhesives are at the forefront of glass labeling, as they can be applied on wet and dry, cold and hot glass surfaces, and yield strong bonds. Furthermore, they have excellent running performance on labeling machines, even at the highest speeds (up to 80 000 bottles per hour or more).

However, as neither the availability nor the commodity price of casein is consistent, in recent years the development of new raw materials has made it possible to more closely approach the aim of developing synthetic adhesives with a machine running performance similar to that of casein-based adhesives. Labeling adhesives based on synthetic polymers have a high spreading capacity and initial adhesion ('tack') and, when compared to casein-based adhesives, savings of adhesive of up to 20% are possible when using them on dry glass bottles. They also have a less intense inherent smell and, owing to their very clean running properties, they prevent soiling of the labeling machines. Waterborne systems are usually applied by means of applicator drums and segments (Figure 8.34).

Whereas, casein-based adhesives are not suited for the labeling of plastic bottles, this does not apply to adhesives based on synthetic polymers (solutions or dispersions of synthetic resins) because their adhesion spectrum is much larger. Depending on the material of the containers, different types of adhesive are used. Special synthetic resins that are internally plasticized by copolymerization are particularly well suited for PE and PP containers. Colloid solutions of synthetic resin can also be used for polyethylene terephthalate (PET) and polycarbonate

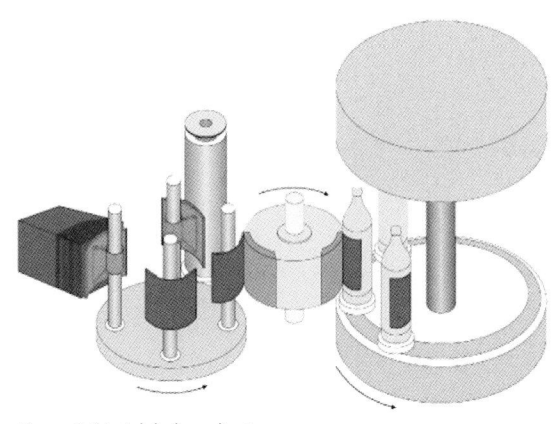

Figure 8.34 A label applicator.

containers. Waterborne adhesives allow both disposable and returnable containers to be reliably labeled.

Another interesting development that has taken place in the waterborne labeling adhesive sector in recent years has been the development of adhesives suited for film labels. Previously, waterborne adhesive use for film labels had failed due to the fact that neither the film nor the container material (e.g. glass or plastic) was able to absorb the moisture contained in the labeling adhesive, so that no reliable labeling was possible. However, newly developed systems allow the reliable labeling of film labels on nonabsorbent surfaces by means of waterborne adhesives. These have a high initial tack (i.e. they safely fix the labels) and are relatively fast setting, even in the case of nonabsorbent substrates. As they set into totally transparent films, they are also suited for 'no label-look' labeling. In addition, they can be used on existing labeling machinery designed for waterborne systems, and they have a good cost–performance ratio. Foil labels provide for high glossiness and resistance to water, and provide excellent clarity behind the label to see the bottle content.

Hot-Melt Labeling Adhesives Hot melts are 100% solids applied in the molten liquid state, either on the label or on the container, where they quickly cool to form durable bonds. Hot-melt labeling adhesives are based on synthetic thermoplastic rubber or EVA copolymers and resins, waxes and oils, with a mixing ratio tailored to the requirements of the desired application. Their open time (i.e. the time available for creating the bond) may be short or very long, as necessary. The hot melts are processed at temperatures between 120 °C and 170 °C, with a trend towards low processing temperatures, particularly for foil labels.

Hot melts have delivered particular performance in the 'wraparound' labeling of bottles and cans. Depending on the container material, different hot melts with different adhesion properties are available. These may be suited to the labeling of glass as well as of metal or plastic containers. One particular strength of hot melts is their excellent 'pick-up' function owing to a high initial adhesion and rapid bonding due to short curing periods. However, care must be taken that the temperature of the

Figure 8.35 A hot-melt applicator drum.

container to be labeled is not too low, in order to prevent the hot melt from cooling and becoming solid too rapidly; hence, it is recommended that the surfaces of the containers are dried under control by using a fan. When wraparound labeling with hot melts, line speeds of up to 60 000 containers per hour can be achieved.

Both, paper and a wide variety of plastic foils are used as labeling materials. Transparent foil labels produced by reverse printing can be processed as accurately as can labels made from foamed plastic with insulating properties. For hot-melt labeling, the bottles coated with hot-melt adhesive draw the cut labels from a magazine stack (instead of a pallet dispenser coated with adhesive, as in the case of labeling with waterborne adhesives). The hot melts are applied either by means of applicator drums (Figure 8.35) or nozzles (Figure 8.36).

Figure 8.36 A hot-melt applicator nozzle.

In the plastic bottle sector, returnable bottles (made from PET or polycarbonate) have become popular in recent years [35, 36]. As with disposable bottles, the adhesives are required not to interfere with the cleaning or recycling process, and consequently different hot melts have been developed for different recycling methods. Frequently, plastic bottles are washed prior to recycling, and any extrinsic material such as labels and adhesive must be removed in this process. Water-soluble or redispersable hot melts meet these requirements and provide for the clean recycling of plastic material.

In recent years, not only the formulation but also the handling of hot-melt adhesives has been improved; at present, granulated hot melts can be conveyed by automatic conveying systems. In the past, however, pressure-sensitive hot melts were packed in siliconized papers, although at present they are supplied in foil packages. With this innovative packaging solution, hot melts need not be unpacked but simply placed into the melting container together with their package wrapping, thus providing easy processing and avoiding the disposal of package waste.

With ever-increasing demands as to the features of the containers, new processing technologies arise again and again. The introduction of new container materials and specifications by marketing and legal authorities has led to many innovations. Often, however, the desired results are beyond what properties of a single adhesive, and hence several adhesives must be combined. For example, tamper-evident seals that must not be water-soluble are bonded to bottle closures by means of hot melts, while waterborne adhesives are used for the labeling.

8.5.5.3 Tamper-Evident Bottle Closures

The purpose of a tamper-evident bottle closure is to comply with statutory regulations and to issue a guarantee for producer bottling and product integrity. Either hot melts or waterborne adhesives can be used, as necessary and depending on the bottle material. Both adhesive systems must meet high demands as to the initial strength, detachability and removability. Bottle closures are made from either plastic or coated metal (e.g. twist-off caps), on the surfaces of which it is much more difficult to obtain adhesion than, for example, on glass surfaces. The surface that is the more difficult to bond (usually the surface of the closure) determines the adhesive used. Furthermore, as the surfaces available generally have a very small area (a few square millimeters to a few square centimeters), in order to provide a secure positioning of the labels and avoid slipping during the curing process, the adhesives must have a high initial strength.

When opening a tamper-evident closure, a bond must be broken, leaving an obvious visual indication that the bottle has indeed been opened. The bonding must therefore not be easy to detach (e.g. by water). A combination of waterborne systems and hot melts may be necessary to bond a paper strip on bottle closures; self-adhesive tamper-evident closures are another option.

In addition to these requirements, tamper-evident closures for returnable bottles must meet another demand – that they be smoothly removable in the bottle-cleaning machine. This is straightforward with casein-based adhesives, but if special requirements with regards to the adhesion properties make it necessary to use synthetic, resin-based adhesives, then care must be taken to select an adhesive that can later be easily removed. For example, if a fast setting rate is required to provide for a rapid and secure fixing of very small tamper-evident labels, then conventional packaging hot melts will not be suited because they are not water-soluble. For this purpose, special hot melts have been formulated that are water-soluble or redispersable so as to be removable during bottle washing.

8.5.5.4 Membrane Bonding

Membrane bonding is a special way of creating a tamper-evident closure for jars. To this end, a membrane made from coated paper or aluminum foil, for example, is bonded to the glass rim. When the jar is opened, the bonding is broken, leaving an obvious visual indication that the jar has indeed been opened. The adhesives are derived from EVA copolymer dispersions mixed with polyvinyl alcohol solutions that allow their viscosity and open time to be adjusted, respectively. The jars are conveyed under topside gluing systems that coat the rim with adhesive. By screwing on the lid, the membrane, which has usually been placed into the lid, is pressed against the adhesive on the rim. Another option is to coat the membrane with hot-melt adhesive that is activated by infrared irradiation following closure of the jars. After cooling, the membranes securely adhere to the rims.

8.5.6
Cigarette Production

One special application of adhesives in the use of paper is the production of cigarettes. At present, cigarette rods are manufactured at a line speed of up to 750 m min^{-1}, during which the cigarette papers are side-seam-bonded using dispersion adhesives and starch-based adhesives are applied by means of discs or nozzles.

One particularly high-demanding application is the bonding of filter tips. To this end, two cigarettes are bonded at a time to a filter rod by means of a paper wrapper. High line speeds allow the bonding of up to 22 000 cigarettes per minute. To adapt the adhesives to ever-increasing production speeds, rheology is a key factor. For example, the rheological parameters of starch-based adhesives can be optimized by enzymatic degradation that modifies their molecular weight. This makes it possible to develop adhesives with a higher solids content, which allows faster setting rate and thus provides for higher production speeds.

Cigarettes are packed either in hardboxes or soft packs. Owing to their high initial adhesion, dextrin adhesives are used for the manufacture of soft packs from cuts that are produced at a rate of 1000 packs per minute. Due to high wet bonding capacity, they are able to overcome the restoring forces prevailing during the bonding process. Hardboxes are usually bonded by means of dispersion adhesives.

8.5.7
Manufacture of Tissue Paper Products

The manufacture of tissue papers is an important sector of the paper industry, as a wide variety of tissue papers is now available for daily use. The adhesive systems and raw materials used for these applications are tailored to the needs of the tissue products and the machinery used to produce them. The adhesive systems are derived from a range of raw material specifically suited to meet the needs of the desired application and machinery.

8.5.7.1 Core Winding

Tissue paper is commonly wound onto cores (board rolls) made from strips of paperboard wound on a horizontal spindle while simultaneously being bonded layer to layer; the result is a continuous core that is then cut to the desired length (Figure 8.37).

The paperboard has a basic weight of $120–160\,g\,m^{-2}$, with two to four (or more) layers being commonly bonded to each other. The winding speeds can reach 140 m of core per minute. Core winding adhesives are often based on dextrin with a solids content of 35–50%, or on polyvinyl acetate homopolymers. In some special cases, animal glue is also employed. The adhesives must provide secure, rapid bonding of the paper strips, allowing the wound cores to be very accurately (tolerance $\pm\,2\,mm$) cut immediately after manufacture. As the cores must be mechanically stable, the adhesives must contribute towards the stiffness of the core.

Polyvinyl acetate homopolymer-derived adhesives are used for machines running at more than $60–80\,m\,min^{-1}$. They have premium adhesion properties, even on difficult papers, and can easily be redispersed during the recycling process. Although adhesives derived from dextrin (usually applied at $50\,°C$) also confer the desired properties to the cores, their slow setting process is a limiting factor for winding speed. When using dextrin for core winding, it is important to know that, during the recycling of the paper, the pulp will contain adhesives and auxiliary products such as borate that are commonly added to dextrin adhesives to increase the setting rate. The borate may react with polyvinyl alcohol from other sources (e.g. lamination) and

Figure 8.37 Core winding.

cause flocculation that may give rise to problems during recycling (e.g. so-called 'fisheyes'). Polyvinyl acetate homopolymer-derived adhesives are always supplied as ready-to-use products, whereas dextrin-based adhesives may be ready for use or supplied as dry powders that are mixed with water prior to use. When using dry dextrin, care must be taken to use the correct mixing ratio in order to ensure the correct viscosity, processing properties and setting speed. Core-winding adhesives are applied to the paperboard either by means of roller or nozzle applicators.

8.5.7.2 Lamination

The lamination process (ply bonding) is crucial for the quality of the finished tissue product. Lamination adhesives join two or three (or even more) tissue plies together. There are several ways in which the plies of embossed products are brought together; one way is to apply the adhesive to the tips of the raised elements of one of the two layers, and those tips come into contact with and are bonded to the recessed areas or floors of the other layer (nested laminating). In pin-to-pin, tip-to-tip or point-to-point laminating, the raised elements of each layer contact each other at the tips, which is where the bonding occurs. The paper typically has a weight of 15–25 g m^{-2} and can be either bleached or unbleached. A typical machine speed is 600–800 m min^{-1}, but even faster equipment is available which runs at speeds of up to 1000 m min^{-1}.

Although, in the past, laminating adhesives were mainly derived from dissolved cellulose, at present they are increasingly being replaced by synthetic polymers, particularly polyvinyl alcohol with its excellent adhesion properties. This is especially the case on tissue papers, which are more difficult to bond due to their wet strength, or because they are impregnated with lotions. The key characteristics of these adhesives are appropriate wet bonding capacity, no or only little foaming, and good processability. Adhesives with fast setting speeds enable the lamination to perform well in the machinery and make it possible to avoid staining on the tissue paper. Laminating adhesives must be flexible enough so as not to impair the softness of the finished product. Water-based products release the majority of their water content into the paper during lamination, which can threaten the embossed structure of the tissue laminate. Recent years have seen the development of hot-melt laminating adhesives that retain the embossing and provide a 'softer feel'. Higher production speeds and a better fixation of the layers (compared to mechanical fixing, particularly on high-speed machinery) are further benefits of hot-melt laminating adhesives. These solid systems are applied by spraying but, owing to their rheological properties (Figure 8.38) and modern application technology, very low application weights of around 2–10 mg per running meter are possible with production speeds of 600–800 m min^{-1} and drum widths of 2.7 to 3.5 m. Hot-melt laminating adhesives have been tested at production speeds of up to 1000 m min^{-1}.

Although, most paper-bonding applications have been dominated by hot-melt adhesives for many years, in the production of tissue paper hot melts have only recently begun to acquire popularity. This is due to the fact that neither an appropriate application technology nor water-soluble or easily redispersable polymers suited for the formulation of hot melts, had been available before. Today, the hot-melt adhesives used for most paper-bonding purposes are commonly based on thermoplastic, non-water-

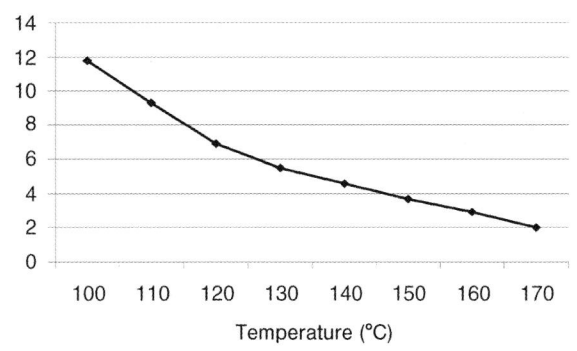

Figure 8.38 The viscosity of hot melts.

soluble polymers (e.g. EVA or block copolymers). To meet the specific needs of the tissue-producing industry, special raw materials had to be developed; at present, sufficiently hydrophilic polyesters and polyurethanes are available that are redispersable or even totally water-soluble. Tissue-laminating hot melts have been successfully developed from these polymers which combine low application quantities with special physical properties that provide for the lamination of soft tissue products, without staining them. These hot melts should be as clear in color as possible (preferably water-white) so as to not to impair the appearance and color of the tissue paper. They must also be stable during the melting process. Water-soluble or redispersable polymers are generally very polar and therefore susceptible to oxidation, which may cause staining of the product. Today's commercial adhesives remain water-white for hours at their processing temperature, and do not stain the tissue papers.

8.5.7.3 Pick-Up Adhesives

Pick-up adhesives, which sometimes are referred to as 'transfer adhesives', start the laminated tissue paper winding onto the paper core. To provide for reliable transfer at high line speeds, they are designed to have a high wet bonding capacity and good processability. Furthermore, they stain neither the core nor the tissue, and are designed to release easily allowing the last sheet on the core to be removed without fiber tear.

Depending on the machine construction and the applicator used, different pick-up adhesives are employed. Most frequently, these are aqueous systems based upon natural or synthetic polymers such as starch derivatives or high-molecular-weight polyethylene glycol (PEG) that can be processed at room temperature. The adhesives are applied to the core either by means of disks or nozzles. Hot melts, such as those based on synthetic resins usually applied by means of disks, are only rarely used as pick-up adhesives for tissue papers.

8.5.7.4 Tail Tie Adhesives

Tail tie adhesives bond tissue paper to tissue paper in order to prevent unwinding during transportation to the packager. They are designed to rapidly wet the paper, to penetrate many tissue plies, and to hold the last sheet securely in place on the roll through the cutting process.

Tail tie adhesives also enable the consumer to start the tissue or towel roll easily and without fiber tear, and are designed not to stain the tissue paper. Application methods can be adjusted to the machine settings using nozzles, drums, wires or rods, and the viscosity of an adhesive must be adjusted accordingly. Tail tie adhesives are aqueous systems based on cellulose derivatives, synthetic water-soluble polymers, or mixtures of synthetic polymers with cellulose derivatives. The choice of new raw materials depends upon the application system and sometimes the inherent strength of the tissue paper. The viscosity and initial adhesion of tail tie adhesives must be designed to suit, for example, the way in which the feeding of the first sheet is adapted to the core in the machinery.

8.5.8
Graphic Products

Products of the print finishing and bookbinding industries have become an integral part of our daily life, and today life without them is simply inconceivable. Books, catalogs, magazines and other graphic products such as advertising mail fulfill our need for information. This situation becomes even more obvious from the fact that, despite the presence of computers and electronic media, worldwide paper consumption is on the increase. Whereas, in 1950 around 50 million tons of paper was produced, by 2005 this had risen to almost 370 million tons. By the year 2010, worldwide consumption is expected to be almost 400 million tons. It is predicted that, by the year 2015 the amount of print products will most likely increase by 60%, with the very large majority of these being manufactured with the aid of adhesives. With such a wide variety of graphic paper products available, a huge number of optimized adhesives, each adapted to a desired use, have been developed during the past decades in order to meet the demands of these different applications [37].

Within the print-finishing industry, all of the adhesive systems created for, and suited to, the bonding of paper are currently employed. Whenever, during the past few years the basic conditions have been changed due to higher machine performances and new paper qualities (e.g. with a high share of coatings or recycled fibers), improved adhesive systems have been developed that can securely handle these difficult tasks, in spite of the increasingly higher production speeds.

8.5.8.1 Unsewn Binding
In terms of quantity, the most important application in the graphic industry is that of unsewn binding ('perfect binding') by means of dispersion adhesives, thermoplastic hot melts, or reactive polyurethane hot melts, depending on the requirements [38]. Previously, animal glue protein was used in this role, but is no longer important in unsewn binding. The decision as to which of these systems can be applied depends, for example, on the materials to be bonded, on the existing machinery, and on the adhesive application systems. The wear and tear to be expected later in the life of the bonded material is another issue worthy of consideration. The most important

Requirement	Glutin adhesives	Dispesion adhesives	EVA hot melts	Reactive polyurethane hot melts
Adhesion	++	+++	++	++++(+)
Low temperature	(+)	++	(+)	++++(+)
High temperature	+	+++	+	++++(+)
Oil resistant	+++	+++	(+)	++++
Durability	(+)	++++	++	++++
Curing velocity	+++	+	++++	+++(+)
Cuttability	++	+	+++	++
Deformability	-	+++	+++	+++(+)
Application properties	+++	+++	+++	+++

Figure 8.39 A comparison of adhesives for unsewn binding.

properties of common adhesive systems used for unsewn binding today are compared in Figure 8.39.

8.5.8.2 Unsewn Binding with Dispersion Adhesives

The dispersion adhesives used for unsewn binding are commonly based on plasticized polyvinyl homopolymers. They are used, for example, when excellent lay-flat behavior is required, or when the risk of printing inks influencing the bond is very high. They are usually applied via rollers. Dispersion adhesives set by absorption or evaporation of water; however, as this process requires a certain time period the capacity of machines working with dispersion adhesives is limited. Drying by means of high-frequency (HF) energy can speed up the process, but this is expensive [39].

8.5.8.3 Unsewn Binding with Hot Melts

Hot melts have gained ground because they provide for high production speeds. EVA copolymer-based hot melts are commonly used for the unsewn binding of magazines and catalogs, although other thermoplastic polymers are also used on occasion. In order to adjust the hot melts to specific properties they are mixed with resins and waxes. As with waterborne adhesives, hot melts are generally applied via roller application.

In recent years, low-temperature hot melts processed at temperatures ranging between 120 °C and 140 °C have become popular in the graphics industry. As with hot-melt packaging adhesives, their main advantages are low thermal stress, improved viscosity and color, longer service life and less equipment maintenance and energy consumption. Because the difference between processing temperature and room temperature is low, setting is commonly very rapid. This property is a major advantage for three-side trimming in the case of short cooling lines and high processing speeds, because it helps to reduce wrinkling, and the knifes will become coated with adhesive. Another advantage of a low application temperature is that local drying out and curling of sensitive paper, or paper that is inappropriately grained, is reduced and the paper more quickly recovers its equilibrium moisture content.

However, when the application temperature is close to the room temperature the adhesive molecules may have less time to wet the paper surface and this may impair adhesion. For papers that are difficult to bond, it is advisable to perform a test prior to production to ensure correct adhesion. Another parameter that must be tested is that of thermal endurance.

8.5.8.4 Unsewn Binding with Reactive Polyurethane Hot Melts

In the 1980s, the development of reactive polyurethane hot-melt adhesives for unsewn binding was an important innovation for the graphics industry and enabled the print finishers to use unsewn binding techniques for products that hitherto had been 'impossible to bond' [40]. Reactive polyurethane hot melts are the adhesive of choice whenever using papers that are difficult to bond, whenever products are exposed to high temperatures (e.g. a road atlas on the hat rack of a car), and whenever high page pull and page flex values are required due to special stress. Polyurethane hot melts are tailor-made polyurethane prepolymers with reactive end groups. Following application, these end groups are subject to a chemical reaction under the effect of air humidity and/or paper moisture, resulting in a large increase in molecular weight. The high adhesive capacity of the prepolymers provides for excellent adhesion to the paper edge. Following chemical reaction, polyurethane films are highly cohesive and provide outstanding durability, making them immune to the migration of mineral oils from printing inks, and giving them a good resistance to aging. At present, the use of reactive polyurethane hot melts is common practice in bookbindery, providing for high machine speeds of more than 10 000 books per hour. It is essential to adjust the machinery to the requirements of reactive polyurethane hot melts to make bookbinding with PU resins successful. For example, coated glue reservoirs are needed so that the sticky adhesives can be easily removed at cleaning. Another key factor is an optimized temperature control, because reactive adhesive systems are clearly more demanding with regards to thermal stress. Recent years have seen the development of special nozzle systems for the application of reactive polyurethane hot melts in unsewn binding.

However, it is not only the applicators that must be modified, but also the melting equipment. As polyurethane adhesives crosslink under the effect of moisture, it is essential to keep them away from air humidity during manufacture, transport and, if possible, also during further processing. One possibility is to deliver them in drums emptied with so-called 'drum melters' (Figure 8.40).

Drum melters provide for a durable sealing against air humidity. 'Melting on demand' makes it possible to limit the thermal stress to which the adhesive is exposed, with only the amount of adhesive actually needed at a certain time being molten. Although the reaction times of the first-generation reactive PU resins were rather long, at present these highly accelerated systems enable unsewn-bound products to be ready for use within a few hours (6–16 h) (Figure 8.41).

These systems stand out for their low processing viscosity, good initial adhesion and, in particular, for their excellent viscosity stability (which is essential for tank melters) during processing.

Figure 8.40 Drum melter.

However, as reactive polyurethane adhesives take advantage of the reactivity of isocyanate groups, free (monomer) isocyanate exposure and emissions must be controlled. Isocyanates are classified as sensitizing to the skin and respiratory tract, and most PU resin adhesives must be categorized as hazardous preparations, indicated by the hazard symbol Xn ('harmful'). As many measurements by professional associations have demonstrated, emissions occur not only during the processing of the adhesives but also when changing or cleaning the drums. By taking precautions, however, conventional products can be processed safely. Nonetheless, alternative options have been developed, brought forth by the ever-increasing safety

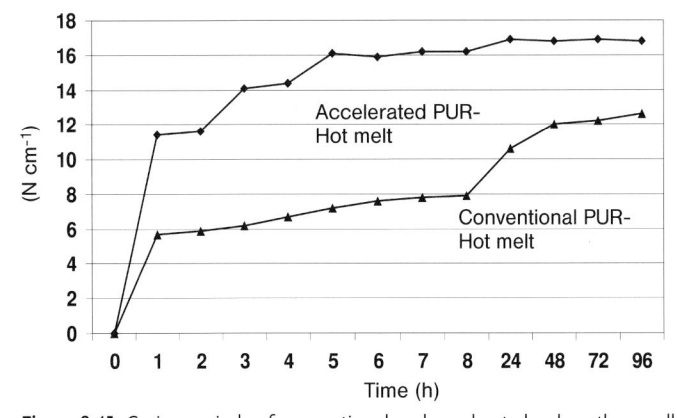

Figure 8.41 Curing periods of conventional and accelerated polyurethane adhesives.

awareness of the bookbinders. New adhesives that contain substantially less (<0.1%) monomer isocyanate than do conventional adhesives have been developed, without loss of any of the positive characteristics of PUs such as low viscosity and low increase in viscosity, high initial strength, and rapid crosslinking mechanism. In these products, the isocyanate concentration remains below the limit value from which the products must be categorized as hazardous preparations according to the regulations stipulated by legal authorities in the EU countries. With good processing practice, these products emit up to 90% less isocyanate vapors than do conventional reactive polyurethane hot melts.

Reactive polyurethane hot melts suitable for the inline manufacturing of rounded, unsewn-bound products have also been developed in recent years. Until then, problems had occurred with the relatively low-molecular-weight prepolymers because, immediately following application, their cohesive strength was not high enough to withstand the rounding process. This demand now can be met with so-called 'dual cure systems'; these are reactive polyurethane hot melts that cure not only under the effect of moisture, but also under irradiation with UV light. Within seconds after application, curing under UV light induces sufficient strength to start the rounding process and to yield high-quality results. The adhesive film is further stabilized by the subsequent reaction of the isocyanate groups with moisture.

8.5.8.5 Multilayer Unsewn Binding

For process improvement in unsewn binding, two or rarely three different adhesives can be applied one after the other (multilayer binding) [41]. Here, one popular technique is the so-called 'two-shot system', which uses two adhesives. The application of two different dispersions for primer and topcoat has a long tradition; they are frequently derived from the same formulation, their viscosities being adapted to the demands of a primer and a topcoat, respectively. These systems can be processed with or without intermediate drying. When using two-shot systems with two dispersion adhesives, the first dispersion optimizes the adhesion to the paper edge, while the second dispersion ensures a safe pick-up of the covers, especially when these are rather heavy. Two-shot systems also optimize the cold flexibility of the product. Two-shot, hot-melt systems have a similar functionality that is beneficial in the manufacture of catalogs, where increased machine capacities and paper qualities with a high share of secondary fibers have transformed the production process. They also provide better page pull and page flex values at consistent production speeds.

Two-shot systems that combine dispersions with hot melts benefit from the advantages of both. A relatively low-viscosity dispersion adhesive is usually used as primer, providing excellent bond strength to the paper edge. The dispersion can be applied in a very thin layer as this adhesive is only responsible for bonding the paper edges. After drying, a hot melt is applied as a topcoat that has been designed to adhere particularly well to the dispersion film. The hot melt provides stability to the back and also for the pick-up of the cover. New combinations of raw materials have better matched up dispersions and hot melts, meeting the requirements perfectly. The low-viscosity primer dispersions show excellent wetting properties, and can be applied either via standard or foam rollers. The hot melts provide very good adhesion to the dispersion film; indeed, when their film properties act together they provide good lay-flat behavior and very good page flex values. A careful selection of two-shot unsewn binding systems leads to an increased product quality compared to conventional unsewn binding with hot melts, and to increased production speeds compared to perfect binding with dispersions. However, since two different adhesives are employed, the binding process requires more expertise.

Another option is the combination of dispersions, applied as primers, with reactive polyurethane hot melts, applied as topcoats. These systems are processed with intermediate drying via rollers, as in conventional processing, and via nozzles in the case of PU resin hot melts, applying the adhesive, for example, to the cover.

The low processing viscosities and rheological properties of both products allow for a minimum layer thickness, leading to an improved lay-flat behavior and, last but not least, to a more economic production. The economic effect is fostered by the fact that these systems set up high strength within a very short period of time, such that the products are ready for use in a few hours.

8.5.8.6 Side Gluing

In unsewn binding, adhesives are needed not only for the bonding of the back, but also for gluing the side of the cover to fix it on the first page (side glues). To this end, either dispersion adhesives or hot melts can be employed. Dispersion adhesives used for side gluing are commonly based on plasticized polyvinyl acetate homopolymers or EVA copolymers. Hot melts are especially suitable for heavy covers and are based on EVA copolymers or synthetic thermoplastic rubbers mixed with resins and oils. The adhesives are applied via rollers or nozzles to the book block (Figure 8.42).

Figure 8.42 Side glue application by wheel.

These products have a very high hot 'tack' – that is, a high initial adhesion which ensures a safe pick-up and bonding of heavy covers. As graphical products are increasingly sophistically designed, conventional adhesive systems are causing more and more problems in side gluing and pasting of endpapers due to weak bonding, oil migration, and so on. Reactive polyurethane hot melts are better suited here and provide safer results. Contacting or noncontacting application to the book block or cover is possible via slot die or round nozzle.

8.5.8.7 Backgluing of Thread-Sewn Book Blocks

Backgluing is used to stabilize the book blocks, and when backgluing thread-sewn book blocks similar adhesive systems are used as for unsewn binding. Softcover books are directly glued into the cover by means of the backgluing adhesive. In the manufacture of book blocks, the backs are then provided with guard or mull strips.

8.5.8.8 Book Manufacture

A wide variety of adhesives are used in the manufacture of books. Book blocks are provided with endpapers and super cloth (guard or mull strips) bonded to the back of the book block with a 10- to 20-mm flap on both sides, extending over and covering the endpapers. These help to stiffen the joints of the endpapers, thus increasing the form stability of unsewn-bound, thread-sealed and thread-sewn book blocks. After gluing, for production of the book, the blocks are trimmed on three sides. Before casing in the book blocks, they can be rounded and provided with an edge coloring and additional parts (ribbon, head and tail band, inlay and hollow).

Ribbons are inserted to make handling of the books easier; they are bonded to the head of the book blocks or, in the case of brochures, on the last page near the back. Ribbon-inserting machines can be integrated in a book line, inserting the ribbon, for example, in the headbanding machine. When used as stand-alone units, a hot-melt gluing system applies a positionable adhesive line onto the back by means of rollers or nozzles, the latter being very well suited for thin book blocks. The headbands – or rather the head and tail bands – are affixed at the ends of the back of the finished book block. Although, nowadays, their purpose is merely decorative, their role is to hide the hollow spine with its guard, lining material and spine inlay of the case. To some extent, a headband helps to stiffen the layers at the upper and lower ends of the back. The purpose of the lining material is to stiffen the back of the book block and to provide a stable and durable form to the back, either rounded or nonrounded. The lining material (i.e. the inlay) is shorter than the height of the book block by few millimeters, and its width goes from one groove to the other (unlike the super cloth, that has two flaps). Lining is a technological prerequisite if the headbands are to be affixed automatically. Lining and affixing of the headbands is performed in a single step. The headband is bonded to the back of the book block in such a way that it spans from the first to the last page, with its roll sitting evenly on the edge. To this end, nozzle-gluing systems apply a small line of dispersion adhesive to the head and end of the previously cut inlay strip that must be absolutely flat in order to make it possible to place it accurately on the back. It must not curl under the influence of the adhesive, and must not be stretchable, because otherwise it will not help to stabilize the back.

Furthermore, the lining material must be impervious to the adhesive so as to ensure that it does not adhere to the spine liner of the case during the casing-in process. A mull cloth can be affixed to the back prior to the affixing of the headbands and the inlay strip, although from a technological viewpoint this is only necessary in exceptional cases.

In the industrial bookbinding sector, it is no longer common to use *hollows*. These are tubes made from thin but very firm paper (kraft sack paper with a density of 50 g m^{-2}), and are mainly used for large-size, heavy products and products with high quality demands (e.g. encyclopedias). They are industrially manufactured by special machines that glue them onto the spine inlay of the case, with the grain orientation in parallel with the book spine. To this end, a dispersion adhesive is applied to the spine liner, and the overlapping side of the hollow is bonded to the inlay and then pressed onto it.

8.5.8.9 Making the Case

Whereas, protein-based adhesives have only minor importance in unsewn bookbinding, they still play an important role in the making of bookcases [42]. Animal glues (see Section 5.9) have proven to be particularly well suited here; these are derived from skin or animal bone collagen and rank among the most ancient adhesives used by mankind. Their sol–gel transition is characterized by a very rapid gelatinization – that is, a change from a liquid to a glassy state – induced by a decrease in temperature of only few degrees below the gelatinization point. This fast-setting feature and their excellent 'tack' properties are taken advantage of for the 'turning-in' (i.e. the folding-over) of the cover cloths around the edges of the two boards of book covers. To this end, the gelatins are heated in hot-water jacketed tubs and applied via rollers at a temperature of around 50–60 °C.

The tubs are filled via premelters provided with a water quench, a stirring unit and a water-circulating pump. When the glues cool to room temperature, they very rapidly build up a high 'tack' which enables them to safely overcome the spring-back forces typical of the turning-in process. High-speed case manufacturing machines produce up to 100 cases per minute. Animal glues are very popular in the traditional bookbinding sector, because they do not affect the paper, can be removed without leaving any residues in the case of a rebinding in book restoration, and can easily be removed from machinery and tools.

8.5.8.10 Casing-In

In soft-back production, books are directly cased-in in the unsewn binding machine, whereas in hard-back production a special machine performs this task. First, the book blocks are compiled via unsewn binding or thread sewing. The endsheets are then bonded to the book block, mainly by means of polyvinyl acetate-based dispersion adhesives. The outer surfaces of the book block are then coated with dispersion adhesives (usually plasticized polyvinyl acetate-based dispersions) and then cased-in by introducing the book block into the cover. It is then pressed in a finishing press that establishes a close contact between the endsheets and the cover. For book covers provided with a hollow, the casing-in machine has a cushion gluing system which is designed to bond the hollow to the back of the book block. Dispersion or hot-melt adhesives are applied to the hollow via a nozzle gluing system prior to joining it with the inlay of the book block.

8.5.8.11 Fold Gluing

Fold gluing is an additional processing step that can be performed at the same time as the folding process itself. A glue line is applied to the fold line via nozzles on the sheet in the sheet folder or on the paper web in the rotary offset printing press in such a way that, after the folding process, the leaves of the sheet are joined with each other in the gutter. Folded sheets, for example flysheets or inserts, are manufactured in this way. Fold-glued sheets can be bound to blocks following compiling, without the need for unsewn spine binding, and this provides for a very high stability. Fold gluing is also used for the manufacture of brochure or flysheet pockets, for envelopes or lottery tickets, or during paper lamination in the production of postcards. The benefits of fold-glued sheets are manifold. First, they are less bulky in the fold than saddle-back-stitched brochures, and trimming, stacking and packaging is easier. Second, their manufacture is more economical. For fold gluing, either plasticized polyvinyl acetate homopolymer or EVA copolymer-based dispersion adhesives are used. A proximity sensor and a control unit activate the fold gluing aggregate that applies a glue line (the length and width of which is accurately positioned) to the sheet. The first-generation fold gluers were contact systems, but since the early 1990s the adhesive has been applied via electric noncontact nozzles that provide for an airtight closure during machine stoppages. As with an inkjet printer, glue dots are applied at a rate of 600 or 1000 per second. Multirow applicators are available for special applications, such as the closure of pockets.

8.5.9
Envelopes and Advertising Mail

Envelopes are manufactured from a wide variety of materials, and in many types and sizes. Not surprisingly, in the manufacture of envelopes production speeds have been ever-increasing, with today's state of the art reaching 1500 units per minute (Figure 8.43).

Although plastic films are used, paper is still the most important raw material for the production of envelopes. The side joints of envelopes made from paper are commonly bonded with plasticized polyvinyl acetate homopolymer-based adhesives, whereas both plasticized polyvinyl acetate homopolymer or EVA copolymer-based adhesives are used for the bonding of windows made from glassine or transparent polystyrene foils. A special feature of the typical envelope is another adhesive application which is used only by the consumer for the purpose of closing the envelope – the water-reactivated gummed flap. These widely known products are manufactured by applying dextrin-based, dextrin/synthetic resin-based or synthetic resin-based adhesives and force-drying them before packing. A number of requirements must be fulfilled: the consumer wants the gum to be quickly reactivated with water, while the producer wants it to dry quickly and to have anti-blocking properties to prevent premature bonding. The humidity in the air must not reactivate the gum in

Figure 8.43 The production of window envelopes.

such a way that bonding takes place. Envelopes can also be closed by means of latex or PSA coatings. Latex is rubber milk which originates from the *Hevea* tree; latex closures require a coating to be placed on both sides of the envelope, with bonding taking place when the two sides are pressed against; at this point a diffusion bonding process occurs between both sides. Currently, PSA-coated envelopes are capturing the market, most likely because the PSA film has simply to be pressed onto the paper surface to close the envelope. PSA coatings must be covered with siliconized papers or plastic foil to prevent premature bonding, and are commonly made from acrylate dispersions. Occasionally, acrylate hot melts or thermoplastic rubbers, mixed with resins or oils according to the desired application, are used.

The production of *advertising mail* is one of the most thriving and prospering sectors of the paper industry. Advertising mail is a short-lived, printed product that frequently contains PSA labels or other adhesive applications, for example for laminating purposes (to reach the thickness required by postcards), or for the resealing (gumming or PSA) of envelopes contained in the advertising mail.

8.5.10
Reuse and Recycling of Paper and Packages

Clearly, natural resources will not last forever, and paying attention to saving resources is a must. The reuse of costly products and consistent recycling on the highest technical level are effective methods to avert the imminent danger of a shortage of raw material. This is especially important in the case of products that have a lifespan of only few days or weeks (as do many packages and graphic products) and should be taken into consideration early in the stage of product conception.

8.5.10.1 Reuse
The longer a product performs its task, the lower the demands for raw material for the manufacture of a replacement. Hence, the decision must be made as to whether it is worth increasing the lifespan of the product, or to reuse it. Returnable systems meet these demands and, as far as the packaging sector is concerned, returnable bottles are

but one of many examples. The adhesives used in the labeling of returnable glass bottles must be formulated in such a way that they provide both trouble-free filling of the bottles and excellent adhesion during bottle use. They must also be easily removable in the bottle-cleaning machine, without leaving any residues or disturbing the process. Since the introduction of returnable PET soft drink bottles and milk containers made from PC, reuse is no longer limited to glass bottles. Although plastic bottles have the advantages that they are lightweight (transport is easier) and unbreakable (no risk of injuries), they place greater demands on the labeling adhesives. However, such demands can now be met by the new synthetic systems that ensure good bonds and, occasionally, a milder cleaning of plastic bottles [1.5% (v/v) NaOH at about 60 °C].

8.5.10.2 Recycling
Even with the best product design and most careful treatment, it is unavoidable that one day a product will no longer fulfill its purpose and must be replaced by a new version. Within the concept of a closed-loop economy, preference is given to material recycling in order to reuse materials and save resources. *Unmixed recycling* offers the best advantage for reusing materials, though the provisions to be taken to obtain unmixed recycled material depend on the recycling method employed for the primary material. In contrast, the influence of adhesives on material recycling can be disregarded for materials that are recycled at high temperatures (e.g. glass or metal). Indeed, recent intense discussions have been conducted with regards to the recycling of paper or plastic materials.

8.5.10.3 Recycling of Paper Products
In Europe, in 2006, the large-scale recycling of paper, board and paperboard containers accounted for more than 58 millions tons of recycled paper products (Figure 8.44) [43].

Although adhesives enter the recycling process along with paper products, they must not disturb the process. The impact of adhesives on recycling operations

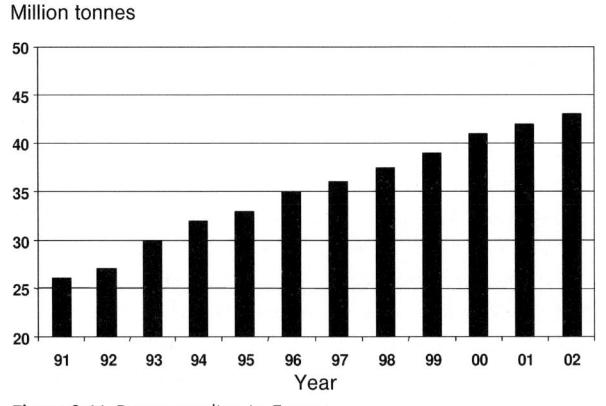

Figure 8.44 Paper recycling in Europe.

```
                          Adhesive films
                    ┌───────────┴───────────┐
            Water-soluble              Not water-soluble
            re-emulgating              not re-emulgating
            ┌───────┴───────┐          ┌───────┴───────┐
      Thermoplastic   Not thermoplastic  Thermoplastic   Not thermoplastic
            │               │                │                │
         for ex.         for ex.          for ex.          for ex.
  Water-soluble polymers Starch adhesives Dispersion adhesives PUR-Hot-melt adhesives
   Dispersion adhesives Dextrin adhesives Hot melt adhesives  Reactive adhesives
   Water-soluble hot-melts Cellulose adhesives
```

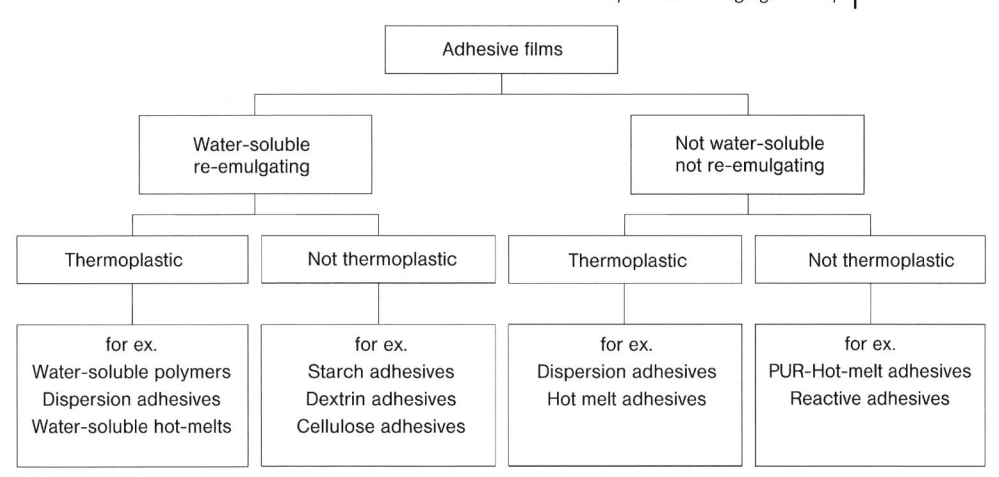

Figure 8.45 Types of adhesive films.

depends on the physico-chemical properties of the adhesive films themselves [44], because recovered paper contains adhesive in the form of a cured film (Figure 8.45). Usually, it is best to sort the adhesive films out as early in the process as possible.

An important property of adhesives with regard to recycling is their water solubility or *redispersability*, because papers are recycled in aqueous media. With few exceptions, hot melts that are basically derived from synthetic thermoplastics are totally insoluble in water. Waterborne adhesives may be either colloidal solutions (e.g. based on animal protein, starch, or polyvinyl alcohol) or dispersions. In the paper sector, dispersion adhesives are the most important group of adhesives used (on a quantity basis). The basic polymers of these synthetic thermoplastics are also insoluble in water; hence, in order to obtain a liquid phase and to enable them to wet a substrate, they must be 'hydrophilized' by means of emulsifiers or protective colloids, allowing them to be dispersed in water. During the setting of an adhesive the primary particles 'flow' into each other to form a solid film; the water-soluble protective sheaths are then incorporated into the film in such a way that dispersion-adhesive films are always somewhat hydrophilic. The composition of dispersion-adhesive films has an influence on solubility and redispersability that strongly depends on the pH value and the temperature of the water. This also holds true for colloidal adhesive solutions [45].

With regards to recycling, it is useful to distinguish between thermoplastic and nonthermoplastic adhesives. In the recycling of paper, the adhesive film is exposed to rather high temperatures (e.g. 40–50 °C in the pulper, 80–90 °C in the disperser, and 80–120 °C in the dryer). The majority of physically setting adhesives are thermoplastic. *Thermoplasticity* is defined as the ability of a material to melt to a liquid when the temperature rises. Whereas, low-molecular-weight substances melt at a defined temperature point, high-molecular-weight materials such as adhesives melt within a temperature range; moreover, as adhesives usually are a mixture of several different polymers this range may be quite large. Whether an adhesive is thermoplastic, or not, depends on its molecular structure. Usually, high-molecular-weight and cross-linked systems are not meltable and are hence nonthermoplastic (e.g. celluloses, crosslinked polyurethanes, epoxy resins). Adhesives may also be nonthermoplastic if they begin to decompose at a temperature which is required for melting (e.g. animal protein, starch). The 'softening' behavior of a thermoplastic substance is due to the molecular mobility that increases with rising temperature, and which also improves its ability to build 'adhesion bridges'. Soft, thermoplastic substances therefore feel sticky to the touch, and their molecular structure defines the temperature at which this effect occurs. In the case of high-molecular-weight thermoplastics (such as many technical plastics) this effect may occur at temperatures of more than 200 °C, whereas other systems (e.g. polyacrylates) – and particularly mixtures – begin to show this effect below a temperature of 0 °C. The T_g-value is a useful parameter for monitoring the 'softness' of a thermoplastic (the temperature range in which the polymer softens) as it describes the temperature at which a thermoplastic 'melts' internally, resulting in a high molecular mobility. High-molecular-weight thermoplastics are still relatively cohesive above their T_g, and still form mechanically stable films. Adhesive films that are sufficiently soft at room temperature to build up proper adhesion are known as 'PSAs' because they wet the surfaces to be bonded when pressure is applied. The 'softness' of the films ensures 'flowing' under pressure, resulting in proper adhesion. With increasing temperature, there will be loss in cohesion until eventually, even high-molecular-weight systems will melt. The above-described phenomena apply to both hot-melt adhesives applied as thermoplastics, and to aqueous systems or solvent-based adhesives following setting. Paper recycling is basically a mechanical process, and therefore the thickness of the adhesive film greatly influences how easily it can be sorted out. Whereas sorting-out is no problem for thick films, thin films are often destroyed and their removal becomes more difficult.

However, thick, compact films generally do not comply with the paper manufacturer's option. If only for cost reasons, the amount of adhesive used should be as low as possible, and for this reason computerized state-of-the-art nozzle systems that provide for high-precision, extremely economical and clean application of adhesives are widely used (Figure 8.46).

This not only saves adhesive and keeps the production process trouble-free, but these systems also have ecological advantages. Closed systems require substantially

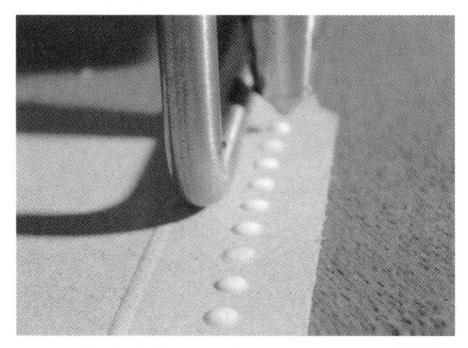

Figure 8.46 The high-precision application of dots of a dispersion adhesive.

less cleaning efforts and produce few waste products or wastewater. Yet, even if these adhesives are mechanically stable and water-insoluble, the films generated by the droplets can be too small to be sorted out in paper recycling facilities. Provided that these aspects are taken into consideration, adhesives can be chosen for the majority of applications in such a way that trouble-free paper recycling becomes possible.

8.5.10.4 Recycling of Plastic Packaging

Today, packages made from plastics are also recycled on a major scale. As with paper recycling, the adhesives used for plastic packaging are required not to disturb the recycling process. As a wide variety of different recycling methods is available for plastics, the adhesives must meet manifold demands. The most established method starts with washing of the plastic packages (the recycling of PET bottles is an example of this technology). The bottles are first shredded, and then cleaned and separated from any secondary products in a water bath. Hence, the adhesives used must be water-soluble or at least dispersible. Whereas, waterborne systems such as colloidal solutions or dispersions do meet this requirement, the development of water-soluble hot melts represented a major challenge for the adhesives industry. Nonetheless, these hot melts are now available commercially and are mainly used for the purpose of wrap-around labeling using plastic labels, providing for an easy recycling of these plastic packages.

8.5.10.5 Compostable Products

Composting represents another option allowing the material loop of products made from organic materials to be closed. *Biological degradation* of the material allows the products to be reconveyed to a form that is useful to Nature. Especially in the packaging sector, this is an attractive option, in particular when packages are contaminated with food residues. With the development of biodegradable materials, for the first time in industrial development a group of materials was conceived that was specifically targeted towards reuse and recyclability. Although natural raw materials are more advantageous in terms of biological degradation, special synthetic polymers can also meet these requirements. Yet, natural raw materials have the advantage of renewability, which saves resources.

To ensure reliable composting, it is not only the basic material of these products which must be biodegradable, but also the processing aids such as adhesives. Adhesives based on plant or animal raw materials easily meet this requirement. With regards to the market in Western Europe, more than 6% of the adhesives used are derived from natural polymers. Raw materials derived from plants comprise polymers such as starch, celluloses, natural latex, proteins and low-molecular-weight substances such as resins (e.g. rosin) and diols. Raw materials derived from animals particularly include proteins such as casein and hide glue.

Waterborne adhesives and hot melts exist that also meet the requirements of the packaging industry with regards to compostability. These systems can be based on natural or fossil biodegradable raw materials, such as natural polyester. Aliphatic polyesters are preferred to their aromatic counterparts because they have a higher degree of biodegradability. As shown in Figure 8.47, the biological degradation of these hot melts is comparable to that of cellulose.

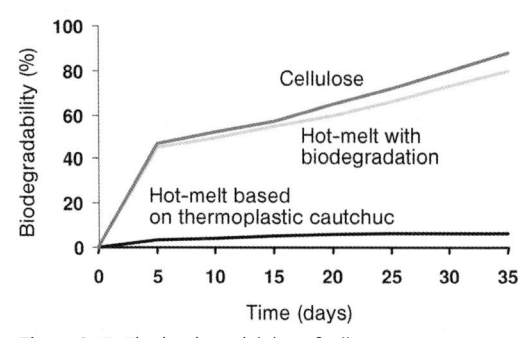

Figure 8.47 The biodegradability of adhesives.

As many biodegradable materials are available as films, it is hardly surprising that biodegradable multilayer films have also been developed that consist of at least two different films laminated by an adhesive. This makes it possible to combine the performances of each single film and also to save weight; moreover, if both films are biodegradable this should also apply to the adhesive. Considerable progress has been made in this field, making available biodegradable adhesives for different types of film.

8.5.11
Regulations Stipulated by Law

Among the multitude of statutory regulations applying to industrial processing, food-law regulations, such as the European Framework Regulation 1935/2004, are of particular importance for packaging adhesives because many packages are used for foodstuffs. A wide array of raw materials is available for the formulation of adhesives that comply with food law requirements, such as the Recommendations of the German Federal Institute for Risk Assessment (Bundesinstitut für Risikobewertung BfR). A European harmonization of the Food and Consumer Goods Laws will bring forth new rules, because European Laws particularly evaluate the migration of substances from consumer goods (e.g. packages) into food, and set out global and specific migration limits. As regulations related to food law are most often related to quantities, it is the responsibility of the manufacturer of the packaging or the bottler of the food to assess whether the statutory regulations have been observed.

8.5.12
Perspectives

Over the past few years, new packaging concepts and materials have brought forth more sophisticated and refined packaging options. A steady increase in the world population and growth in industrial production will have a further growing market impact on the packaging sector that will become even more complex owing to active and intelligent packaging options:

- *Active packaging* is packaging which is specifically designed to change the chemical, biological or physical parameters of the packed products, for instance by using new materials.
- *Intelligent packaging* is packaging designed to convey appropriate information to the user about, for example, the history of the product and its quality during transport and storage by means of external or internal tags or other indicators.

Intelligent packaging with extended EAN (European Article Numbering) technology or RFID (Radio Frequency Identification System) technology is also under discussion. One technique being developed is that of 'plastic chips'; these are made from conductive or semiconductive organic polymers and can be printed onto paper or plastic film and then affixed to packages as labels. An increase in production speed and ecological aspects should also provide a major impetus for further developments. The avoidance of waste by reuse and recycling will lead to new systems in the packaging sector. At this stage, many packaging adhesives have already been tailored to later reuse and recycling of the packaging material. Yet, adhesives will continue to play an important role to help packages to perform their tasks during transport, storage and presentation on the shelf.

In the graphics sector, personalized products are set to grow – 'print on demand' and subsequent processing ('bond on demand') will bring forth special adhesive systems that provide for an economic manufacture of the smallest print runs. Advertising mail will become ever more sophisticated so as to allow delineation from advertising in an electronic medium, and adhesives will continue to play an important role also in this sector.

8.6
Small-Scale Industries and Handicraft Trades

Adhesive bonding is gaining ground in modern production processes in small-scale industries, handicraft trades and in the do-it-yourself sector. A wide range of products has become available for the adhesive bonding of most different materials under most different conditions and impacts.

A wide spectrum of pressure-sensitive rubber-, acrylate- and silicone-based adhesives for the purposes of protecting, splicing, masking, damping and labeling enable the customer to identify innovative solutions, to increase flexibility, and to make production more efficient.

8.6.1
Joining

Adhesives and double-sided adhesive tapes provide for a firm connection between dissimilar materials. With one-part and two-part adhesives, structural bond strengths can easily be achieved at room temperature, even on low-energy surfaces such as PE and PP. The adhesives perform tasks that previously were reserved for mechanical joining techniques, such as joining by screws, riveting or spot welding.

In time, mechanical fasteners may be replaced not only by adhesives but also by high-performance adhesive tapes that fulfill the tasks of, for example, joining by screws, riveting, clipping and spot-welding. These provide for a secure, reliable and performing joint because they allow most different materials to be joined, have excellent tack, allow immediate processing, avoid the generation of stresses when joining dissimilar materials, have excellent resistance to elevated temperature, and are able to meet a customer's specific needs when delivered as die-cuts. *Spray adhesives* or *reclosable fasteners* can be applied where quick joining or temporary joining is essential. Screen-printed adhesives can be placed directly and exactly where they are needed. New systems are available that rapidly cure by UV-light at room temperature without harming the bonded surfaces.

8.6.1.1 Structural Adhesives

One-part and two-part structural adhesives are solvent-free and cured under the effect of heat (one-part adhesives) or at room temperature (two-part adhesives). The curing process of two-part adhesives can be accelerated by heat (e.g. a temperature of 65 °C for 120 min). Structural bond strength can be achieved with versatile high-performance epoxy resin, polyurethane and acrylate adhesives that have been designed for the bonding of metals (e.g. steel, aluminum, copper, brass) and plastics (e.g. polycarbonates, PVC, acryl-butadiene–styrene (ABS), polymethyl methacrylate), as well as glass, ceramics and wood. The manufacturer tailors the flexibility of the adhesives to the desired application, ranging from hard to rubber-like and flexible.

In the world of plastics, PE and PP are versatile materials. However, a major challenge is encountered when attempts are made to bond them to each other, or to other materials. In both industry and handicraft trades, PE and PP are used on a large scale (as commodity plastics), although until recently structural adhesive bonding was only possible following elaborate surface preparation.

Whilst these thermoplastics can indeed be welded to each other, due to their low-surface energies the structural bonding requires previous chemical or physical surface preparation, such as corona or plasma pretreatment. Thus, mechanical solutions were always the first choice in industry when a simple method was sought to join these materials.

Recently new rubber-like, two-part structural acrylic adhesives have made it possible to create an easy, quick – and, above all, high-strength – bonding between materials such as PE or PP, without the need for any surface preparation. This new generation of adhesives is also suitable for bondings that involve thermoplastic elastomers (TPE).

As soon as both parts of the adhesive are mixed, a chemical reaction starts on the wetted surface that provides for an adhesion-friendly surface condition with a higher energetic state. The newly created surfaces can also easily be bonded to other materials, as the pot-life is approximately 2.5–3 min and the joined materials can be processed further at only 30–60 min after adhesive application. These new

Figure 8.48 Application systems for two-part adhesives.

adhesive systems are solvent-free and cure at room temperature, with the final adhesive strength taking place within 8 to 24 h at room temperature, or within 30 min when heated to 70–75 °C. The structural bond strength of the joint is comparable to the inherent strength of the low-energy materials (PE and PP).

Structural adhesives are either manually or automatically applied by means of knife coating, extrusion or injection. Special equipment allows the two-part adhesives to be processed in an economical, rapid, clean and accurate manner (Figure 8.48), for example with hand-held applicators and air-spray guns with double-chamber cartridges and static mixer nozzles. The adhesive is applied as dots or beads, with its metering, mixing and application being performed in a single step.

8.6.1.2 Cyanoacrylate Adhesives

Cyanoacrylate adhesives are colorless, solvent-free, rapidly polymerizing and cold-setting one-part structural adhesives. The moisture contained in the air initiates curing, while light pressure contact generates a thin film that sets within seconds, making these adhesives particularly well suited for small and plane surfaces. Activators can be employed to accelerate the curing process or to enable bonding on critical surfaces.

Very good bond strengths are achieved with cyanoacrylate adhesives on a wide variety of materials, including metals, plastics (e.g. ABS, PVC, polyamide, PC) and elastomers [e.g. ethylene propylene diene rubber (EPDM), styrene butadiene rubber (SBR) and fluorinated rubber], as well as glass, ceramics, wood, leather and cork. These adhesives can be classified according to their suitability for a specific type of material (elastomers, plastics, metals) or according to their viscosity (ranging from very good to nonflowing), and have a good resistance to aging, temperature and chemicals. They are applied dropwise, either manually or by means of metering devices.

According to specific fields of application, cyanoacrylate adhesive products can be classified into the following groups:

- Bonding of profiles, round cords, and so on, made from solid or foamed rubber (elastomers that are difficult to bond such as EPDM, and fluoroelastomers such as Fluorel™ and Viton).

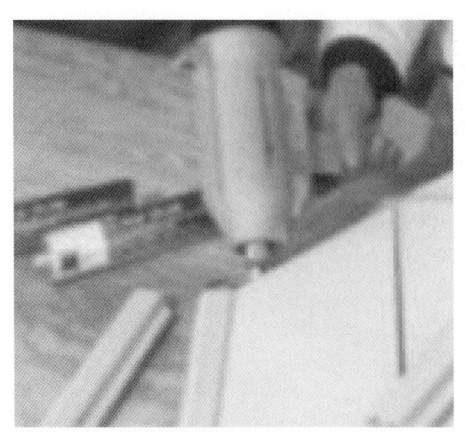

Figure 8.49 Application of postcuring polyurethane hot melts.

- Joining of plastic materials, elastomers as well as porous materials such as wood, cork and leather.
- Bonding of metals such as iron, steel, nonferrous heavy metals and light metals.

Activators for cyanoacrylate adhesives accelerate curing and make it possible to bond materials that are less adhesive-friendly and that give an acid reaction, such as wood and cork. The use of activators also prevents the exudation of plastic parts.

8.6.1.3 Postcuring Hot Melts

Postcuring polyurethane hot melts are solvent-free, one-part, moisture-setting adhesives with a solids content of 100%. They rapidly build up strength resulting in a high initial tack (as do conventional hot melts), and this is followed by a postcuring that results in a high structural strength. These adhesives offer good flexibility and resistance to impacts as well as a good resistance to plasticizers. They have different open times (1–10 min) that provide for rapid processing and rework and/or long adhesive beads. Different viscosities make it possible to create very thin or gap-filling bond-lines.

Polyurethane hot melts are employed to bond wood or plastic materials with themselves or with other materials such as rubber, leather, textiles, aluminum, glass or painted metals, resulting in strong and loadable bonds. Excess adhesive can be removed easily and cleanly during the wax phase (within about 20 min after application). Polyurethane hot melts are applied by means of special cartridge guns (Figure 8.49), automatic extrusion or spray applicators.

8.6.1.4 Hot-Melt Adhesive Systems

Hot-melt adhesive systems comprise versatile performance hot melts and applicators with auxiliary equipment for continuous use in industry (Figure 8.50). Hot-melt adhesives with 100% solids content have been developed from solvent-free thermoplastic hot-melt resins. In the molten state, they are capable of wetting a surface and of developing a strong bond by means of heat dissipation, without chemical reactions, to

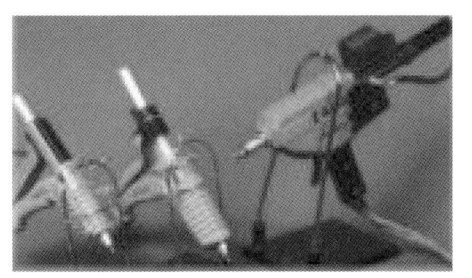

Figure 8.50 Application systems for hot melts.

many different materials such as wood, board, paper, cork, leather, rubber, felt, textiles, plastics and foamed material, ceramics, glass and metals. They are not only well suited to the joining of different materials with themselves or with each other, but also for the sealing and filling of cavities and gaps. No clamping devices are necessary, and the final strength is achieved after only a few seconds. Performing applicators are available for processing.

Low-melt adhesives are applied at a relatively low melting point (120–130 °C), and perform as standard systems. They are especially suitable for materials that are sensitive to heat, such as thermoplastic foils or Styropor®, avoiding staining, shrinkage and warping. Furthermore, the danger of the materials to be burned, and the development of smoke or odors, are reduced to a minimum.

8.6.1.5 Aerosol Adhesives

Aerosol adhesives are applied at the push of a button in an easy, rapid, portable and clean manner, without the need for auxiliary equipment such as brushes and knifes, and having to clean these afterwards. Aerosol cans serve as transport protection, storage container and applicator simultaneously.

Aerosol adhesives are versatile systems that are suited to the temporary or permanent bonding of paper, board, textiles, foils, felt, cork and leather with themselves or with wood, metal, glass, plastics, rubber, and so on. They are used in a wide field of applications, for example in the bonding of insulating material, Styropor®, foamed material, material for decoration, stencils and labels.

8.6.1.6 Dispersion Adhesives

The characteristics and fields of applications of polychloroprene and acrylic dispersion adhesives are similar to those of solvent-containing adhesives. However, the adhesive itself is dispersed rather than dissolved in water. In contrast to solvent-containing adhesives, dispersion adhesives cannot be reactivated.. When delivered, dispersion adhesives are nonflammable, which is an advantage for both storage and production. Owing to their high solids content, dispersion adhesives have a high productivity. The drying time can be shortened by means of heat supply (infrared, hot air).

Dispersion adhesives can be used with a wide variety of materials, including wood, paper, board, cork, leather, felt, textiles, rubber, plastics, foamed material, glass, ceramics and coated metals.

Most frequently, the adhesive is applied to both adherents, which are joined under pressure after drying (the color changes from white to transparent).

8.6.1.7 Solvent-Based Adhesives

Solvent-based adhesives have a good wetting behavior and can be applied easily in different ways. Good bond strength is obtained rapidly after evaporation of the solvent. According to the adhesive basis, there is a range of performances available. Solvent-based adhesives are suited to a wide array of different materials such as metals, plastics, rubber, leather, cork, felt, wood, board, textiles, glass, ceramics and concrete ('all-purpose adhesive').

Most frequently, the adhesive is applied to both adherents, which are joined under pressure after evaporation of the solvent. If at least one material is porous, the adhesive can be applied to one substrate only (wet bonding). The adherents can also be first coated/dried and then later bonded after reactivation by a solvent or by heat.

8.6.1.8 High-Performance Acrylic PSA Tapes

In many applications, acrylic-based high-performance PSA foam tapes have replaced mechanical fastening methods such as joining by screws, riveting, clipping and spot-welding. Over the past 20 years they have become widely accepted in numerous applications. Owing to their viscoelastic behavior, they absorb dynamic forces and vibrations to a much higher extent than do mechanical joints. They can also compensate for differences in the linear expansion of the joined materials up to three times their own thickness, making it possible to save both weight and production costs. A safe, reliable and performing joint is created owing to high tack, immediate processability, stress-free joining of dissimilar materials, high resistance to elevated temperature, and the availability of customized die-cuts (Table 8.3).

High-performance foamed adhesive tapes are a virtually inseparable unit consisting of an acrylic closed-cell adhesive core and two functional adhesive surfaces.

Table 8.3 Advantages of high-performance PSA tapes (foamed).

Quality	Cost	Flexibility
• Permanent, safe joint • Rapid, easy and clean assembly • No disturbing screw or rivet heads • Smooth, unblemished surfaces • Improved design • Reduced vibrations and noise • Excellent sealing properties • Resistance to UV radiation • Resistance to chemicals	• Fast, effective joining technique • Fewer rejects • Saves weight (thinner and lighter materials can be joined) • No investment needed (e.g. for machinery) • Easy processing: no subsequent treatment needed	• Easy joining of metals, plastic materials, glass, etc. • Strain-free bonding of dissimilar materials • Expanded design possibilities • Prepacking possible • Customized die-cuts on demand

Unlike the case of conventional foamed adhesive tapes, the continuous viscoelastic adhesive allows a permanent, stress-free joint to be created. These systems are based on acrylic copolymers containing C–C single bonds that have a very high resistance to energy in the form of heat or UV radiation, and excellent resistance to chemical attacks. This property makes them particularly well suited to the production of conveyor belts that are exposed to high temperatures, for example in shrink tunnels in the food industry. These adhesive tapes have permanent resistance to elevated temperatures; indeed, some of them can shortly resist temperatures of up to 260 °C.

With acrylic-based high-performance PSA foam tapes, a wide variety of materials with different surface properties can be joined, with the forces always being distributed evenly. Unlike mechanical joints, adhesive tapes can absorb shear forces owing to their viscoelastic behavior. When the temperature increases by 50 °C, the linear expansion of aluminum is 1.15 mm (relative to 1 m) and that of steel 0.85 mm; this results in a difference of 0.3 mm, which is easily compensated by the tape.

Adhesive tapes also have excellent adhesive capacity. On steel, for example, they have a peel strength of 120–450 N per 100 mm and a tensile shear strength of 42–69 N cm^{-2}.

Acrylic high-performance PSA film tapes are transparent; foam tapes are available in black, white, gray or transparent, with thicknesses ranging from 0.05 to 3 mm and widths from 6 to 1200 mm. Transparent tapes play an important role in new production techniques, notably in the glass industry. *Solar modules* are also produced with the help of acrylic-based high-performance PSA tapes; an example is the solar modules (which were installed to walk on) in front of the glass dome of the Reichstag Parliament Building in Berlin. The transparency was tested in an accelerated aging test by exposing them to high temperatures and intense UV radiation for 3000 h, during which time the transparency decreased by only 1% (from 88.2 to 87.3).

Innovative solutions are also made possible by the use of high-performance PSA foam tapes in other industry sectors. For example, they are used for roof structures of trailers, in ceilings with integrated cooling systems, or in façade construction (Figures 8.51 and 8.52). At the Jumeirah Beach Hotel in Dubai, a total of 60 000 m of PSA tape was used when constructing the building. In truck construction, the roof panel is often directly bonded to a load-bearing structure without screws, rivets or spot-welding and, as there are no boreholes, there are no points of attack for leakage or corrosion. In this way, when bonding the stiffener profiles of side

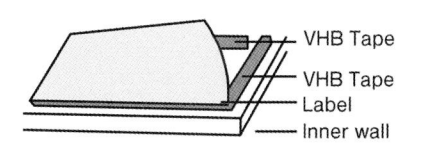

Figure 8.51 The fastening of a label in a railway carriage (VHB® tape; 3M).

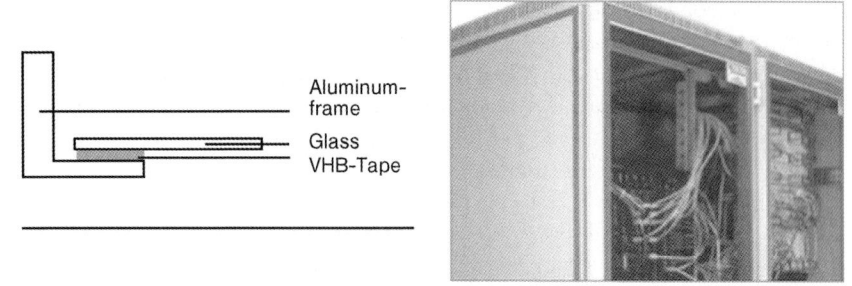

Figure 8.52 Adhesive bonding of glass windows (VHB® tape; 3M).

panels, for example of trucks, a uniform undamaged surface is created which is ideal for the application of decorative PSA films as there are no awkward screws or rivets. In such applications PSA tapes also help to reduce noise.

Today's ever-smaller components do not offer enough space for conventional fastening methods. For this purpose, customized die-cuts are available that have shock-absorbing and sealing properties for later use. They can also be used, for example, in the sealing of switch cabinets or underwater cameras.

8.6.1.9 Reclosable Fasteners

When it necessary to create a safe and removable assembly, reclosable fasteners (dual-lock fasteners or 'hook-and-loop' fasteners) offer outstanding advantages and an almost unlimited variety of applications. In the wall and roof panels of trains, trams and buses, reclosable high-performance fasteners replace screws, thereby avoiding corrosion owing to a lack of holes, and allowing access for maintenance purposes in a matter of seconds.

The interlocking of the stems of *dual-lock fasteners* is signaled by a distinct noise. The tensile strength of these fasteners is sufficiently high so as to replace nonesthetic mechanical fasteners in many applications. In terms of reliability, dual-lock fasteners can easily be opened and closed many thousands of times (Figure 8.53). These systems are not only reclosable, but also have excellent tensile strength, damping characteristics with regard to noise and vibrations, and resistance to both solvents and temperatures ranging from $-30\,°C$ to $+95\,°C$.

Hook-and-loop fasteners are ideal for products that require thousands of easy openings and closings; often, they are used to hold in place the cushions of motorcycle, bicycle or safety helmets. In operation, the tiny, stiff hooks mesh with the soft, pliable loops following the application of slight pressure to provide a quick fastening. Moreover, their production is both quick and cheap.

Figure 8.53 The set-up of a reclosable fastener (dual-lock system).

8.6.1.10 Screen-Printable Adhesives

Screen-printable adhesives can be positioned, but only where they are needed. The new systems rapidly cure with UV light, without harming the surfaces involved. As these products can be applied locally and with high precision, they will in time replace the fastening systems that use double liner tapes (that used pre-cut films) in the production of switches, instruments or light-emitting diodes.

New systems are exclusively derived from liquid acrylates, with up to 40% higher yield and greater flexibility than conventional, dispersion-based, screen-printed adhesives. As they contain neither solvents nor water, the printing results are excellent, providing for sharp edges. In addition, even during breaks in production the screen does not suffer clogging, and there is no need for intermediate cleaning or replacement. Consequently, material savings of up to 30% can be achieved simply by using up the adhesive after breaks. However, although the very rapid UV curing saves production time, the precutting of films leads to the production of large quantities of waste compared to die-cut production (up to 80% more).

Screen-printable adhesives are available which have thicknesses after printing and curing that range from 0.025 to 0.075 mm, and a resistance to temperatures between $-40\,°C$ and $+80\,°C$.

8.6.2 Protecting, Splicing and Masking

During the early 20th century, the 3M Company (St. Paul, Minnesota, USA) developed the world's first adhesive tape designed specifically for masking purposes in automobile body shops. Today's applications (masking, fastening, bundling, decorating) of most different varieties of adhesive tape are as manifold as the material combinations and resulting properties. The wide spectrum of adhesive tapes features a variety of backings (paper, polyester, polyurethane, coated fabrics, cellophane, PTFE, aluminum or PVC; see Table 8.4), incorporating many adhesives, colors, widths and lengths.

Three different types of adhesive are employed in industrial adhesive tapes, each with its own advantages and disadvantages (Table 8.5).

Adhesive Tapes with Paper Backings This group of products comprises adhesive tapes basically used for reinforcing, closing, splicing, masking during painting, and for belt applications.

Adhesive Tapes with PVC Backings Adhesive tapes with PVC backings have a wide field of applications owing to their high performance. Indeed, they are resistant to solvents, possess a high quality of color, can be stretched, and are resistant to abrasion. They are used for closing, marking and labeling, protecting, masking, bundling, fastening/tacking, and for many other purposes.

Adhesive Tapes with Polyester Backings Tapes with polyester backings are used for fastening, processing web materials, protecting and covering. They possess high

Table 8.4 Properties of different tape backing materials.

Tape backing material	Advantages	Disadvantages
PVC	• Good deformability • Resistance to acids	• Recovers after stretch • Poor flexibility at low temperatures • Temperature resistance up to +75 °C
Polyethylene	• Good deformability • Resistance to acids and solvents • Economical	• No dimensional stability • Temperature resistance up to +75 °C
Polyester	• Good abrasion resistance • Tensile strength • Resistance to solvents • Temperature resistance up to +150 °C	• Tears easily when the edges are damaged • Difficult application due to thinness and electrostatic charging
Metal foil	• Excellent moisture and dust barrier • Good heat-reflective properties • High temperature resistance (up to +315 °C for aluminum; up to +150 °C for lead foil) • Flame resistance • Outdoor weathering resistance • Resistance to acids (lead foil)	• Poor flexibility • Low resistance to acids (aluminum foil)
PTFE	• Excellent anti-stick properties • High temperature resistance (up to +250 °C continuously, up to +275 °C intermittently) • Resistant to acids and solvents • Flexibility at low temperatures • Excellent moisture barrier properties	• Relatively high price
Glass cloth	• Excellent tensile strength • Good heat insulation barrier • High temperature resistance up to +205 °C for several hours • Flame resistance	• Poor moisture barrier • Poor conformability

resistance to aging, high tensile strength, and excellent resistance to chemicals and abrasion. They are available as colored or transparent tapes, and have a temperature performance between $-50\,°C$ and $+180\,°C$.

Adhesive Tapes with Metal Backings There is no easier, faster or cleaner way to process metals. Metal backings possess excellent resistance to aging, and resistance to high – if not the highest – temperatures. Furthermore, they are very resistant to solvents, show low water-vapor permeability, and are suitable for die cutting.

Table 8.5 Properties of different PSA adhesives.

Adhesive	Advantages	Disadvantages
Rubber-resin adhesive	• Low price • High initial tack • Relatively good peel strength on steel	• Moderate temperature resistance (up to +95 °C) • Poor low-temperature resistance
Acrylic adhesive	• Extremely long service life • High temperature resistance up to +150 °C • Excellent peel strength on steel • Moderately priced • Transparency • Sunlight resistance	
Silicone adhesive	• Excellent high temperature resistance up to +205 °C (continuously) or higher (short-term) • Long service life • Excellent low-temperature performance • Good adhesion to silicone	• Relatively high price • Generally low initial tack

Sliding Tapes Sliding tapes are based upon PTFE and ultra-high molecular weight (UHMW) PE, and have special properties. Both offer a high resistance to temperatures and the lowest coefficient of friction. UHMW PE tapes were specifically designed to have a high abrasion resistance. PTFE tapes have the following properties:

- low coefficient of friction (the 'self-lubrication effect' of PTFE adhesive films improves the processability of web materials)
- excellent heat resistance (PTFE glass cloth tapes with high resistance to elevated temperature provide longlasting performance on many heat-sealing machines)
- anti-stick/release effect (facilitates clean-up of hot plastic extrusion coating by not allowing plastic build-up on rolls)
- resistance to chemicals (PTFE tapes provide a chemical barrier in production sectors with high exposure to chemicals)
- conformability (PTFE tapes provide for easy use and effective service of rolls)

The important properties of UHMW PE tapes are:

- abrasion resistance (UHMW PE tapes are used to protect the plastic or metal surfaces of conveyor belts, guide rails and containers from wear)
- noise damping (efficient damping of noise and vibrations in many fields of application)
- anti-stick/release effect (provide for easy clean-up of components/surfaces because sticky materials such as paints, adhesives or varnishes do not adhere to UHMW PE tapes)

Cloth Adhesive Tapes Cloth adhesive tapes and special tapes made from aluminum and/or glass cloth have a wide field of application owing to their coated cloth backing, in combination with adhesive varieties. They offer high tensile strength (glass cloth backings have an unmatched tensile strength that is seven times higher than that of polyester backings) and high resistance to elevated temperatures.

Adhesive Tapes with Polyurethane Backings Polyurethane tapes comprise a performing polyurethane backing coated with different types of adhesives on one side. Polyurethane tapes coated with acrylic adhesive are suitable for permanent bonding applications, and have very good resistance to solvents and UV radiation. Polyurethane tapes coated with rubber-resin adhesives are employed for temporary bonding applications. The outstanding properties of such tapes include excellent abrasion resistance, optimal protection against erosion, excellent tensile and disruptive strength, and good printability.

Adhesive Tapes with Cellulose Acetate and Cellophane Backings These tapes possess high transparency and resistance to aging, and are used mainly for the splicing of film footage, for closing, covering and protecting, tacking and marking purposes.

8.6.3
Damping

Elastic PSA buffers (bump-on products) are available to improve design and esthetics, to confer anti-slipping properties, or as spacers and stoppers. They absorb vibrations and noise and adhere to a wide variety of surfaces. They have outstanding anti-slipping properties and a high abrasion resistance, and do not leave any marks or traces (Figure 8.54).

These small PSA buffers not only reduce noise on doors, windows, furniture, glass, porcelain, ceramics, and so on, but they also reduce vibrations, for example on domestic appliances or computers, and absorb shocks on façade elements or metal cabinets. In addition, they prevent sensitive surfaces from being scratched and objects from shifting on smooth surfaces.

Figure 8.54 Pressure-sensitive adhesive buffers.

Adhesive tapes also allow noise to be reduced, particularly in the case of sensitive components. Tapes are available with foamed material backings in combination with a highly stressable acrylic adhesive.

8.6.4
Labeling, Marking and Coding

Barcode labels, nameplates or warning labels not only communicate important information, but also convey the manufacturer's image. Therefore, it is of utmost importance that they adhere reliably to the substrates.

A wide variety of pressure-sensitive films provides for product identification under the most severe conditions. Indeed, every day, label manufacturers are developing new customized identification systems, for example for the automotive (Figure 8.55) or electronics industry, that defy most adverse conditions caused by chemicals, temperatures or difficult surfaces.

Special recyclable labels are available that have been tailored to the recycling needs of popular plastic materials [e.g. PC, ABS, PC-ABS, PS, high-strength polystyrene (HIPS)] used for the manufacture of casings for information and entertainment electronics equipment, avoiding the time-consuming and costly removal of labels before starting the recycling process.

In industry, there is a trend towards ever-increasing rates of turnover with regards to products of all kinds, and this has immediate impact on processes such as sorting and counting, controlling, storing and commissioning. To this end, *retroreflecting* materials are available that make it possible to mark products or their transport containers for identification by photoelectrical scanning systems (Figure 8.56). According to the requirements, these materials are attached directly to the product or its transport container in the form of a single sign, or a combination of signs (code).

Laser engraving allows all types of fonts and logos (high-density barcodes included) to be produced by removing the upper layer of a double-layer acrylic film by means of a laser beam, exposing a different colored material layer below as information

Figure 8.55 A retroreflecting license plate.

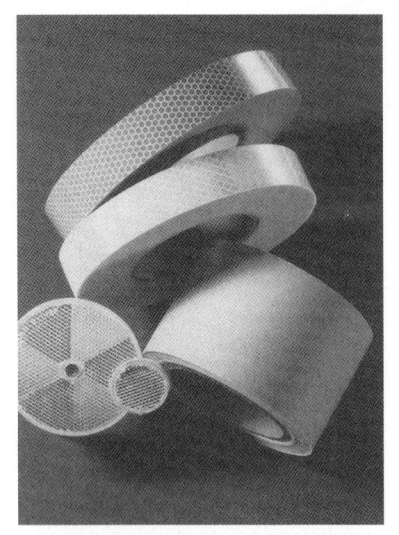

Figure 8.56 Retroreflecting labeling tapes.

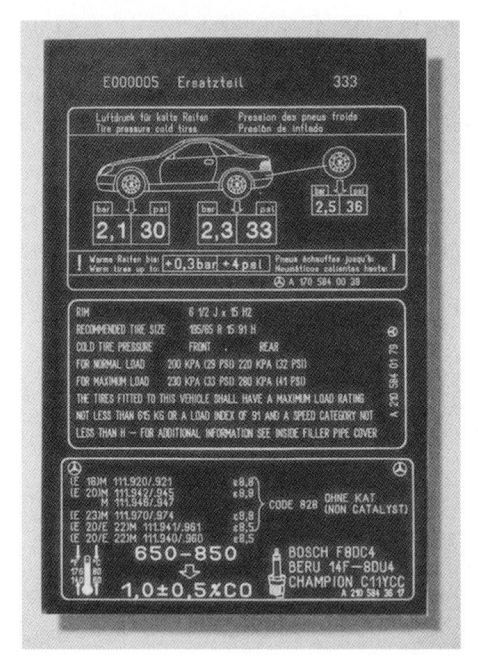

Figure 8.57 Information label in cars.

(Figure 8.57). At the same time, the laser beam can cut any label form, making die-cuts redundant. Laser engraving offers the following benefits:

- noise-reduced, flexible processing
- easy format change

- permanent, nonremovable printing
- no preprinting and precutting required
- suitable for small print runs
- perfectly suited for high-resolution barcodes

These films are employed for labeling purposes in the automotive industry, for transfer-evident labeling, and for high-performance labeling.

The special features of safety labeling films only become evident in the case of a manipulation attempt. Indeed, when trying to remove a label once bonded, all hopes of being able to transfer the label intact are dashed because the label is destroyed. These special films detect whether a manipulation attempt has been made.

8.7
Electronics Industry

In the electronics industry, adhesives are employed for the attachment and joining of components, for energy management, product safety, protection and identification, and for shielding purposes.

The task of the adhesive is not only to join the components, but also to provide for electrical conductivity, insulation, heat dissipation, sealing, protection and noise reduction. These tasks can be illustrated by the bonding of components to *printed circuit boards*. In a first step, the adhesive must provide for an easy and rapid application. Small and smallest components are frequently bonded by means of accurate metering devices; for this purpose, the adhesive must be resistant to thermal stress, particularly during soldering, and must level out any possible irregularities of the surface and the stress generated by the joining of dissimilar materials.

8.7.1
Thermally Conductive Adhesives

Today, electronic devices are becoming smaller and smaller, their performance is becoming higher and higher, and the mounting density levels in electronic components are growing. Hence, effective solutions for the dissipation of thermal energy are vital. *Heat sinks, heat spreaders* and other cooling devices are frequently bonded directly to the electronic board to achieve a heat management which is as effective as possible. To this end, the adhesives must not only fulfill the usual demands, for example with regard to a fast and clean processing, but also meet the following requirements:

- good thermal conductivity to provide for a heat transfer between the components
- good dielectric strength to avoid undesired current flows or short circuits
- good vibration damping
- excellent gap-filling features to avoid entrapping of air and to provide for optimal heat transfer

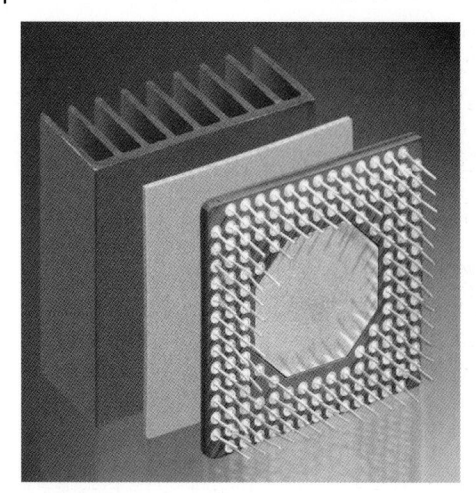

Figure 8.58 Bonding of heat sinks to active components.

- good relaxation behavior to compensate for stress generated by different thermal expansions
- low shrinkage

Thermally conductive products are available as adhesive films, elastomer pads and adhesives. They frequently contain thermally conductive, ceramics-based fillers that substantially improve the heat dissipation from electronic components and, at the same time, have an electrically insulating effect. In addition, they level out any possible irregularities of the surface and have a good initial tack and final bond strength to provide for a correct fastening of the components.

Today, thermally conductive adhesive films (Figure 8.58) have partly replaced thermal conductivity pastes, eliminating the need for mechanical fasteners such as clamps or screws. They are capable of leveling out any possible irregularities of up to 20% of their own thickness. A typical application is the bonding of heat sinks to computer central processing units and to flexible or rigid circuit boards. Thick adhesive films are also suited for heat dissipation in modern plasma panels. Weakly adhering, thermally conductive elastomer pads have special gap-bridging properties capable of leveling out differences of up to 2.5 mm, although additional mechanical fasteners are required. Thermally conductive adhesives combine low thermal impedance with high mechanical strength, and also have excellent wetting and gap-filling properties.

8.7.2
Electrically Conductive Adhesives

A most versatile option to make an electrical interconnection between two or more contact points is to use electrically conductive adhesive films that contain graphite fiber nonwovens or metal/metallized particles. These not only allow the use of lighter materials and flat design, while improving the reliability of the finished products

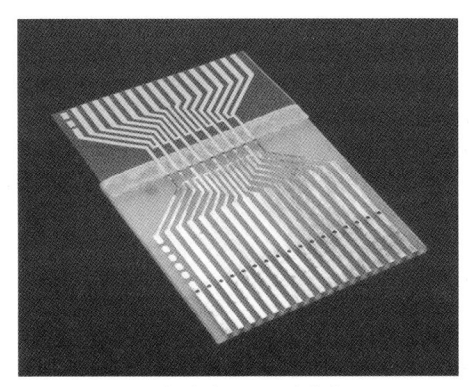

Figure 8.59 A polyethylene terephthalate (PET) circuit module bonded to a printed circuit board.

(Figure 8.59), but also help to reduce costs with regards to materials and processes. Depending on the current flow, a distinction must be made between *isotropic* and *anisotropic* electrical conductivity.

Isotropic adhesives are XYZ-axes electrically conductive; that is, the current flows not only through the adhesive thickness (Z-axis) but also in the plane of the adhesive (XY-axes). Isotropic adhesives contain, for example, electrically conductive fibers that interlock in such a way as to allow interconnection in the X, Y and Z directions.

Anisotropic adhesives allow interconnection through the adhesive thickness only (Z-axis); that is, the current only flows in one direction. Anisotropic adhesives are filled with conductive particles, for example high-temperature curable adhesive films filled with silver, nickel or gold particles. The adhesive acts as an insulator between the particles.

First-generation anisotropic adhesive films were specifically designed to make the manufacture of liquid crystal displays (LCDs) easier because the 'fine' electrical contacts of the flexible circuit boards could no longer be soldered. Another typical application is the bonding of flexible silver-paste printed polyester circuit modules to rigid plates. Adhesive films are also suited to the bonding of flexible circuit modules to flexible membrane switches. Products designed for this purpose also offer the benefit of a fast processing because they are slightly pressure-sensitive before crosslinking at 130 °C. However, the majority of anisotropic films require special cooling during storage and transportation below 0 °C because they are very sensitive to temperatures.

Depending on the desired application, anisotropic adhesive films can contain either hard or soft particles. Hard particles are suited for the interconnection of flexible to flexible or flexible to rigid plates, and are embedded into the conductive path by means of pressure; this increases the contact area and obtains good electrical properties. Soft particles are used for flex-to-glass interconnections – that is, the bonding of flexible circuit modules to glass substrates (LCDs). When they come into contact with hard surfaces, they deform and improve the electrical properties. Selection of the correct system requires a consideration to be made of the type of adhesive and the size and concentration of the particles, as these parameters have an influence on the electrical and mechanical characteristics, the bonding process, and the fields of application. Put simply – the more particles in the matrix, the better the conductivity; the less particles, the better the adhesion.

Thermoplastic anisotropic adhesive films have special features allowing smart cards to be converted into multifunction cards. Besides conventional chip bonding, they allow the integration of radiofrequency identification (RFID) antennas, finger print sensors and displays into smart cards. As the processing time of the adhesive films is very short (maximum 3 s), roll-to-roll lamination is possible. In addition, unlike high-temperature curable adhesive films, these adhesives do not require any cooling during storage and transportation.

8.7.3
Underfill Materials

The electrical characteristics of electronic components with solder bumps such as flip chips or ball grid arrays (BGAs) must remain stable in the long term. However, heat cycles, vibrations or direct mechanical stress used to have a negative effect on the durability. When these components are soldered onto a circuit plate, any gaps will remain between the bumps. So as to ensure as stable an interconnection between the components as possible, so-called 'underfill materials', which are based on epoxy adhesive technology, are used to completely fill the space between the bumps and to protect the solder joint. Two methods are available:

- No-flow process: Application of the underfill adhesive before the soldering process. When the components are applied, the adhesive is pressed into the gaps using a paste-like material (e.g. a one-part epoxy adhesive), followed by a thermal curing of the adhesive.
- Capillary process: The material (e.g. two-part epoxy adhesive) is injected into the gaps between the bumps by means of a metering needle following the soldering process, and then cured.

The adhesives are required to build up good adhesion to the substrates, and to be shock and vibration resistant, respectively.

8.7.4
Functional PSA Tapes

Transfer tapes with different functionalities are frequently used in the electronics industry to bond dissimilar surfaces (Figure 8.60). Acrylic-based PSA tapes offer exceptional holding forces as well as excellent durability, sealing properties and shock resistance. They are used, for example, in the assembly of circuit boards and for the bonding and sealing of underwater cameras.

Transfer tapes with very low concentrations of monomers, organic acids and organometallic compounds are available for applications that are very sensitive to contaminations. These products are employed, for example, in the bonding of electronic components in clean room environments and for the assembly of hard disc drives. Optically clean transfer tapes with more than 99% light transmission are

Figure 8.60 Circuit boards bonded to aluminum sheets.

used for the bonding of displays and touch screens to ensure excellent transparency and accurate colors.

8.7.5
Spacers

Spacers composed of PSAs coated on polymer films allow a certain width to be maintained between electronic components. Figure 8.61 shows a spacer for foil keypads. The spacers are required to be flexible and to compensate for external stresses and strains.

8.7.6
Labeling

Electronic components or circuit boards are labeled by means of products that are resistant to high temperatures and that are printable by thermotransfer. The labeling

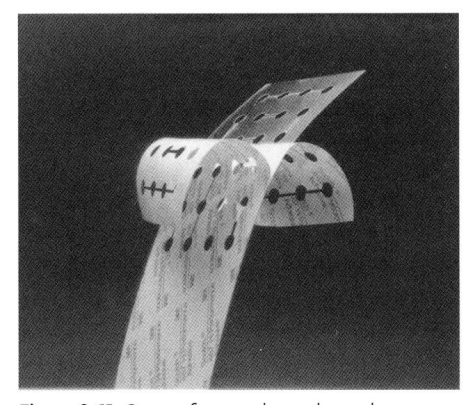

Figure 8.61 Spacer for membrane keypads.

is required to be maintained even after processing of the components, particularly after the soldering process. Whereas, conventional soldering was performed at temperatures of 230 °C, the components and labels must now withstand temperatures of up to 260 °C with the new lead-free solder. Smart labels have integrated chips and an antenna providing for a contact-free labeling, for example, as RFIDs.

8.8
Optical Industry

In the optical manufacturing industry, adhesives are employed for three different tasks that are completely independent from each other: (i) the temporary blocking of glass elements to allow processing; (ii) optical cementing; and (iii) the bonding of optical elements into mounts. According to the desired application, optical adhesives or cements must fulfill different requirements.

8.8.1
General Requirements for Adhesives in Optical Industry

Adhesives used to temporarily fixture glass elements to allow processing are required to cure within seconds, without shrinking. After processing, they are required to debond quickly, without harming the surfaces. Whilst conventional adhesives are not suited to this task, hot melts do meet the requirements of this application, provided that no strains are produced in the optical elements when the adhesive cools to room temperature after having been heated for the bonding process.

For *optical cementing*, the adhesive must have optical clarity and a matching refractive index, without causing any strains. Optical cements are used to bond the majority of devices used for optical observing, in optical measuring instruments and in space research, in medical devices required to be designated 'sterile', and in sensor technology. According to the relevant application, the requirements for optical clarity and resistance to aging may differ considerably. In optical components, the cement must compensate for the different coefficients of thermal expansion of the optical glasses, and the cement layer will differ depending on the refractive index and thickness of the adhesive. Black or dark adhesives are well suited for the bonding of optical elements into mechanical mounts because they reduce light scatter. *Flexibility* is crucial for these adhesives in order to prevent strains from being induced in the optical elements in as wide a temperature range as possible. The elements to be bonded generally have differences in thermal expansion. When the correct adhesive is chosen, its flexibility – as well as a predefined thickness of the bond-line – can help to minimize the strains induced. Bubble-free application is necessary to provide for imperviousness and resistance to moisture-induced aging. In order to ensure correct adhesion of the cement to the glass surface, the surface must be cleaned prior to bonding. The mounting adhesives must be resistant to fungal attack and also exhibit low outgassing figures. Thus, the cements used for mounting optical elements include epoxy, polysulfide,

8.8.2
Optical Adhesives for the Temporary Fixture of Optical Elements

When fixing the glass elements, they must be fixed in the desired position as rapidly as possible. Hot melts (also known as 'optical pitch', 'sealing wax' or simply 'wax') are very popular owing to their high curing rate and solubility in organic solvents.

In optical components, strains are induced by shrinkage of the adhesive during solidification, as well as by the different coefficients of thermal expansion of the materials during cooling. Inorganic fillers such as chalk powder or quartz filler can reduce shrinkage of the adhesive, while a flexible resin matrix can help to minimize the strains caused by differences in thermal expansion during cooling. The flexibility of the matrix is limited, however, due to the creeping behavior of the resins, which may have an effect on the alignment of the optical elements during the curing process. Due to the undesired creeping behavior of soft thermoplastic resins, hot melts cannot fully meet the requirement of placement accuracy and minimization of strains in optical components.

Reactive adhesive systems also allow the components to be fixed at room temperature, although after curing they are required to have a low resistance to solvents. Besides acrylics that crosslink at room temperature, modified acrylics which set under the effect of UV light are used (Figure 8.62).

After processing, the adhesive must be decemented rapidly, without causing any harm to the sensitive optical surfaces (see Section 8.17.5), a requirement that cannot be fulfilled by conventional adhesives. For this purpose, UV-setting adhesives are available that have been modified in such a way as to debond as soon as they are heated in an aqueous soap solution, or as soon as they are immersed in organic

Figure 8.62 A bonded beamsplitting cube. The prisms are glued together with optical cement, and the cube is bonded to the mount. (Photo courtesy of Swissoptic AG).

solvents. The resistance to aqueous solutions, however, must be high enough so as to withstand exposure to the aqueous emulsions used in the manufacturing process. Therefore, the use of UV-setting adhesives is a very sensitive process.

8.8.3
Optical Cementing

Optical systems have a higher resistance to the environment and better long-term stability when the surfaces in the ray path are kept small. When the optical cement is applied in a very thin layer, the elements can be placed with very high accuracy. An *optical cement* is a highly transparent adhesive that bonds the optical surfaces of two optical components. Unlike air gaps, the optical cement influences the optical path of the devices in service, due to its refractive index. This adhesive must therefore be highly transparent and free from particles. Usually, the refractive index of the optical cement is similar to that of the glass used. *Decementing* may occur due to thermal strains or moisture aging that may induce internal reflections and reduced transmission, leading to a complete failure of the system. It is therefore crucial that measures are taken to reliably prevent decementing.

When replacing compact mounts with optical cements, the measuring accuracy and the accuracy of image are improved. Loss of light will be reduced, and the system will be lightweight. Compared to structural adhesives, optical cements must meet totally different requirements because they are applied within the optical path. The key characteristics of optical cements are:

- spectral transmission
- refractive index and dispersion
- viscosity
- lack of strains and flexibility
- build-up of adhesion without primer
- resistance to UV light
- fluorescence

Special optical cements are available for devices used in space research, in medical devices required to be designated 'sterile', and in sensor technology. The optical cleanliness and resistance to aging of these cements may differ considerably. Selecting the proper optical cement requires a consideration of the temperature and moisture conditions to which it will be exposed. As glass surfaces are subject to moisture aging due to their chemical structure, immediately before cementing the glass components are cleaned in aqueous ultrasonic equipment. The removal of any surface contaminants is a prerequisite for uniform wetting and adhesion. The resistance to moisture aging at elevated temperatures depends largely on the chemical composition of the optical glasses used; for example, glasses with a high resistance to moisture, acids and bases will have a higher stability with regard to moisture aging. Frequently, the glass surfaces to be cemented are optically coated. Usually, the optical cement adheres better to the hydrophilic coating than to non-coated glass surfaces.

Owing to its flexibility, the optical cement acts as a buffer when bonding optical glass elements with different coefficients of expansion. When cementing different types of glass, the optical cement layers can be applied in variable thicknesses, depending on the viscosity of the cement and the pressure applied when joining the components. Viscous and more flexible optical cements are used when cementing large optical elements made from different types of glass with different coefficients of thermal expansion. However, in practice, thicker cement layers induce undesired wedge errors.

In industrial production, UV-setting acrylics, urethanes or epoxy cements are mainly used owing to their easy metering and rapid curing characteristics. Besides flexible two-part epoxy systems, two-part silicone systems are employed for special requirements, particularly for components that are very sensitive to strains. These cements are also used whenever a high spectral transmission is required below 400 nm and/or whenever radiation is very intense.

8.8.4
Bonding Optical Elements to Mounts

It is of utmost importance that the adhesive has good mechanical properties when optical lenses and prisms are bonded to mechanical fixtures and mounts. As the bonding area is usually smaller than in the case of optical cementing, the mechanical strains are much higher. The dimensions of the elements, the temperature range and the coefficients of thermal expansion of the substrates to be bonded determine how flexible the adhesive must be. A key characteristic of these adhesives is elasticity to prevent strains from occurring within the optical component over as large a temperature range as possible, without impairing placement accuracy through creeping effects of the adhesive. Strains caused by thermal expansion can be minimized not only by a highly elastic adhesive, but also by sticking to a predefined thickness of the bond-line. The optical system must be designed to prevent any liquid adhesive from penetrating into thin centering gaps. At a bond-line thickness of a few micrometers, even elastic adhesives will cause strains in the optical components at temperature changes.

Black or dark adhesives are well suited for the mounting of lenses and prisms when light scatter must be prevented. Bubble-free application of the adhesive is a key requirement for process stability, as it improves the tightness and moisture aging stability of the bond.

To allow for good adhesion of the adhesive on the glass surface, the hydrated glass surface must be cleaned immediately prior to bonding. For industrial applications, the glass elements are cleaned with aqueous tenside solutions in special ultrasound equipment, where alkali metal ions as well as particles and impurities are reliably removed from the surface. In spite of this cleaning process, following moisture aging, the adhesive does not adhere sufficiently to the glass surface, and this results in an inadequate long-term stability of the bonding. Special glass primers, mainly derived from silanes, sustainably improve the resistance to moisture aging.

Besides the capillary forces that play a pivotal role, a key characteristic of automatically applied adhesives in the bonding of optical components is the *viscosity*. Selecting the correct viscosity of the adhesive and choosing the correct design allows the bond-line thickness to be kept constant, resulting in a tight, strain-free bonding.

When UV-setting adhesives are employed, an optical alignment tool can be used to accurately place the optical components, and the mechanical tolerances of the elements can be leveled out by a variation of the bond-line thickness. However, as it is not always possible to irradiate the optical elements from every side after joining, a combination of setting mechanisms (hybrid of UV light and heat, or UV light and anaerobic conditions) offers a solution to this problem. These systems, however, have limited temperature stability due to the high coefficients of expansion of the flexible adhesive systems. Owing to thin bond-lines, the placement accuracy of optical microcomponents is better than that of optical components with larger dimensions.

The key requirements of adhesives used for the bonding of lenses and prisms to fixtures and mounts are resistance to fungal attack and low outgassing figures, and this makes epoxies, polysulfides, polyurethanes, urethane-modified acrylics and silicones (for special applications, e.g. in space) the first-choice adhesives. Unfortunately, silicones have a poor resistance to fungi, and the surfaces may be contaminated by dimethyl siloxane spreading out onto the surface; the surface can then no longer be cemented, bonded or coated unless a special surface preparation is carried out. Hence, the processing of silicone-based adhesives is subject to special guidelines on the handling of this material to prevent surface contaminations.

8.9
Mechanical Engineering and Process Equipment Construction

In mechanical engineering and process equipment construction, welding and joining by screws are still the first-choice joining techniques. Adhesive bonding is suitable for small series production because only minor equipment is required for this technique. Fast-setting adhesives that can be applied without costly devices (such as autoclaves and/or mixing and metering units) are preferred. The prerequisites for reliable adhesive bonds are a correct design which meets the demands of bonding, a reliable surface preparation of the adherents, and an extensive knowledge of the long-term behavior of adhesive bonds.

8.9.1
Mechanical Engineering

Many adhesives are suited for the bonding of metals used in mechanical engineering. Epoxy resins and acrylics are particularly well suited. Bonding is mainly used for the production of composite materials, where welding cannot be used. Plain steel, low-alloy steel and high-alloy steel, as well as different aluminum and casting alloys, are the most important materials bonded in mechanical engineering.

Cold-setting adhesives with high reactivity, such as two-part adhesives, methyl methacrylates, anaerobic adhesives and cyanoacrylates, are the most frequently used adhesives. It is not only essential to conceive the design so as to meet the demands of adhesive bonding, but also to be knowledgeable about the long-term behavior of bondings under prevailing conditions. These are classified into 'exposure to chemicals' (e.g. oil or water vapors) and 'physical requirements' (e.g. creep and vibration). The specification of the requirements usually allows an optimal adhesive to be found for a specific application.

Bonding also offers advantages in the manufacture of cutting tools such as machine reamers or end mills. In both cases, hard-metal (carbides pressed into a metal matrix) cutting tips are bonded to steel flutes using a hot-setting epoxy resin adhesive. Originally, these tools were manufactured by brazing, but this induced distortion and softening of the steel, microcracking of the cutting material, and incomplete joining of hard metal and steel. The key advantages of adhesive bonding are: (i) the joining of hard metal and steel over the entire surface, without harming the materials; (ii) the prevention of crevice corrosion; (iii) the even distribution of strains; and (iv) the damping of vibrations [46].

8.9.2
Domestic Appliance Industry

Painted and coated surfaces are bonded with high-performance adhesives. Many functional parts of domestic appliances are bonded; for example, the cooling coils at the back of a refrigerator are bonded with heat-conducting adhesives, while the cover of a top-loading washing machine, which seals the drum, is bonded with polyurethane adhesives [47].

Double-sided, high-performance adhesive tapes are used to produce high-strength joints of powder-coated metal parts, also in combination with other materials such as glass, and without the need for surface preparation. The adhesive tapes adjust to any form in an optimal way such that even very thin or structured surfaces can be joined over the entire area, without producing any strains. Further advantages are high impact strength at low temperatures, a high initial tack, absolute tightness, and the absorption of noise and vibrations.

8.9.3
Manufacture of Internal Combustion Engines and Transmission Manufacturing

For securing or locking, joining and sealing in the manufacture of internal combustion engines, and in transmission manufacture, the use of anaerobic diacrylate-based adhesives is virtually self-evident today. Anaerobic adhesives remain liquid as long as they remain in contact with oxygen from the air (see Section 5.7).

When a screw is tightened in a thread, the adhesive sets via a radical mechanism initiated by metal ions. Excess adhesive escaping at the screw thread end can easily be removed. Screw locking by using anaerobic adhesives provides very good resistance to vibrations and reliable sealing against different media such as oils or fuels. Extensive tables on the resistance of anaerobic adhesives to solvents are available that allow the reliable selection of a correct adhesive [48].

Fast-setting anaerobic adhesives also allow the caps of core holes of motor and cylinder head closings to be reliably sealed, so that the heads can be subjected to an on-road test shortly after bonding. When replacing a flywheel, high functional strength is rapidly achieved by using a screw-locking anaerobic adhesive. In tests performed with a setting time of 20 min, M10 screws stressed with 50 Nm withstood 10^4 loadings [47].

Another example is the bonding of *camshafts*. Rather than using forged parts, the cams are bonded to tubular shafts, allowing the free choice of materials for both parts. The same applies to crankshafts that can be bonded instead of being forged or cast. Sometimes, this can be less expensive or even mandatory in terms of structural design. The bonding of shaft-hub connections has been thoroughly investigated in a joint project funded by the former German Federal Ministry of Research and Technology BMFT (now Federal German Ministry of Education and Research BMBF) [49].

Bonding cams made from gray cast iron, or sintered material, and applied onto hollow steel shafts, require a shear strength of $10\,\text{N}\,\text{mm}^{-2}$, or more. The key parameters of investigation were defined as: (i) a torsional fatigue strength during 15 years of service; (ii) resistance to a temperature of up to 160 °C; and (iii) resistance to motor oil containing fuel, combustion residue and water. Another important parameter was creeping strength at a temperature of 80 °C or more during several hours because, when a motor is very hot and then switched off, it is always assumed that a cam opens a valve, and the adhesive joint is continuously exposed to stress. The cam then must not change position on the shaft at all.

These investigations showed that anaerobic adhesives are well suited for shaft-hub connections for the following reasons: (i) they are easily applied; (ii) nonset adhesive residues can be removed without the need to rework the camshaft; (iii) the setting of thin adhesive layers takes place rapidly and completely at room temperature; and (iv) the thermosetting adhesives have good resistance to media and elevated temperatures.

Shaft-hub connections (Figure 8.63) have higher fatigue resistance than keyway assemblies or gearings. Therefore, the large-dimensioned, long-life drive trains of steel mills can only be manufactured using bonded shaft-hub connections.

Shrinkage is a well-known, reliable method of joining elements. The strength obtained with the shrinkage of differential ring gears used in the automobile industry, however, does not allow exposure to elevated temperatures. Shrink-bonding by means of anaerobic adhesives is the solution, providing a strength which is three- to fourfold as high as that of joints produced by shrinkage alone. At the same time, contact and fretting corrosion is avoided. The increase in strength is mainly due to the fact that the contact area increases from 30% to 100% of the real surface owing to material joining, provided that the application and setting of the adhesive are accurately adjusted to the prevailing manufacturing conditions. Shrink-bonding is performed at around 170 °C.

Figure 8.63 Application of anaerobic adhesive for fixing the shaft in a small direct current (DC) motor. (Photo courtesy of Dorel Verlag).

8.9.4
Stainless Steel Structures

Special types of steel are useful as structural materials for virtually all industrial production sectors owing to their particular material characteristics, especially their anti-corrosion properties. Historically, stainless steel has been used in the chemical and food industry, but it is also becoming popular in the building industry because their surfaces are metallically clean and visually flawless, without further surface preparation. Today, it is not only the good resistance but also the mechanical characteristics of stainless steels that are increasingly appreciated. For example, in the automobile industry, special types of steel are used for highly stressed parts to increase the rigidity and crash resistance of cars (Figure 8.64).

To take full advantage of the characteristics of special steels in an overall structure, a joining technique is required that does not impair their good anti-corrosion properties and mechanical characteristics. Generally speaking, material joining techniques are preferred to mechanical ones. *Welding* is possible, but this entails the risk of coarse grain formation (leading to material embrittlement) and sensitization by precipitation of chromium carbides along the grain boundaries of the structure. Furthermore, with spot/tack welding and step-by-step welding, fissures are induced

Figure 8.64 A bonded stainless steel door frame in a truck container. (Photo courtesy of Schmitz Cargobull).

Figure 8.65 Corrosion induced by chloride ions in a bonded joint of stainless steel (1.4376 after 5 months VDA test. Adhesive: hot curing, one-part epoxy).

that are not allowed in food-related equipment for reasons of hygiene [50]. Strength tests performed on bonded stainless steel have shown that the quality obtained with structurally bonded stainless steel is comparable to that obtained with bonded body steel [51]. The crash-optimized, one-part epoxy resin adhesives tested have a high contamination tolerance, for example towards common auxiliary material used for forming purposes, so that costly surface preparation can be dispensed with.

It must also be considered that, with stainless steel joints, the service life expectation under climatic exposure is determined by other failure mechanisms than with the metal adherents (see also Section 7.5.1). When exposed to a warm humid environment for a prolonged time, the strength of bondings of adhesives with low resistance to corrosion was impaired by bond-line corrosion which started at the cut edges, whereas the strength of bonded Nirosta® 4301 (X5CrNi18-10) and Nirosta® H400 (X5CrMnNi19-7-3) was impaired by damages at the interface between the polymer and the metal.

Furthermore, it should be noted that, under the influence of media containing chlorides, electrochemical concentration elements are sometimes formed that may impair the anti-corrosion properties in the area of bonding (Figure 8.65). In this case, the use of stainless steels with a higher resistance to corrosion may be the solution, or perhaps passivation such as etching or electropolishing.

8.10
Textile Industry

8.10.1
The Early History of Fabric-Reinforced Rubber Materials

Today, so much of the modern world relies upon articles made from polymer–textile (mainly rubber–textile) composites; in fact, it is difficult to imagine how we could survive without these goods in modern life. Products such as transportation systems (rubber–textile tires), material-handling systems (conveyor belts) and mechanical driving systems (belts) reflect the importance of these materials. Although textiles have been produced and used for many thousands of years, rubber (gum) was introduced into Europe only 500 years ago [52], and only during the past 200 years has it been really utilized, notably in combination with textiles. Since then, however, a substantial development has taken place in the design and use of these materials.

Over the past 85 years, synthetic products have replaced natural materials (i.e. rubber and cotton) in the production of textiles and polymers, with the result being a wide variety of versatile, composite materials.

From the skills of the ancient Mesoamericans, who were very knowledgeable about the characteristics of native rubber, rubber processing progressed towards applications such as the manufacture of waterproof coatings of sails, hoses and bags. All of these products, however, had a limited use performance. The final landmark in the early history of the rubber industry was the discovery of *vulcanization* at the start of the 19th century; this made it possible to produce pneumatic tires [53, 54] that consisted of an inner tube and an outer tire. The outer tire was produced from several layers of plain-woven cotton fabric and rubber, and provided with bead wires to hold the tire in place on the rim. Although, in 1915, the replacement of tarpaulin with cord fabrics led to a major improvement in tire performance, the latter was still limited by the rubber material used at the time. Subsequently, the introduction of *carbon black* doubled the service life of tires from 3000 to 6000 km, while the fabrics 'caught up' with a relatively new fiber called 'rayon', a regenerated cellulose. Unfortunately, this created a new problem since, for the first time, a fiber other than cotton was used. Until then, the combination of cotton with rubber was perfect because the fine cotton fiber ends could be incorporated into rubber, resulting in a mechanical bonding of the rubber to the textile reinforcement. Synthetic fibers, however, are composed of continuous filaments with only a few fiber ends. So, to make synthetic textiles bondable to rubber, an adhesive solution was developed that was originally based on natural rubber and casein, and used to treat textiles by helping them to bond well to rubber. Shortly afterwards, a resorcinol–formaldehyde-based resin replaced the casein component. As a consequence, with every newly developed fiber a matching adhesive system was developed with regards to continuing the performance of the composite material, such that the development of rubber and textile technologies ran in parallel. Today, a wide array of composite materials is available for whatever requirements are to be met under most adverse conditions, whether in space, in the deep sea, or at arctic temperatures [55].

8.10.2
How Textile Fibers Contribute to the Living Standards of the Industrial World

The bonding of textiles is a key application in the manufacture and use of many products in virtually all domains of daily life. Examples are clothing (not only up-to-date fashion, but also functional and protective clothing), home textiles (carpets, wallpapers) and textile cleaning products (cleaning rags, extractor hood filters). The extent to which synthetic fibers have added – and continue to add – to everyday life is illustrated in the following sections.

8.10.2.1 Technical Textiles

In recent years, synthetic fibers have increasingly been used for products involving technical textiles. The technical advances resulting thereof directly or indirectly add to many domains of daily life without being recognized as 'involving textiles'.

Few people know, for example, that the safety and stability of dams, dyke reinforcements, roads, streets and freeways is achieved owing to specifically developed textile reinforcement fabrics, so-called 'geotextiles', in the form of spunbonded nonwovens, layings, cloths or knittings manufactured from high-strength synthetic fibers designed to achieve filtration, media separation, soil reinforcement and stabilization of the soil or even in rockets used for aerospace purposes. Nobody cares about the 'heart' of fiber-reinforced cement, for example in bridge piers made from concrete construction elements provided with prestressed high-tech filaments that prevent the generation of dangerous cracks, and neither do bituminous spunbonded nonwoven fabrics that replace galvanized sheets as roofing membranes attract much attention. In so many other domains, textile materials reflect technical progress in structural engineering, from artificial grass and outdoor textile floorings or canvas covers, tents and air-supported halls to textile architecture of high artistic value (e.g. the convertible roof of the open-air theater of the abbey ruins of Bad Hersfeld, Germany).

Modern high-tech fibers and specifically developed textures have translated into highly advanced technical components, such as in aerospace engineering. For example, the ergometric layout of the textile reinforcement of plastic components that have a lower specific weight than aluminum can accurately be tailored to the coordinate system of the applied forces, resulting in extraordinary aerodynamic performance [56].

All of these examples have in common that technical textiles are 'bonded' to themselves or to other components at least at one process stage or the other, with a wide variety of adhesive systems being used.

Although all textiles mentioned here are 'bonded' (hot-melt powders being particularly used for clothing textiles), only textile-reinforced rubber goods will be discussed in detail in the following paragraphs.

8.10.2.2 Textiles as Rubber Reinforcements

Today, a modern medium-class vehicle contains around 250 kg of chemical materials, with plastics alone contributing 125 kg. Currently, approximately 900 000 tons of polyurethanes are employed in automobile production, mainly in the manufacture of seats, fenders and side fairings. Textiles and carpets provide comfort in the interior, while paints and underbody coatings provide an appealing appearance to the precious vehicle and protect against corrosion. The second most important material, on a quantity basis, is rubber. With textile reinforcement, this elastomer has high-performance characteristics. In modern cars, examples of rubber parts containing threads and yarns include the air-conditioning, power steering, and the brake, hydraulic and fuel hoses. Toothed belts and other drive belts contain cords (i.e. plied yarns).

Belts and hoses are *flexible joints* which transfer high forces and/or convey media under high pressure. Technical fabrics are tailored to the customer requirements to

Table 8.6 Properties of textile reinforced composites.

Properties determined by the rubber compound	Properties determined by the reinforcement
• Resistance to solvents • Flexibility at low temperatures • Resistance to ozone • Water- and gas-proof • Aging properties (thermal and chemical) • Resistance to abrasion • Flammability • Pressure loss	• Adhesion • Deformation during use • Tensile strength and elongation • Resistance to pressure or vacuum • Bursting strength

become refined products that meet current market needs. Important fields of applications are air-spring assemblies, conveyor belts, membranes, the above-mentioned tires, and special fabrics.

Textile-reinforced rubber components take advantage of the properties of both, rubber and textiles (Table 8.6), which is reflected by a synergy of both.

Modern high-performance yarns and fabrics must meet the highest requirements as to physical parameters such as elongation, tenacity, shrinkage and resistance to temperatures and chemicals; moreover, they must also adhere to all rubber mixtures. The specific requirements of the different applications met by different types of yarn are summarized in Table 8.7.

8.10.3
General Properties of Elastomers (Rubber)

The properties of elastomers make them very different from other materials, in particular with regards to entropy elasticity, viscoelasticity and chemical structure, which is a macro network.

Due to a three-dimensional network, elastomers do not flow and are insoluble, but swellable. When the temperature increases, they will not melt until degradation occurs.

Table 8.7 Fiber types and characteristic applications.

Article	Specific requirement(s)	Material(s)
Air-conditioning hose	Resistance to chemicals, high pressure	Aramid
Brake hose	Low elongation, minimal heat shrinkage	Rayon
Power-steering hose	High elongation	Nylon
Hydraulic hose	High tenacity	PES, Aramid, PVA
Fuel hose	Antistatic	Modal, steel
V-belt and toothed belt	High tenacity, high shrinkage force, low elongation	PES, glass

PES = polyethylene–styrene; PVA = polyvinyl acetate.

When conceiving a rubber formulation, its components can carefully be selected to obtain a wide variety of properties. The T_g determines their viscoelastic properties, and active fillers induce typical nonlinearities (dependence of deformation). Owing to a chemical network, elastomers have high failure tolerance with regard to solvents and peak temperatures.

Low strength, however, is a disadvantage of elastomers, as is sometimes their low modulus, but reinforcing elastomers with steel, textile or glass fibers can solve this problem. During the forming process, chemical crosslinking takes place that allows chemical bonding to take place between the elastomer and different substrates (fibers, wires, metal parts). When rigid materials are used as reinforcements, anisotropic components can be produced – that is, components that show high strength and elastic behavior in one or several directions, and high deformability and viscoelastic behavior in other directions. This unique combination of properties is taken advantage of in the manufacture of a wide variety of products that contribute 90% of the total amount of textile-reinforced products manufactured. Composites are mainly subject to dynamic stress and corrosive conditions, requiring strong and durable adhesion between the elastomer and the reinforcement to withstand the mechanical and chemical conditions, which can only be achieved by chemical bonds.

8.10.4
Treatment of Textiles for Rubber Reinforcement

Fiber-reinforced rubber composites rely upon three systems: (i) reinforcement textiles; (ii) adhesive systems used to treat textiles to make them bondable to rubber; and (iii) rubber.

8.10.4.1 Reinforcement Fibers

Inorganic fibers, organic fibers and metal fibers are well suited reinforcements for elastomer composites, and may be either threads or planelike structures. *Filament yarn* is used for power transmission in one preferred direction (e.g. textile reinforcement in V-belts), whereas planelike structures (fabrics, knittings, metal parts) are employed for two-dimensional power transmission (conveyor belts, hoses) or to confine elongation (air-spring assemblies, tires).

Fibers and yarns are sensitive to compression and shear forces, and shall only be subjected to tensile loading. To this end, covulcanization of the reinforcement with the matrix takes place under prestressed conditions.

8.10.4.2 Adhesive Systems Used to Treat Textiles

To resist to dynamic loading and corrosive conditions, the fibers and the elastomer matrix must build up chemical bonds. This is achieved by treating textile fibers with an adhesion promoter in a separate processing step. Mechanical anchorage of the fiber is sufficient for products subjected to minor stress (low-pressure hoses, the typical 'garden hose'). No adhesion promoter is needed for certain articles with open-weave nettings owing to 'strike-through' adhesion – that is, the matrix is bonded through the spaces between the textile reinforcement. Short-fibril fibers such as cotton yarns are anchored in the elastomer via anchorage fibrils.

The following production steps are required for treatment with an adhesion promoter to enhance the adhesion of rubber to textile fibers: The textiles are unrolled under predefined tension, dipped (put through an impregnating bath) with the depth of penetration depending on the tension, the direction of the twist, and the viscosity of the adhesion promoter (dry uptake: tire cord: 4–6%, conveyor belt fabrics: 6–8%, glass cord (toothed belt): up to 20%), and then dried (convection/hot air, infrared, microwave, HF). This is followed by condensation of the adhesion-promoting resin as well as heatsetting and winding of the yarn onto precision winders.

The following treatments are possible to have reinforcements build up chemical bonds to elastomers [57]:

- soaking in solutions containing bifunctional adhesion promoters (cement)
- impregnation of yarns with aqueous resin adhesion promoters (RFL systems)
- addition of the adhesion promoter to the rubber mixture (direct adhesion)

Polyfunctional Adhesion Promoters Di- and triisocyanates are used as adhesion promoters for fibers that have only a few reactive groups, or that have only low reactivity such as PET (polyester) or polyphenylene terephthalamide (aramid). Adhesion promoters are applied either by soaking the fibers or the fabrics in a solution, or in the form of a paste. Due to high reactivity, fibers or fabrics coated with isocyanates must be processed rapidly or covered with a rubber layer. For polyester, epoxy resins are also suitable as adhesion promoters.

Resorcinol Formaldehyde Latex (RFL) Dip Yarns and fabrics are usually impregnated with a mixture composed of latex and a condensate of resorcinol and formaldehyde. Generally, styrene vinylpyridine–butadiene terpolymer or chloroprene latex is used. Natural latex has a high tack and is only rarely employed. The latex component is added to resorcinol–formaldehyde precondensate; occasionally, waxes or tackifiers are also contained in RFL dips.

The reaction product of these substances determines the specific characteristics of the dip, such as viscosity, durability, adhesive strength and flexibility. The latex in the RFL is selected for good bonding to the solid rubber that will be used in the final product. Details of the compatibility between the rubber types and latex types are listed in Table 8.8.

Table 8.8 Compatibility between rubber and latex polymers.

Solid rubber polymer	Latex polymer
Styrene–butadiene rubber (SBR); isoprene rubber (IR)	Vinyl pyridine rubber (VP and/or SBR)
Natural rubber (NR); butyl rubber (BR)	VP
Chloroprene rubber (CR)	CR
Acryl-nitrile butadiene rubber (NBR)	NBR
Ethylene propylene diene rubber (EPDM)	No universal solution to date

RFL is produced in a two-step process, starting with the generation of a resorcinol–formaldehyde precondensate composed of mono-, di- and trihydroxy-methylated resorcinol, followed by the addition of latex. Alternatively, the precondensate can be obtained commercially. The rubber component is a blend of SBR latex with vinyl pyridine latex. After a waiting period of approximately 6 h, the aqueous dip is applied to the fibers, predried at 100–130 °C, and finally allowed to react with the fibers in the main dryer at temperatures of 150–230 °C. The binding reaction takes place as a condensation reaction between the methylol groups of the resin and active hydrogen atoms of the fiber. At the same time, the resin reacts with the latex rubber component. Binding to the matrix takes place by covulcanization with the latex component. Nitrile rubber lattices are well suited for producing polar rubbers.

Polyester fibers have only a few reactive OH groups, and are predipped at manufacture; they can then also be coated with RFL dip. Predips for polyester filaments are composed of a water-soluble epoxy resin (e.g. glycerol polyglycidyl ether) and a blocked isocyanate (e.g. phenol-blocked methylenebis(4-phenyl isocyanate)) that only reacts at elevated temperatures. The components are applied as dispersion and allowed to react at approximately 220 °C.

Aramid filaments are also slow-reacting, and are provided with a predip composed of epoxy resin and ε-caprolactam-blocked isocyanate. Glass fibers are typically coated with an adhesion promoter that is already applied during manufacture.

8.10.4.3 Effect of the Fiber Type on the Adhesive System

Although the treatment of textile fibers to enhance their adhesion to rubber is a typical process, some particularities must be noted. First, the adhesive system must be matched to the type of the fiber, and the basic formulation is frequently modified to some extent to achieve optimum enhancement with regard to special elastomers. The application of adhesive is generally combined with heatsetting.

Rayon With the advance in manufacture processes aiming at the production of high-strength rayon fibers, it was found that SBR lattices did not confer sufficiently adhesive properties to rayon; hence, vinyl pyridine (VP) lattices were developed. The typical formulations of RFL systems with VP contents up to 80% for the preparation of high-strength rayon fibers are shown in Table 8.9.

Table 8.9 Typical formulation of RFL systems (parts per thousand by wet weight).

Component	Resol resin system	Novolak resin system
Resorcinol	18.8	—
Novolak (75%)	—	35.8
Sodium hydroxide solution (50%)	2.8	1.6
Formaldehyde (37%)	27.6	18.6
Ammonia water (28%)	—	9.8
Latex	424	405
Water	526.8	529.2
Total	1000	1000

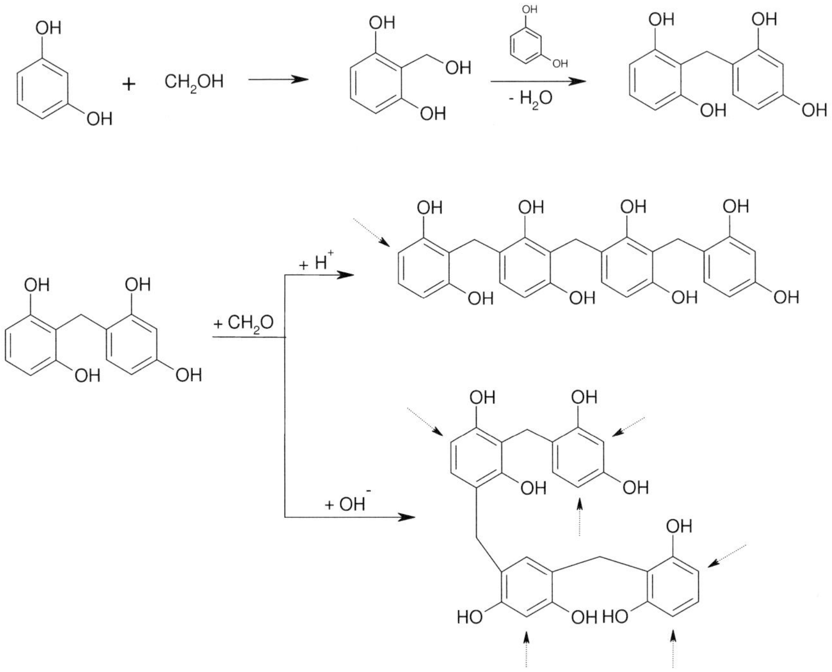

Figure 8.66 Acid condensation produces linear novolaks; basic condensation produces three-dimensionally crosslinked resol.

These formulations can contain either resol resin produced by basic catalysis, or novolak resin produced by acid precondensation. As acid catalysis activates *ortho* protons, novolak is a one-dimensional, noncrosslinked resin, whereas resol contains crosslinks (both activated in *ortho* and *para* positions; see Figure 8.66). Therefore, in order to allow crosslinking a novolak resin-based formulation must contain a higher amount of formalin compared to resol resin-based formulations (see Section 5.4).

Although, usually, comparable results are obtained with these formulations, for articles exposed to high dynamic stress, formulations without sodium hydroxide solution as the basic catalyst are better suited because the sodium hydroxide solution increases the modulus of the resulting film, which impairs dynamic performance.

Polyester As mentioned above, standard RFL systems do not confer satisfactory adhesive properties to polyester. In view of the environmental aspects and fire protection, aqueous rather than solvent-containing systems have therefore been developed which are based on dispersed blocked isocyanates that split off at heat (ca. 230 °C) to yield free isocyanate groups that are allowed to react. The solvent-containing systems are applied as a predip before the RFL dipping and allowed to react. The adhesion-promoting effect is almost as good as with solvent-containing systems. Alternatively, the predip is added to the RFL dip. Both systems are used widely for soft cords and conveyor belts. In contrast, stiff cords for the belt industry (raw-edge belts) must not fray at the cut edges, and therefore a high, interfilamentary adhesion must be produced. To date, this has only been achieved by using solvent-containing isocyanate systems. Owing to their high penetration capacity, they build up a high-strength polymer matrix between the filaments. To date, no waterborne system could be established commercially, despite intense developmental activities.

Aramid Although aramid (aromatic polyamide) is chemically closely related to nylon (aliphatic polyamide), no satisfactory results have been obtained with standard RFL systems. Whilst the established methods applied to the preparation of polyethylene–styrene (PES) can be used here, they sometimes impair the dynamic resistance of textiles. When using dispersed glycerol polyglycidyl ether as predip, followed by standard RFL dip in a second stage, it must be noted that optimum results are only obtained when the elevated temperature is maintained over a sufficiently long period (for about 60 s at 240 °C).

Special adhesion-promoted aramid fibers that no longer depend on a two-stage process are also available commercially, at least from one supplier. However, for complex systems, a two-bath dipping process performed in-house usually gives better results. As with PES, minor modifications in the rubber mixture can have major effects on adhesion, although in some particular cases the addition of carbon black to the dip can level this out.

Particularly with aramid, the latex component of the dip should be carefully adjusted to the rubber with regards to miscibility, but this problem does not occur with commercially available latex components (CR, NBR, etc.). Care should be taken, however, when using butyl rubber (IIR) or EPDM, for example. In this case, only polymer emulsions are available, which may not produce the same results.

8.10.4.4 Direct Adhesion Through Addition of the Adhesion Promoter to the Rubber Mixture

The labor-intensive dip process can be avoided when the components of the adhesive system are added directly to the rubber mixture. Highly dispersed silica is used to activate the reaction, and hexamethylenetetramine or hydroxyl methyl derivatives are employed as formaldehyde donors. The reaction between fiber and matrix takes place according to the mechanism described above. This method has the disadvantage that the adhesion promoter will be contained throughout the matrix, which may impair mechanical and dynamic properties.

When high adhesion performance is required, a combination of adhesion-promoted textiles and adhesion-promoted rubber mixtures has proven to be very useful, provided that other desired properties, such as dynamic resistance or resistance to aging, are not too much impaired [58].

8.10.5
Adhesion Mechanisms

With regards to the adhesion mechanisms involved in textile reinforcement of rubbers, a distinction must be made between adhesion at the dip/rubber interface and adhesion at the dip/textile interface.

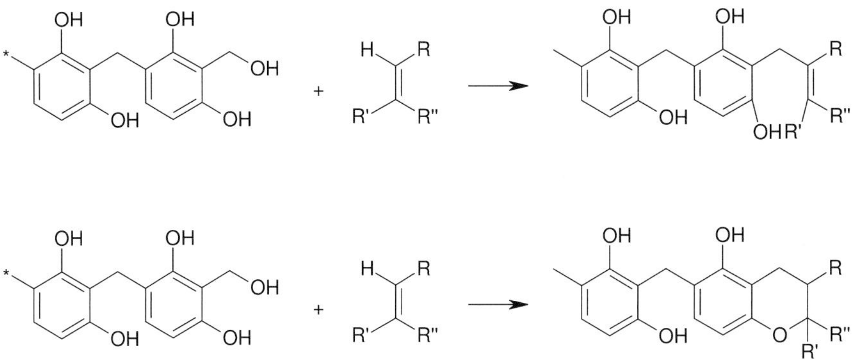

Figure 8.67 Possible condensation or addition reaction resols with latex double bonds.

Adhesion at the dip/rubber interface is obtained by the direct crosslinking of dip components (double bonds, hydroxyl groups, etc.) with the rubber matrix, in particular with its sulfur components. This explains the fact that crosslinking agents have a direct effect on adhesive strength, and that systems containing low amounts of sulfur or sulfur-free systems can hardly reach high adhesion levels.

The resin system used is only partly responsible for adhesion, which is obtained either via reactions of active H-atoms or via the formation of chroman structures (Figure 8.67). This effect, however, has only limited importance due to the relatively low amount of resin contained in the system (ca. 20% related to the solids content) and the slow reaction rates. The formation of an interpenetrating network is thought to be mainly responsible for adhesion.

Still fewer scientifically proven explanations are available for the dip/textile interface. Indeed, there is hardly any convincing and up-to-date argument in publicly available literature explaining the reasons for the underlying adhesion phenomenon. It is generally accepted that the adhesion is due partly to mechanical anchorage of the resin that penetrates into the spaces between the filaments, although microscopic investigations have shown that the dip only penetrates into two or three filament layers of the yarn (Figure 8.68).

This may be plausible with regard to the surface of rayon fibers that are microscopically quite rough, but it seems not to apply to 'purely' synthetic fibers such as nylon, PES or aramid, which are very 'smooth'. As far as chemical bonds are concerned, Figures 8.69 and 8.70 show that, in the case of cellulosic fibers and possibly nylon, covalent bonds would be a possible explanation, given the fact that 'simple' condensation reactions between the methoxy groups of the resin and active H-atoms or amide groups in the polymer chain are possible.

In the case of PES or aramid, however, these explanations are not acceptable. These chemically resistant polymers have only a few free OH groups (PES) or no chemical bonds at all, which would make it possible to react with the resin under the prevailing conditions. A possible approach for the explanation of the adhesion-promoting effect of the predip is the formation of polyurethane structures on the

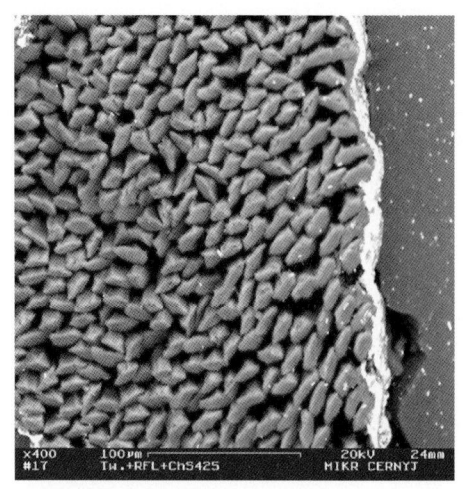

Figure 8.68 Microscopic section illustrating the thickness of a RFL layer (Aramid, dtex 840x1).

surface of the fibers that may provide a diffusion mechanism due to similar solubility parameters [59].

8.10.6
Aging of RFL-Dipped Textiles

Due to the reactivity of RFL systems, RFL-dipped textiles must be stored in a cool, dark and dry place (common storage conditions for such textiles) to preserve the adhesion-promoting effect until processing. Occasionally, correctly stored textiles preserve their properties for an extraordinarily long period of more than 24 months [60]. If, however, the textiles are exposed to UV light and sunlight, ozone and moisture, for example, the durability of the adhesion-promoting effect is rapidly and severely impaired (already after a few hours) because the RFL system is adversely

Figure 8.69 Possible reactions of precondensed resin with cellulosic structures in rayon.

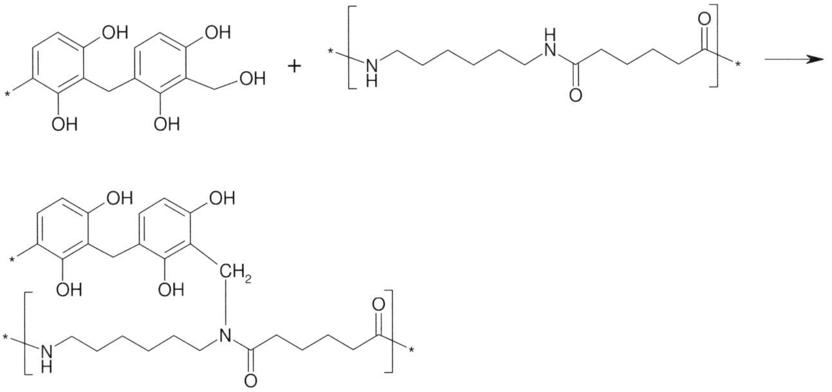

Figure 8.70 Possible reactions of precondensed resin with acid H-atoms in the amide group of PA 6.6.

affected. Investigations performed by the author have shown that, following irradiation with daylight for 100 h, the adhesion-promoting effect decreased by 60%, while 'normal' storage without protective black foil for 100 h induced a decrease of 50%.

Based on results obtained with investigations performed with ozone and UV light in other laboratories [61], it can be deduced that the dramatic adverse effect was due to a degradation of double bonds on the surface of the dip film. During vulcanization with the rubber matrix, due to a lack of double bonds, curing was drastically reduced or made impossible. The failure pattern changed from 'adhesive' failure to cohesive failure.

Moisture itself does not have any significant influence on the degradation of the adhesion-promoting effect over time, but it does increase the adverse effect of ozone. If textiles are too moist, problems may arise due to bubble formation during vulcanization, which may impair the quality of the product, but not adhesion itself. Due to high moisture uptake (up to 13%), problems are most significant with rayon, followed by nylon with a moisture uptake of up to 6%. Half of the uptake occurs very rapidly (within 30 min after leaving the stove) and is completed after about 15 h.

In the case of excess storage, if tests reveal a decrease in adhesion due to oxidative degradation, it is possible to re-dip the textiles to avoid economic loss. The original adhesion-promoting effect can be restored because there is still a sufficient amount of reactive components in the resin.

Textile finishing for the manufacture of fabric-reinforced rubber goods has been an established process for many decades, and indeed all technical fields of application of high-performance fibers rely upon this process. However, the phenomena of adhesion that underlie this process have not yet been fully identified, most likely due to the complexity of the reaction components used. Details of standard formulations and RFL systems, as well as the corresponding rubber mixtures, are publicly available, although the formulations of mixtures actually used in practice are kept as 'trade secrets' because they play a decisive role in the high performance of rubber–textile composites, and hence in the economic success of the involved businesses. Continuous development aiming at solving existing problems will doubtless meet and exceed the growing need for technical performance of reinforced rubber systems. With environmental protection in mind, one of the priorities today is to eliminate potentially harmful chemicals contained in the formulations, such as formaldehyde.

8.11
Footwear Industry

From the early 19th century until the Second World War, footwear was exclusively bonded with nitrocellulose solutions. Polychloroprene rubber solutions were introduced in about 1949, and offered the possibility of bonding chemically dissimilar materials such as rubber and synthetics. Until then, only leather had been used in footwear production, and the introduction of a new adhesive resulted in an extremely dynamic development of new, adhesion-based technologies and special machinery that made production processes less difficult and more efficient.

Around 1970, the bonding technologies in industrial footwear manufacturing took a further development with the introduction of polyurethane adhesives, caused by the need for bonding new sole materials (mainly PVC).

Today, a wide variety of mostly solvent-based adhesives are employed in the manufacture of footwear. However, since 2004, the European Directive for Volatile Organic Compounds (VOC) has made solvent use more difficult and more expensive, and nowadays the REACH regulation continues to urge the industry to find organic solvent-free adhesives in shoe manufacturing. Large-scale footwear manufacturers are seeking waterborne and/or solvent-free alternatives to stay abreast of changes in the regulations stipulated by law, to consolidate their reputation as environmentally friendly companies, to save costs, and to optimize their processes.

The adhesives used in footwear manufacture are solvent-based systems derived from rubber, SBS or SIS block copolymers, dispersions of polyacrylates and polyvinyl acetates, lattices, PSAs and hot melts. Two-part contact adhesives or HF-weldable adhesives meet higher strength requirements, for example when sewn borders are replaced. EVA hot glues are used for inner uppers and stiffeners, while polyamide hot glues are employed for folds and corner joints. The parts to be joined, inner uppers, interlinings of the upper and other reinforcements, toe paddings and stiffeners can be primed with adhesives and joined by heat sealing.

In footwear manufacture, *lasting* – the fixing of the upper onto and around the rim of the insole – is the most demanding process with regards to structural requirements. The upper is exposed to high stress during this process, but usually this does not involve any additional sewing or stitching. A typical lasting system is a combination of pulling-over and fixation with automatic injection of hot glue, typically a rapidly solidifying polyester or a flexible polyamide; both are used in the form of sticks or granulated. For special designs only, polychloroprene contact adhesives or, for light shoes, natural rubber solutions or lattices are also employed.

The soles are bonded with solvent-based adhesives derived from polychloroprene or polyurethane. Different methods for the preparation of surfaces to be bonded are available. Substrates with pores on their surfaces are primed with a thin adhesive film, while vulcanized rubber and thermoplastic rubber soles are halogenated (treatment with bleach or organic solvent solutions of trichloroisocyanuric) or sulfonated (treatment with sulfuric acid, also called cyclization) [62, 63]. Investigations have been performed with physical surface preparations (corona discharge, low- and high-pressure plasma, UV radiation) of rubber to enhance adhesion [64]. Nylon and other synthetic polymers are primed with a primer coat [65, 66].

One-part contact adhesives are applied as a thin layer on the sole. The coated soles are then stored in a dust-free place until needed for the manufacturing process. When two-part isocyanate adhesives with hardeners are used, care must be taken not to exceed the open time. Bonding by heat activation is achieved with IR irradiation of coated soles for several seconds until the temperature of the surface is between 55 °C and 80 °C. As the temperature of the sole itself remains much lower, a strong bond is obtained within a relatively short period of time. Soft or flexible soles that tend to deform can be bonded with adhesives requiring low activation energy at room temperature.

Other manufacturing processes involved are the *direct injection molding* of thermoplastic rubber or PVC, vulcanization of rubber, and low-pressure reaction molding of integral skin foam polyurethane. In this case, the fixed frame is coated with a two-part polyurethane adhesive, to which a latent hardener is frequently added, or with a self-vulcanizing synthetic rubber solution [67].

The application of hot glues derived from EVA is limited to the bonding of light footwear or textile uppers.

Adhesives are also used for *shoe repair*. One-part, polychloroprene-derived solvent-containing adhesives are universal adhesives that work with most different types of shoes, materials and material combinations. PVC and plastic materials are bonded with polyurethane adhesives.

Bonding strength in the footwear industry is defined by the European standard DIN EN 1392:2006 *Adhesives for leather and footwear materials – Solvent-based and dispersion adhesives – Testing of bond strength under specified conditions*, as well as the *Specification and Testing of Adhesives and Materials for Shoe Production*, as accredited by the European Association of Adhesives Manufacturers (FEICA).

8.12
Road and Highway Engineering

Within the EU, more than 40 000 lives are lost in traffic accidents each year, and some 9000 people are injured each day. Traffic accidents are the most common cause of death for those aged below 45 years, and cause more casualties than either heart disease or cancer. The financial burden is estimated at €160 billion every year, which is more than 2% of the gross domestic product of the EU.

During the 1930s, at a customer's suggestion, Harry Heltzer, the future President of the US company 3M, was assigned to a new project to develop a self-adhesive reflective material for use on highways to make the center marking on highways more reflective at night than was possible with the standard white or yellow paint. The

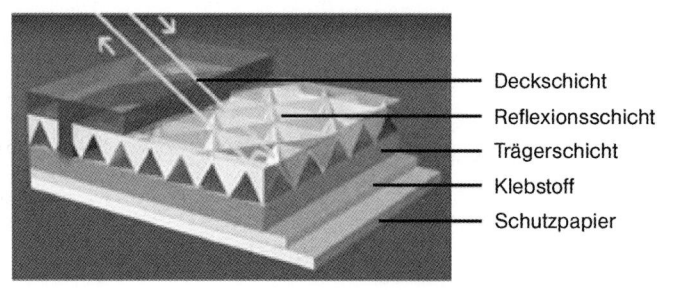

Figure 8.71 Structure of a self-adhesive reflecting foil made from microprisms.

solution was a pavement-marking tape coated with tiny glass beads that was also suited for use on traffic signs. The first retroreflecting traffic signs with the new reflective sheeting were installed in Minneapolis (Minnesota) in 1939.

Subsequently, a wide variety of self-adhesive retroreflecting products were developed that rapidly conquered the market. With the increase in mobility and private car use throughout the world, traffic safety needs also increased, and for many decades retroreflecting materials have contributed substantially to making the roads a safer place. Shortly afterwards, not only traffic signs but also license plates, town signs and street name signs, were provided with self-adhesive reflective sheeting. Today, retroreflecting contour marking represents a new trend in the rear and side marking of trucks.

For many years, 'be seen, be safe' has also been the motto in the leisure and professional clothing sector. From schoolchildren to sportspersons, roadmen or firefighters, reflective elements applied to clothes and accessories have become widely accepted as enhancing not only safety, but also design and visual appearance.

In extremely dangerous situations, for example in the case of accidents at sea, visibility is a matter of life and death. The provision of lifesaving equipment with reflective material substantially increases the chances of being rescued in case of wreckage, even at night and regardless of the weather.

The development of *microreplication*, a trendsetting technology, has further enhanced the performance of reflective materials in recent years. Here, a structured surface composed of tiny, identical microprisms with substantially higher reflection values than traditional reflective materials, replaces the glass beads (Figure 8.71).

8.12.1
Traffic Signs

Traffic signs and devices must be highly visible and legible to clearly alert and inform drivers in time. High-quality traffic signs can contribute to meeting the higher road safety needs arising from a traffic volume that has drastically increased in recent years. Especially in urban traffic conditions, traffic participants are blinded or distracted by extraneous lights, and the number of elderly traffic participants with visual impairment has steadily increased.

Self-adhesive retroreflective sheeting allows information to be safely communicated in time, even in adverse weather conditions, owing to high retroreflective

Figure 8.72 Traffic signs with reflecting foils.

performance (Figure 8.72). Scientific studies and national and international field tests have shown that carefully placed high-quality traffic signs also significantly enhance traffic safety.

Fluorescent yellow traffic signs are eye-catchers and help to avoid accidents. In Germany, each day, many traffic accidents occur at poorly marked danger spots. For example, over half of the total 25 000 railroad crossings are marked only by a conventional St. Andrew's cross; this is supposed to warn the road users but it does not – one in four of the accidents that occur at railroad crossings leads to fatalities (74 in the year 2000 alone). But many dramatic accidents also happen at construction sites and schools. In Germany in 2001, 106 schoolchildren were killed outside their school building. National and international studies have demonstrated that fluorescent yellow sheeting significantly increase the visibility of traffic signs, and the German automobile club, ADAC, has called for their use at railroad crossings.

Self-adhesive reflective sheeting for traffic signs are applied by rubber rollers, print rolls or knives to aluminum alloy or plastic surfaces. When using plastics as application surfaces, the applications should be evaluated to ensure acceptable adhesion. Although adhesion promotion by primers is not necessary, it is advantageous to slightly roughen the sheets such that their surfaces are absolutely clean, dry and free of grease immediately prior to bonding, which is performed at a temperature of at least 18 °C. Raising the temperature to 25 °C leads to an increase in the initial adhesion capacity of the sheeting.

8.12.2
Pavement Markings

In the dark, 40% of roadway markings are poorly visible, while in dark and rainy conditions even the majority of markings cannot be seen. As a consequence, they will no longer serve their purpose, which is to route traffic. Pavement markings with high night-time visibility in rainy conditions definitely enhance road safety and should be used everywhere.

High-quality pavement markings are clearly visible, retain reflectivity even at night and in rainy conditions, and are nonslippery for their period of use. Durable, high-performance self-adhesive reflective pavement marking tapes are available that cope with virtually all conditions with regards to traffic needs, weather conditions and types of road.

Self-adhesive permanent marking tapes with high retroreflective values, nonslippery characteristics and a long life span are composed of glass beads or ceramic particles embedded in a firm polyurethane surface coating that covers an elastic rubber layer, as well as a PSA layer which is adapted to the pavement.

Temporary Markings At construction sites, when roads become narrow the drivers become stressed and must pay close attention to traffic signs and speed limits, to trucks nearby, and also to oncoming traffic which may be only a line's width away. During daytime conditions this situation demands the utmost concentration and driving skills, but at night time and in rainy conditions it is even worse.

Pavement markings provide visual guidance and help drivers to stay in their lanes, guiding them clearly and safely through the dangerous construction site.

Poorly visible or missing lane lines are an important risk factor at construction sites, but this risk can be greatly reduced and even eliminated by using high-quality pavement markings (Figure 8.73).

Reflective, soft, aluminum-based, removable pavement marking tapes are clearly visible and suited for temporary marking at construction sites, deviations, parking lots, workshops, or for permanent marking. These tapes adjust to all types of unevenness and can be driven on immediately after application, as with all types of pavement markings.

Removable tapes with a rubber support layer and an embedded reinforcement network can be used for longer-lasting, temporary marking of construction sites and winter construction sites.

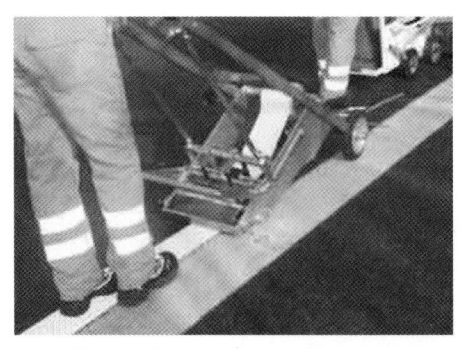

Figure 8.73 Application of removable pavement markings at a construction site.

New fluorescent yellow products provide excellent visibility during the daytime, nonslippery characteristics, and enhanced visibility at night, even under rainy conditions, to provide safe visual guidance.

8.12.3
License Plates

License plates, which were conceived with the birth of the automobile, began as simple color-painted plates but have rapidly evolved into high-tech car identification labels. One landmark in this development was the replacement of color painting by self-adhesive reflective sheeting (binding since 1989). Later, in 1995, European vehicle registration plate graphics were introduced (binding in Germany since 2000).

8.13
Surface Design

8.13.1
Surface Design and Surface Protection

Self-adhesive films for the design or protection of surfaces are specially conceived for use on flat or uneven substrates, with or without rivets and recesses, and adjust to any unevenness without too-high a restoring force. This is achieved by careful adjustment of the rheology of the system, which comprises the film, the surface coating and a PSA.

Modern self-adhesive films are easily and quickly applied, without producing any bubbles or wrinkles; this is due to the presence of microscopic air release channels within the adhesive, through which air can escape during application. Tiny glass beads within the adhesive layer allow even large films to be accurately positioned; consequently, even untrained users can easily apply films without producing flaws and significantly reducing the number of rejects.

Special graffiti-resistant films have been developed to protect the outer surfaces of buses or railroad carriages. Markings made with spray-color paint, felt-tip pens or shoe polish can be easily and quickly removed using commercial cleaners. Unlike the situation with paints, no shadow is left behind and there is no need for time-consuming and costly surface preparation. Another advantage is that, when the sprayers find that a plastic film has been applied they go elsewhere to spray on difficult-to-clean, painted surfaces.

These films confer a clean appearance not only to outdoor applications but also to interior surfaces. Today, in many public transport vehicles the windows are deliberately and needlessly scratched on the inside; applying a scratch-protecting films would prevent such scratching and improve the visibility for passengers.

8.13.2
Fleet Graphics

Although today, many companies have fleets of service vehicles (passenger cars, transporters, trucks) which offer a huge potential for brand promotion, this approach is rarely used. Yet, by using the advertising space provided by these vehicles, whether by using a simple logo or complex graphics, it is possible to create an excellent product image.

In Germany – and indeed throughout Europe – because a large proportion of heavy road traffic runs at night, increasing numbers of accidents are occurring, due mainly to the poorer visibility. This problem could be eased by providing trucks with self-adhesive retroreflecting films that allow them to be seen clearly at night. An investigation conducted by TU (Technical University) Darmstadt, Germany, suggested that by using these retroreflecting films, night-time accidents involving trucks could be reduced by more than 90%. Yet, these films are used on only 1% of all trucks. At night, car drivers perceive only about 5% of the information that they do during daytime; hence, at night an additional visual stimulus is required, and this rear-marking of trucks would make them clearly visible from a distance of at least 100 m.

In a self-adhesive retroreflective sheeting, glass beads or prisms are embedded into a reflection layer that allows the entering rays of light to be returned to their source at the same angle. In this way, reflective sheeting may be used not only to enhance safety but also to transform vehicles into an attractive advertising medium (Figure 8.74). In fact, graphics applied to the sides of trucks may serve as highly memorable and impressive eyecatchers. Moreover, these advertising campaigns are visible around the clock, reach far more people, and also enhance traffic safety.

8.13.3
Public Transport Advertising

A survey conducted by Infratest, Germany, has shown that traffic advertising reaches a target of 60% of the population. In this respect, mobile outdoor advertising is recognized as being highly effective, with 73% of drivers and passengers remembering it overall, and 31% remembering it in detail. The main target groups are buses and street or rail cars, with creatively designed vehicles contributing to the animation

Figure 8.74 Reflective advertising graphics on a truck.

Figure 8.75 Modern traffic advertising.

and entertainment of the street scene, particularly when combined with conventional billboard campaigns.

Often, a self-adhesive film is used to cover the entire side of a bus, windows included (Figure 8.75). In this case the inside window surface is provided with a special film that does not disrupt the passengers' view, whereas people on the outside see a uniform picture.

Traffic advertising is especially suited to short-term advertising campaigns because it is inexpensive. Particularly in city centers, where large surfaces are only available to a limited extent; indeed, studies have shown that in this way target groups can be particularly well reached, notably the mobile, young and dynamic.

Today, aircraft graphics are also becoming more interesting (Figure 8.76). Where else could an advertisement be placed so eye-catchingly in an international environment? A spectacular highlight of the World's Fair Expo 2000 was a Lufthansa Boeing 747-400 decorated with a 800 m^2 Expo logo, which flew round the globe campaigning for the first World's Fair taking place in Germany. Working around the clock for two-and-a-half days, 30 Lufthansa employees assembled and applied the graphic from 270 film pieces.

The special graphic film was provided with a microperforation, was extremely weather-resistant, and could be stretched horizontally and vertically to prevent cracks from being generated during flight. The graphics were not applied to highly exposed components such as the wings, tail edges and areas where leakage of fuel or hydraulic fluid may have occurred. The film was a digital-printed graphic sealed with a screen-printed, two-part clear coat. Laminated film was not employed

Figure 8.76 Aircraft graphics.

because lamination would close the pores. A protective coat was applied manually to the film edges in order to avoid detachment that might represent a safety risk.

Compared to painting, modern aircraft graphics technology has several advantages. Currently, as the majority of airlines lease rather than buy their aircraft, graphic films allow the aircraft to be easily adapted to the corporate design of an airline. Hence, the visual implementation of short-term marketing campaigns is not a problem. Another benefit is that, unlike painting, bonding does not require a dust-free environment and need not be performed in specially prepared workshops. The time factor is another advantage – the bonding of a film is much faster than painting, and significantly reduces the aircraft's expensive downtime.

8.13.4
Building and Floor Graphics

The bigger a building, the more options for eye-catching facade design. If a building is highly visible to thousands of people who drive past it every day, it can be converted into a spectacular billboard for promotions and outdoor advertising.

Some years ago, a mammoth image featuring Edgar Davids, a Dutch soccer player, dominated the skyline of Rotterdam as a mind-boggling graphic converted the highest building in the Netherlands into an impressive, giant advertising pillar (Figure 8.77). At a height of 150 m, the 'Delftse Port', headquarters of the Nationale-Nederlanden insurance company, is taller than all other surrounding office buildings. The huge poster – the biggest building decoration ever produced – was manufactured from a special perforated film that provided an impressive view, without disrupting the view from inside.

Figure 8.77 Decoration of a façade.

To do this, from 4500 self-adhesive pieces weighing a total of 3000 kg, Europe's leading specialists in large-size digital printing had to assemble a giant puzzle. A so-called 'window marking film' was bonded to the front and reverse side of the taller of the two office buildings, as well as on the smaller of the two towers. It took 25 days to print the colorful graphic in Reading, UK. To achieve a perfect assembly, a team was sent to Rotterdam to produce accurate drawings of the building. Each piece was numbered and provided with a code corresponding to the windows of the building, allowing each single piece to be placed on the correct window. In the city center of Rotterdam, a team of eight worked for a total of 30 days to install the graphic that when complete, covered an area of 9500 m^2 and stood more than 60 times larger than life. The team worked in pairs, using window-washing platforms to travel up and down the sides of the building. The working conditions were exceptional, with the team working in all weathers, and managing to install the graphic without using conventional tools such as knives so as not to harm the glass façade, which may have put the structure of the building at risk. The stunning graphic, which was the size of the Statue of Liberty, caught the eye from a distance of several kilometers and created a furore all over Europe. This exceptional project showed what innovative film technology and modern PSAs could achieve – no job is too large or too colorful, and the resulting graphic simply looked great.

Consumer goods giant Procter & Gamble adopted another special method of outdoor advertising when introducing 'Swiffer', a new cleaning cloth. Besides conventional media such as television, print and radio, the branded company banked on floor graphics. A special nonslippery self-adhesive film – so-called 'floorminder' – was used as a huge eye-catcher on the platforms, stairs and pillars of German urban railway stations. The colorful graphics stood out from conventional advertising and left a lasting mark.

These graphics can also be walked on, as they are suited to virtually all types of flooring and can be removed without leaving any residues when the advertising campaign is over. Special films for outdoor applications can be used not only in trade areas, but also for walk-on advertising areas during target group-specific music and sport events.

8.14
Tamper-Indicating Labels

From TV crime thrillers, we are all familiar with the paper sealing labels that detectives use to secure doors against unauthorized opening. These are bonded to the door and the frame and will tear when the door is opened.

More sensitive tamper-indicating labels which signal that the seal has been tampered with by an attempt to remove it or to open the door, are becoming increasingly popular (Figure 8.78). Here, when only slightly disturbed or peeled off, the strip produces an irreversible VOID message.

Tamper-indicating labels are, for example, multilayer strips that have a thin transparent upper layer, which is coated on one side with an adhesion promoter,

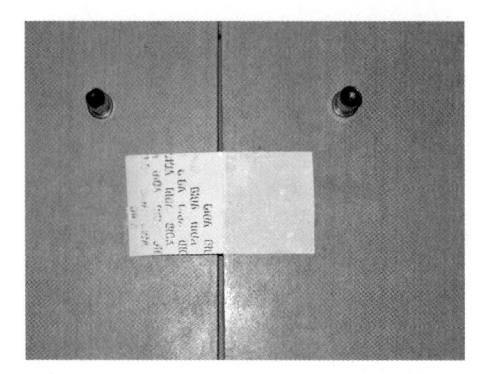

Figure 8.78 Tamper-indicating label at a cabinet door; after application (top), after opening the right-hand door (bottom).

followed by a thin pigmented intermediate layer, and finally a lower layer with a pressure-sensitive acrylic adhesive that is provided with an easily removable protective layer. In the area of the VOID message, the transparent layer is coated with an anti-adhesion layer by means of a special printing method prior to application of the adhesion promoter. When the bonded strip is disrupted, or an attempt is made to remove it, the bonding will detach in the area of the anti-adhesion layer first. The pigmented layer and the acrylic adhesive will remain on the item to be protected. The transparent upper layer exposes the VOID message, which appears on the item with the pigmented layer. Translucent colors can be printed on the noncoated side of the transparent upper layer.

This type of tamper-indicating labels is used not only for sealing doors but also as packaging security devices for sensitive products. Similar systems are employed against counterfeit.

Care must be taken not to damage the tamper-indicating layer when removing the protective layer. This applies particularly to the sealing of doors, which is done manually under conditions that may be difficult to control. Furthermore, the bonding area should be thoroughly cleaned to enable the acrylic layer to adhere well to the surface and allow the VOID message to be exposed when attempts are made to break the seal.

8.15
Medical Sector

In the medical sector, adhesives are used in a wide range of applications where they must meet the highest demands with regards to quality and reliability. Medical applications can be divided into three groups according to the period of contact between the medical device and the human body, namely short-term contact, contact for several days, and long-term contact.

8.15.1
Medical Devices with Short-Term, Superficial Body Contact

Medical instruments or products with short-term contact include syringes, cannulas, catheters, valves, filters, respiratory masks, keyhole surgery instruments, endoscopes and orthodontic aids. These are composed of very different materials that sometimes are difficult to bond, such as metal alloys of aluminum, medical steel, titanium, nickel (Nitinol), or plastics such as polyethylene (PE), polypropylene (PP), polyetheretherketone (PEEK), polyphenylene sulfone (PPSU), polycarbonate (PC), polyvinyl chloride (PVC), or oxidic materials such as glasses or ceramics. Structural adhesive bonding is sometimes the only joining technique suitable for this type of material.

Acrylic systems are used for the manufacture of *sterile disposable products* such as catheters, infusion systems and disposable syringes, and for the production of nonsterile reusable products such as rehabilitation aids and orthopedic devices. Acrylic-based systems set rapidly under the effect of moisture, catalysts, heat, UV light or visible light. Cyanoacrylates usually polymerize and set in the presence of moisture on the substrates.

Epoxy adhesives are used for *sterilizable, reusable products* such as surgery instruments and endoscopes. The materials and adhesives used for manufacture must be biologically inert and resistant to corrosion and temperature. The adhesive bond must have high durability and resistance to cleansing materials, disinfectants and steam sterilization, and a defined strength to ensure correct functionality after the cleansing and sterilization process [68, 69].

8.15.2
Medical Devices with Several (up to 30) Days Body Contact

Self-adhesive products such as 'band-aids' are the most ancient and popular applications of adhesives in the medical sector. PSA tapes come into contact with the human body for several days, while acrylic spray-on adhesives are widely used on cleansed wounds where they replace band-aids. Newly developed *transparent wound dressings* made from polyurethane film and acrylic adhesive may remain on the injection site of central venous catheters for up to seven days. These allow the progress of wound healing to be observed and are permeable to water vapor and oxygen, but impermeable to microorganisms and water [70].

8.15.3
Medical Devices with Long-Term (Over 30 Days) Body Contact

In *dental treatment*, adhesives are in permanent contact with the human body. The methacrylate adhesives which are used as filling materials set rapidly under the effect of light, which is the only setting procedure suited to this field of application. Glass ionomer cements also set under the effect of light, and are used for orthodontic treatment with molar bands and brackets [71].

Methyl methacrylate bone cements have been used for many years to fix endoprostheses (e.g. hip joint). Medically approved polyurethanes and neutral cure silicones are used for implantable devices [69].

Tissue adhesives, which also remain in contact with the body over several months, must be resistant to humid environments and compatible with endogenous tissues. Tissue adhesives are derived from natural and synthetic materials.

Medically approved cyanoacrylates (2-octyl cyanoacrylate, *n*-butyl-2-cyanoacrylate) are used to replace conventional sutures of flesh wounds. They are also employed in plastic surgery [72, 73].

Histoacryl Blue is a topical skin adhesive and consists of *n*-butyl-2-cyanoacrylate which, in the presence of moisture on the skin, polymerizes in seconds. The adhesive contains a blue dye in order to make it easier to see when it is being applied (Figure 8.79) [74].

Cyanoacrylates are also used to repair the middle ear membrane, to prevent embolism in case of arteriovenous anomaly of the brain, to repair the retina, for bone stabilization, and to repair perforations of the sclera and cornea. Other areas of applications are the treatment of liver wounds, herniae (inguinal tissue rupture), perineotomies and trocar incision sites. By hydrolysis, the adhesive is primarily metabolized into formaldehyde and alkyl cyanates, which are later excreted. Surgeons use special ampoules or electronic pumps to apply cyanoacrylates, which form a strong tissue bond within 60 s [71, 75, 76].

Fibrin is an endogenous adhesive formed during coagulation by combination of fibrinogen and thrombin in the presence of factor XIII and calcium. Factor XIII is contained in fibrinogen, and mixing both substances directly initiates the blood coagulation cascade. The fibrinogen and thrombin components are typically delivered

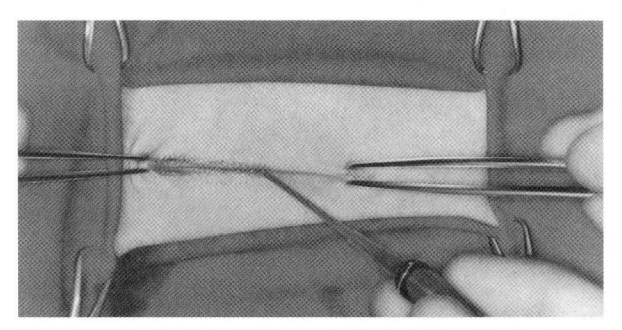

Figure 8.79 Application of Histoacryl, a topical skin adhesive. (Illustration courtesy of Braun).

in lyophilized form and reconstituted later; they are applied separately to the wound using a double syringe with blunt cannula. After application, the sealant becomes viscous, adheres to the wound or incision, and immediately forms a fibrin network. Larger areas are sealed using a spray-on application of the adhesive, while severe hemorrhages are sealed with a sponge or fleece. Fibrin is used in thoracic and cardiovascular surgery (sealing of lung leaks and esophageal leaks or splenic, liver and kidney lesions), in neurology [sealing of cerebrospinal fluid (CSF) leaks], and in ophthalmology [75, 77–79].

Gelatin-resorcinol formaldehyde (GRF) tissue adhesive is produced from gelatin, resorcinol and water in the presence of formaldehyde and glutaraldehyde. GRF was used for aortic dissection repair [80] and enhances hemostasis in hepatic or renal vein resections [75].

Gelatin-resorcinol pentanedial ethanedial (GR-DIAL) adhesive is produced from gelatin, resorcinol and water in the presence of pentanedial and ethanedial; the latter are hardeners which determine the setting rate of the adhesive. GR-DIAL has been used for aortic dissection repair [80].

A *synthetic absorbable hydrogel* derived from polyethylene glycol (PEG) is used for the sealing of vessels in a wide range of indications, including peripheral bypass grafts, dialysis grafts, aortic aneurysms [81] and air leaks in lung surgery [76].

Collagen gel seals all forms of bleedings in the heart, blood vessels and spine surgery. In combination with endogenous plasma, collagen and thrombin suspensions are used for hemostasis [72]. A mixture of collagen, thrombin and endogenous fibrinogen seals CSF leaks in cranial and spine surgery [81].

Adhesive *protein plaque* is found on the tips of byssal threads formed by mussels (a mussel forms around 50–100 adhesive threads). The main component of these proteins is an adhesion molecule called *Mytilus edulis* foot protein (Mefp-I; Figure 8.80; see also Section 8.15) [82]. The biodegradable protein is composed of 17 amino acids, and can be used for the fixture of chondrocytes and osteoblasts [75].

Prolamin gel is also a biodegradable protein, and is used for the intravascular sealing of arteries in tumor resection and for the temporary sealing of the pancreas during transplantation [75].

Figure 8.80 Mussel (*Mytulis edulis*) with byssus threads and foot on an aquarium glass; the mussel diameter is approximately 1.5 cm. (Photo courtesy of IFAM).

In combination with glutaraldehyde, bovine serum albumin (BSA) seals CSF leaks in the surgery of pituitary gland tumors and hemorrhages associated with the use of ventricular devices [83]. Another field of application for BSA is in aortic dissection surgery [79].

Many medical adhesives and sealants are currently under intense development, and novel products are being continuously approved for use in the medical sector.

8.16
Adhesion in Nature

8.16.1
Design Principles of Biological Attachment Devices

In Nature, attachment devices serve as functional systems, the purpose of which is either temporary or permanent mechanical attachment of an organism to a substrate surface, to another organism, or the temporary interconnection of body parts within an organism. Many species of animals and plants are supplied with diverse attachment devices, and there is no doubt that many functional solutions have evolved independently in different lineages. The morphology depends on the species biology and the particular function in which the attachment is involved. The evolutionary background and the habits influence the specific composition of attachment systems in each particular species [84].

There are eight fundamental classes of attachment principles (Figure 8.81): (1) hooks; (2) lock or snap; (3) clamp; (4) spacer; (5) suction; (6) expansion anchor;

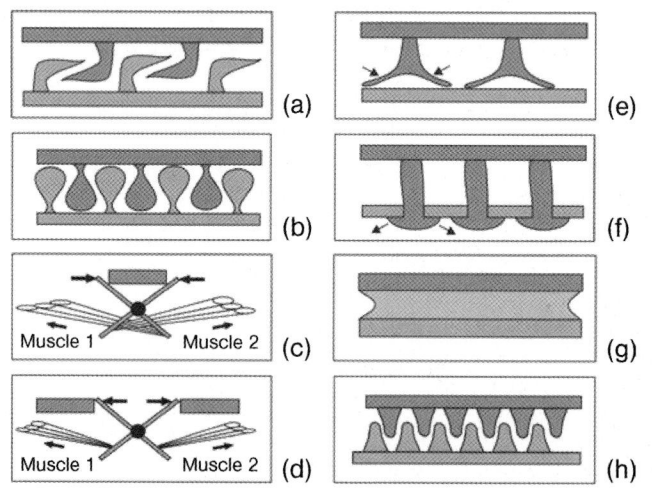

Figure 8.81 Fundamental principles of attachment found in Nature. (a) hooks; (b) lock or snap; (c) clamp; (d) spacer; (e) suction; (f) expansion anchor; (g) adhesive secretions (glue); (h) friction.

(7) adhesive secretions (glue); and (8) friction. However, different combinations of these principles also occur.

Adhesion developed by biological adhesives can be due to the use of different basic physical forces, with the adhesive effect often being enhanced by particular contact geometry and mechanical interlocking between the adhesive and irregularities of the contact surface.

8.16.2
Gluing Under Water

Many marine organisms are able to secrete glues, which can act in aquatic environment (Figure 8.82 and Table 8.10). Flatworms (turbellarians), annelids and gastrotrichs (worm-like animals) use glandular organs composed of three cell types: two distinct gland types and specialized epidermal cells for secretion discharge (Figure 8.82) [86]. The dual function of these glands leads to the term 'duo-glands'.

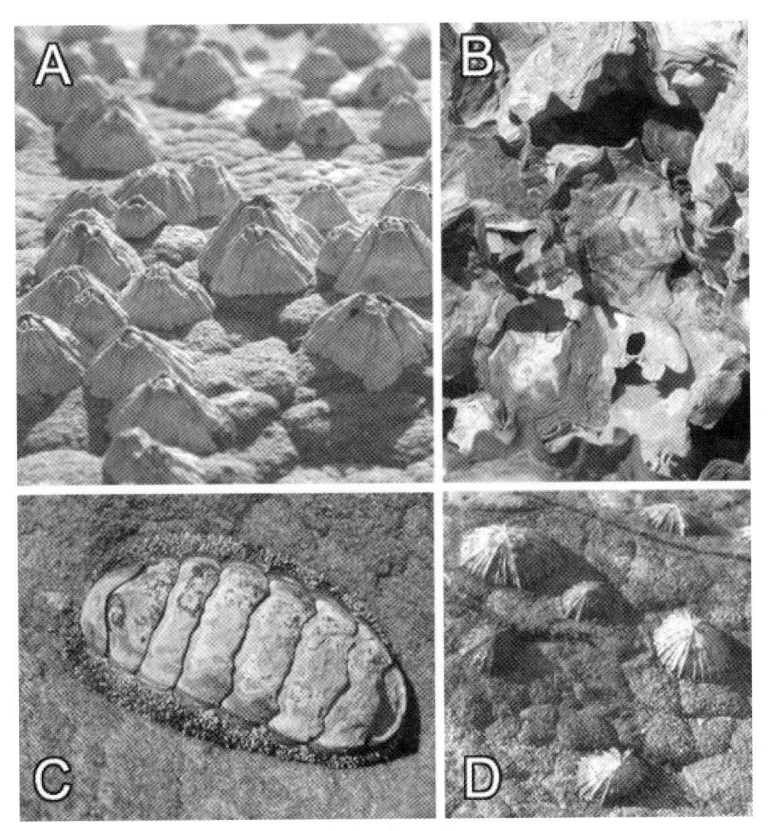

Figure 8.82 Gluing under water. (A) barnacles and (B–D) mollusks on the rock substrate. (A) barnacles *Tesseropora rosea*; (B) rock oyster *Saccostrea glomerata*; (C) chiton *Acanthopleura gaimardi*; (D) common limpet *Cellana tramoserica*.

Table 8.10 Tenacity of some marine invertebrates on smooth glass [85].

Taxa	Structure involved in adhesion	Type of adhesion	Tenacity ($\times 10^5$ N m^{-2})
Cnidaria	Foot	Transitory	0.43
Anthozoa	Foot	Transitory	0.19
Mollusca, Gastropoda	Foot	Transitory	2.28
Mollusca, Bivalvia	Byssus plaque	Permanent	3.16–7.50
Crustacea, Cirripedia	antennules	Temporary	0.98
Crustacea, Cirripedia	Base	Permanent	2.30
Crustacea, Cirripedia	Base	Permanent	5.20
Echinodermata, Asteroidea	Podia	Temporary	1.70
Echinodermata, Holothuroidea	Cuvierian tubules	Permanent	6.00

This triplet of structures, usually operating with three distinctive functions of adhesion, de-adhesion and tension bearing, is recognized in most marine organisms.

It has also been suggested that marine organisms adhere by the secretion of polycationic substances, which bind to substrates, especially to anionic sites [87]. The actual mechanism of adhesion depends on the chemistry of secretion. The attachment systems of polychaete larvae use lectins as adhesives, whereas the majority of other systems bear proteins and glycans (glycoproteins, protoglycans, polysaccharide protein complexes). Systems adapted for permanent adhesion, such as those found in mussels and barnacles, bear proteins and polyphenols (Figure 8.82) [88–92]. Polyphenols play a role in tanning proteinaceous components of the secretion. The variety of adhesive systems, based on two-component adhesion, shows that there is no universal duo-gland mechanism [86].

In systems using adhesive substances as a form of viscoelastic lubricant for locomotion, glycans are the main component (Figure 8.83). These adhesives, which have been reported in molluskan pedals and echinoderm tube feet, are known as 'mucus'. Due to its viscoelastic properties, mucus behaves as an elastic solid under rapidly applied forces (adhesive function), but as a flowing fluid at slow velocities (lubricating function) [93, 94].

8.16.2.1 Byssus Adhesion in Mollusks

Marine bivalve mollusks are probably the most studied adhesive marine organisms. Their adhesive capacity is advantageous, because it allows mollusks to exploit nutrient-rich turbulent zones of the sea.

Oysters are permanently fixed to the substrate by secretion of cement glands (see Figure 8.82), whereas byssated mollusks adhere but are not necessarily permanently attached. They are able to abandon an old attachment site and move to a new one. The *byssus* is an extracellular structure consisting of a thread (0.05–0.20 mm in diameter and 2–4 cm long) attached to the animal at one end and to a substrate at the other (Figure 8.84) [95, 96]. An additional advantage of the byssus are its mechanical properties, allowing better damping of the wave impact.

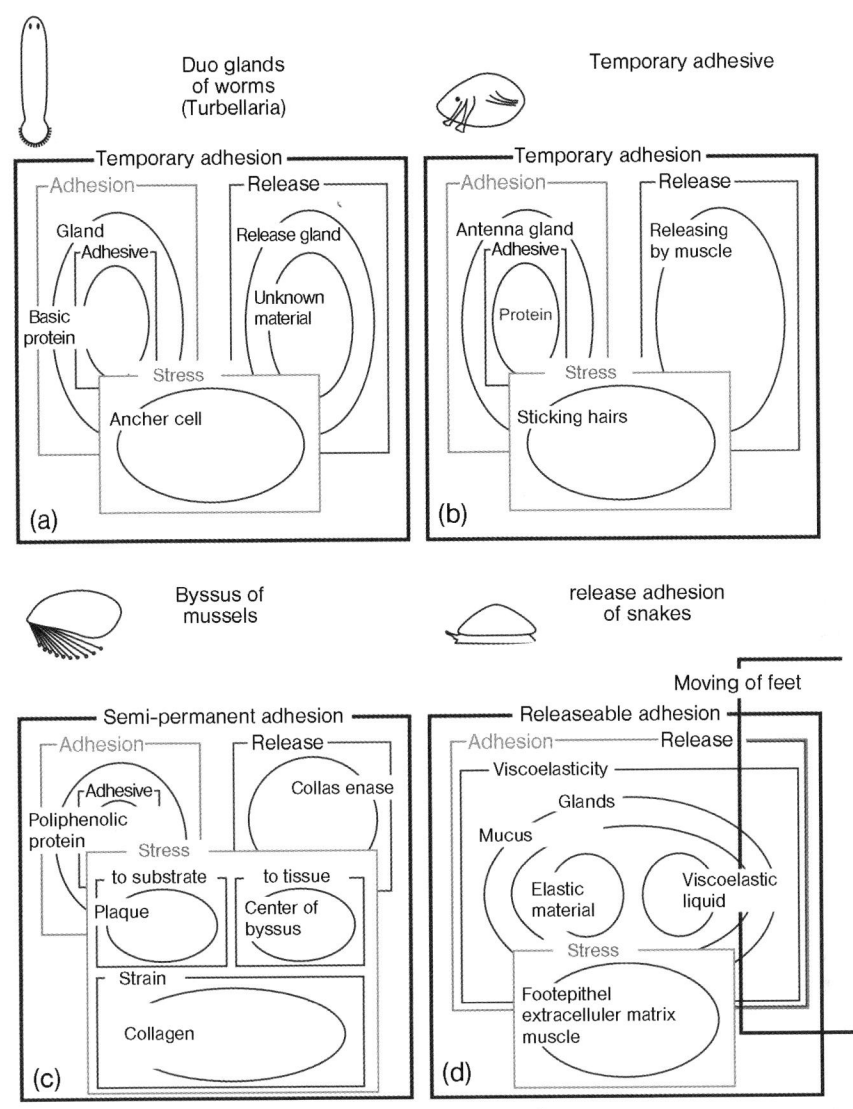

Figure 8.83 Schematic comparison of functional elements of adhesive organs in a tubellarian (a), barnacle (b), mussel (c) and gastropod (d). The functions are indicated by boxes, and the structures which perform them as ovals.

The foot produces the byssus, and the attachment to the animal is at the ventral base of the foot. The threads end at the stem, a root-like process embedded deeply in the base of the foot. The distal ends of the threads are attached to oval plaques (2.6 mm long and 2.0 mm wide).

The average attachment strength of the whole composition of threads of an animal ranges from 10–12 N in *Mytilus edulis* to 90 N in *Septifer bifurcatus* [95]. However, such

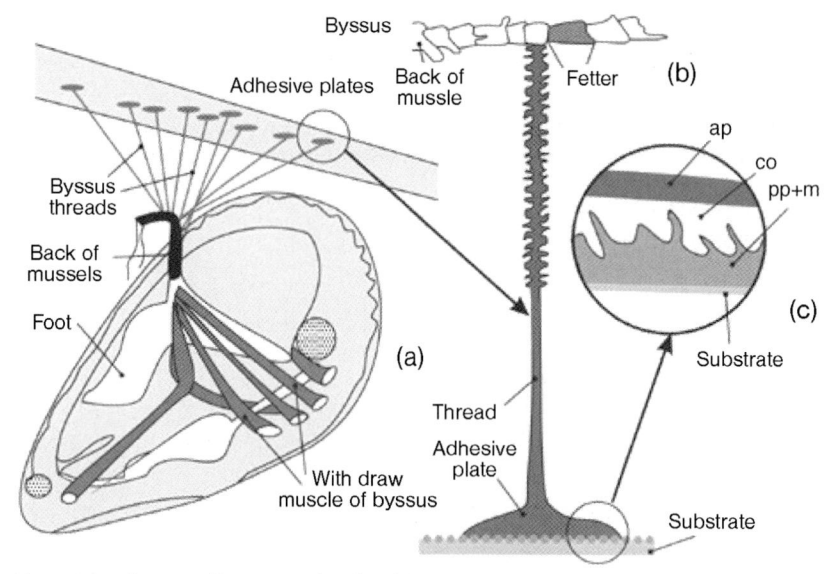

Figure 8.84 The mussel byssus is a bundle of threads consisting of extracellular material. By drawing itself up on these tendinous threads, the mussel can control the tension of its attachment. The distal tips of the threads end in adhesive pads called plaques.
(a) Schematic view of a mussel suspended by a byssus;
(b) schematic of the byssus; (c) distribution of mucosubstance (m), collagen (co) and polyphenolic protein (pp).
ap = adhesive pad.

a force is summed over all structural elements securing animals against external forces. It includes an adhesive bond of the plaque to the substrate, the thread attachment strength to the mussel, and the tensile strength of the threads. Usually, the breaking strength of threads is greater than that of the attachment sites. The proximal region and adhesive plaque are common sites of structural failure [97].

Byssal threads consist of highly ordered crystalline polymers (collagen and b-type protein), with the collagen fibers packed inside a b-type protein sheath [98]. Occasionally, the byssal threads consist of striated fibers.

Five groups of cells or glands participate in byssus production. In most species, the byssus consists of at least four essential components: acid mucopolysaccharides; adhesive protein; fibrous protein; and an oxidative enzyme, polyphenoloxidase. Many chemical details of byssus secretion remain unknown.

The sequence of events during the formation of the attachment plaque has been previously reviewed [95]:

1. With a contact angle of about 6°, a dispersion consisting of phenol granules and a mucoid substance is secreted onto the substrate surface. This dispersion is presumably mixed and applied to the surface in films 50 nm thick by use of the paddle-shaped cilia. Such a mixture results in a polyphenolic protein forming the continuous bonding phase.

2. The secretion becomes increasingly heterogeneous, with collagen added via the longitudinal ducts as the third component. The architecture of this heterogeneous material is cellular (trabecular): spaces between the trabeculae of phenolic material are filled with mucoid secretion. The collagen interacts with phenolic material and penetrates the trabecular architecture to a depth of about 3 µm.

3. The outer coverage of the thread and plaque is a dense substance, probably b-type protein.

As a result of these events, the byssal attachment plaque joins two dissimilar materials, a surface and the fibrous material of the threads.

8.16.2.2 Tube Feet of Starfish

Echinodermata (starfish, see urchins, sea cucumbers) bear tube feet or podia that are used to adhere to and release from substrates by chemical interactions (Figure 8.85).

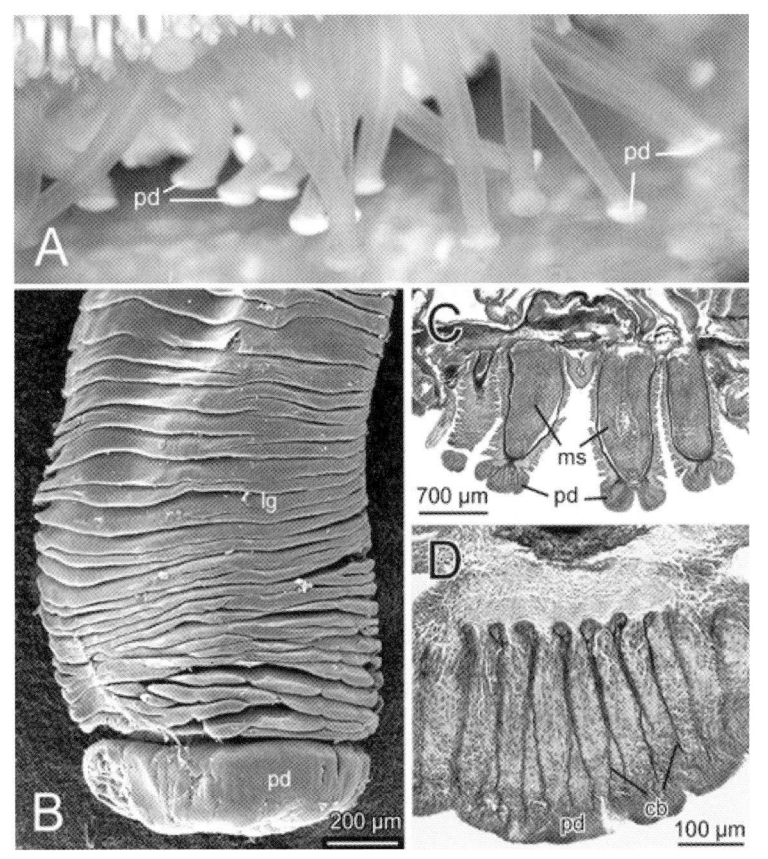

Figure 8.85 Podium of the starfish, *Asterias rubens*. (A) Starfish adhering to glass; (B) side aspect as viewed in the scanning electron microscope; (C,D) longitudinal sections through the podium. cb = collagen bundles; lg = leg; ms = muscles; pd = podium.

This mechanism allows them to move [99, 100] and burrow on substrates. The podia are external appendages of the hydraulic (ambulacral) system, and are driven partly by muscles and partly by hydraulic pressure [101]. The following types of podia have been identified: descending, ramified, knob ending, lamellate and digitate [102].

The histological structure of podia is relatively constant, there being four tissue layers: an inner and outer epithelium, with a connective tissue layer and a nerve plexus between them. The connective tissue layer is made from an amorphous substance that encloses bundles of collagen fibers and various types of cell. The layer consists of a diffuse external layer and a more compact internal layer within which the collagen fibers are arranged in circular fashion. In starfish and sea cucumbers, the external connective layer may contain one or several types of embedded calcareous needle (spicules) [103, 104]. Externally, a multilayered (three to five sublayers) cuticle covers the epidermis, consisting of fibrous and, sometimes, granular material [102].

Only the distal parts of the tube feet (knobs, discs) are involved in attachment. Adhesive organs operate according to mechanical and chemical mechanisms; the podia possess an adhesive system with two types of secretory cell – those that release an adhesive secretion and those that release a de-adhesive secretion. In the podia epidermis there are three main types of specialized cell: secretory, neurosecretory-like, and sensory. In the handling podia the cells are always closely associated, forming sensory-secretory complexes (Figure 8.85) [102], but in the locomotory podia three cell types form a homogeneous cellular layer together with support cells (Figure 8.86). Such a cellular organization of the adhesive surface correlates well with the functions of the podia; a large adhesive area correspondingly provides a stronger attachment during locomotion, whereas discrete adhesive areas seem to be more efficient for the handling of small substrate particles.

Adhesive and de-adhesive secretions are produced by secretory and neurosecretory-like cells, respectively. In most species, the secretory cells contain acid mucopolysaccharides, although in several species proteins rather than mucopolysaccharides have been revealed in these cells. The third group of species possesses both

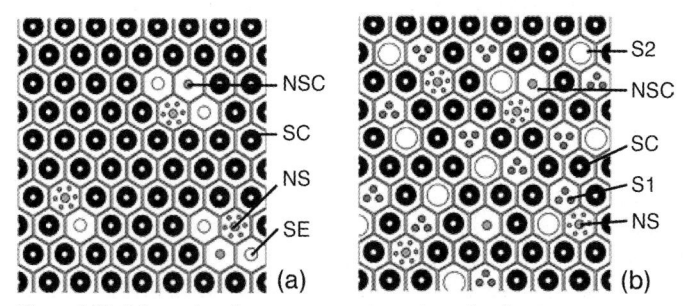

Figure 8.86 Schematics of transverse sections through adhesive areas of echinoderm podia. (a) Handling podia of *Asteronyx loveni* (Ophiuroidea); (b) locomotory podia of *Holothuria forskali* (Holothuroidea). NS = neurosecretory-like cells; NSC = sensory cells; S1, S2 and SE = secretory cells; SC = support cell.

substances. Presumably, different functions of podia require different adhesive strengths and, therefore, adhesive secretions with different composition. Thus, a coarse molecular model for adhesion in echinoderms is a protein–polysaccharide complex, the composition of which varies from one taxon to another [102].

Interestingly, neurosecretory-like cells, which secrete de-adhesive substances, do not stain with any of the classical histochemical dyes. Although the mechanism of detachment is not fully understood, two hypotheses for its explanation have been proposed:

- De-adhesive material interferes with the electrostatic bonds between the adhesive layer and external layer of the cuticle.
- The material works as an enzyme, releasing an external layer of the cuticle from the underlying cuticular layers.

The mechanical mechanism of attachment is restricted to the sucker-mediated attachment. This hypothesis presumes that descending podia can deform the disc shape into a form of sucker that is additionally supplemented by chemical adhesion. The levator muscle in the podial disc is a probable candidate for suction generation. Recent experimental data on the mechanics of the tube feet show very soft properties of their terminal ends.

8.16.2.3 Adhesion in Barnacles

There are two types of adhesion occurring in these crustaceans: (i) temporary adhesion, which occurs in larval barnacles; and (ii) permanent adult adhesion. The antennular attachment organ holds the larva on the substrate while it explores, prior to permanent settlement. The bond allows the larva to remain attached against moving sea waves, but still provides some movements.

The so-called 'antennulary sucker' cannot produce suction; rather, attachment is due to the presence of secretion on the attachment disc. The secretion is produced from the numerous (about 20) glands located in the second antennal segment. These gland cells open individually, in two concentric rings on the surface of the attachment disc [105]. Additionally, the attachment discs demonstrate high deformability, allowing them to adapt to the substrate profile.

There are two cell types responsible for permanent adhesion: (i) the a-type cell, which gives a positive reaction to histochemical tests for proteins, phenols and polyphenoloxidase; and (ii) the b-type cell, which gives a positive reaction only for proteins. In the larva, two cement glands are located just posterior to the compound eyes. From each of these glands a collecting duct leads into the single cement duct, which passes down an antennule and opens out at several points on the attachment disc surface. Although the cement is in a fluid state when secreted, its tanning begins immediately and the liquid cement fills in the microscopic depressions of the surface. The tanning process is gradual, during which the attachment force, when plotted versus time from settlement, shows an asymptotic increase. This natural adhesive apparently meets the criteria for good bonding agents for use in medicine, although further investigations are required in this respect [106].

8.16.2.4 Temporary Adhesion in Cladoceran Crustaceans

Small cladoceran crustaceans, such as *Sida crystallina*, attach themselves to the substrate by adhesive anchors connected to the cuticle. *S. crystallina* bears three anchor organs at the dorsal side of the head. An animal attaches to an anterior organ first, and then additionally attaches to two movable posterior organs. The attachment is presumably due to a cementing secretion, because fine threadlike structures remain attached to the substrate even after the animal frees itself [107]. However, the nature of this secretion and the strength of attachment remain unknown.

8.16.2.5 Adhesion in Larval Fish and Ascidians

The larvae of cichlid fish can attach themselves at the location of emergence during the course of several days after emergence by using the secretions of an adhesive apparatus, which is very variable among the species and allows the larvae to remain under parental defense for several days. As a rule, this apparatus consists of three pairs of cup-shaped glands which are located on the dorsal surface of the larval head and produce a filamentous secretion. Later, the larvae leave and their attachment organs disappear. The secretion gives a positive histochemical reaction for polysaccharides [108–111].

In larval ascidias, the anterior end of the cephalenteron bears three adhesive papillae which enable the larva to attach to the substrate just prior to metamorphosis. In early studies, these have often been referred to as 'suckers', but more recent investigations [112] have provided evidence concerning the adhesive nature of these organs. The papillae bear two types of cell (A and B); the first type is a typical secretory cell, while the second is most likely sensory in nature [113]. The A-cells produce proteins and mucopolysaccharides, which often occur in marine organisms as those providing attachment.

8.16.3
Adhesion in Terrestrial Insects

8.16.3.1 Permanent Adhesion in Insects and Spiders

The coxal vesicles in representatives of the family Machilidae (Archaeognatha) serve to cement the old cuticle during skinning (ecdysis). The exact composition of the cement remains unknown, but it cannot be removed from the surface by the use of water, saline or acetone.

Parasitic insects such as lice or dipterans from the family Oestridae (Hypodermatinae) possess specialized glands to glue their eggs to the host hairs. Hypodermatin eggs also have specialized attachment organs which consist of a flexible petiole and a prominent clasper. The clasper, which is responsible for affixing the egg on a host hair, consists of an attachment groove filled with adhesive substance and a pair of adhesive-coated lateral flanges [114, 115]. The host hair is locked within the groove, and the adhesive solidifies around it. Many other species of insects also use glue-like substances to attach their eggs to each other or/and to the substrate (Figure 8.87).

Adhesive substances are also used for prey capture by orb-web spiders. These spiders use the secretion of the aggregate glands (the essential components of which are glycoproteins) to cover a so-called 'sticky spiral' [86, 116]. Various types of spider

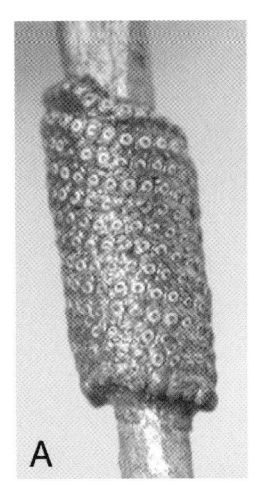

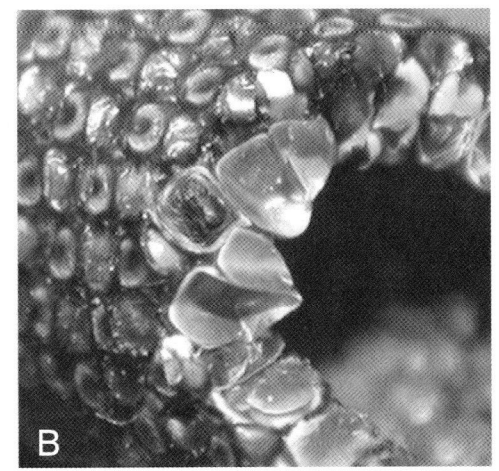

Figure 8.87 Egg aggregation in the moth *Malacosoma neustria* (family Lasiocampidae). (A) Eggs attached to the tree branch; (B) eggs glued to each other.

silk have different tenacity under dry, moist and dusty conditions [117]. The silk of the spiders *Filistata insidiatrix* and *Tegenaria atrica* demonstrated especially high tenacity (2.3–2.7 mN cm^{-2}). Dusty conditions usually reduce tenacity.

8.16.4
Adhesion in Plants

Whilst all of the above examples are from the animal world, adhesion is also used by many plant species to fulfill diverse functions. For example, it is strongly involved in mechanisms for the dispersal of seeds and fruits. Typically, adhesive fruits attach to a passing animal and are carried until they drop off, or the animal detects and removes them. There are two distinct attachment mechanisms found in plants: mechanical interlock or hooking (84%); and gluing of viscid fruits (13%) [118]. Among viscid fruits, three main types can be recognized:

- fruits that become sticky only when wet
- tarry fruits
- fruits with hairs that do not need moisture to become sticky

Diaspores can be viscid in five ways: sticky cuticle (calyces); viscid (glandular) hairs (bracts); viscid seed bodies; viscid drupes/berries; and viscid diaspores with drupes.

The sticky diaspores may be dispersed in three ways. First, the seeds may be attached to the plumage of birds consuming the fruit pulp. Second, sticky seeds may be carried on the feathers after they have passed through the gut of a frugivorous bird. Third, the seeds remain sticky for up to 10 min after fruit explosion and can attach to birds foraging in the vicinity. Adhesion, based on sticky secretions, is used by many plants to seal damages, or by carnivorous plants to catch insects (Figure 8.88).

Figure 8.88 Carnivorous plant spoon-leaf sundew *Drosera spatulata* with sticky hairs covered with adhesive secretion used for trapping insects. The insects are digested and used as additional nutrients for plants growing on poor soils.

The adhesive properties as well as chemical composition of plant glues require further investigation.

8.16.5
Surface-Replicating Mechanism Based on Growth

There is an interesting example of attachment due to replication of surface profile. An important attachment-related property of living tissues is its growth ability. Growing plant tissues can exactly replicate a surface profile by growing into surface depressions. The living tissues, after filling up the surface profile, become hard by thickening of the cell wall. Such a mechanism is difficult to define as either purely adhesive or frictional since, after hardening, the plant interface resembles a mechanical grip firmly attached to the surface. To some extent this is analogous to the glue principle. By using such a mechanism some lianas can climb up and spread over other plants and rocks (Figure 8.89).

8.17
Little-Known Applications

8.17.1
Foils for the Rotors of Wind Turbines

Today, electricity is being produced increasingly from wind energy. However, with rotor blade tip speeds of up to $300\,\text{km}\,\text{h}^{-1}$, blade erosion can occur in wind power plants due to rain, dusts, insects or other particles contained in the air. Innovative PSA foil combinations provide protection and enhance performance.

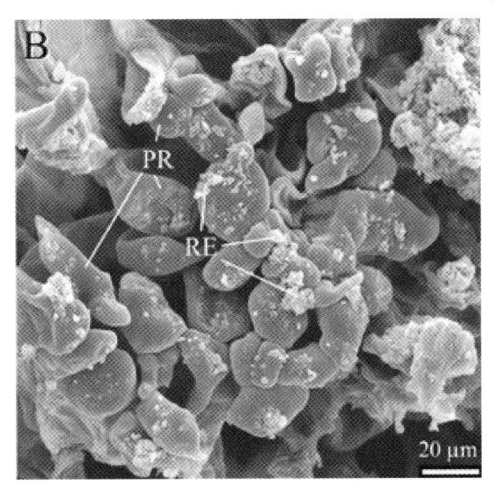

Figure 8.89 Attachment pads of the liana *Parthenocissus tricuspidata*. (A) Attachment devices as they are when connected to a surface; (B) pad removed from the wall of a building. RE = remaining stone material; PR = plant roots grown into the irregularities of hard, stone-like material.

8.17.1.1 Erosion-Protection

The largest mass-produced rotor blade in the world has a length of 37.5 m and is manufactured with leading-edge production methods. Both, the manufacturer's personnel and the materials used must meet the highest demands with regard to quality. As a matter of principle, materials are subjected to in-house incoming inspection and additional intermediate tests before entering the production process.

The quality of the finished product has also continuously been improved such that, at present, rotor blades are provided with so-called 'erosion-protection foils' – special self-adhesive polyurethane films which are applied during the final stage of production. The rotor blades consist of two shells made from glass fiber and epoxy resins, which are assembled and painted. An initial long-term test completed in 1998 showed that the application of a protective foil allowed the maintenance intervals of the bonded surface to be extended to at least four years. By comparison, nonprotected rotor blades had a maintenance interval of two to two-and-a-half years due to erosion problems occurring much earlier. Although erosion-protection is an additional cost-factor, it allows the manufacturers to achieve a quality standard with which they can be associated. Indeed, erosion-protection not only extends the maintenance interval but also improves the public image of the manufacturer of the wind power plant.

At present, all leading German manufacturers fit their rotor blades with protection foils. Typically, a polyurethane film 300 mm wide is applied to the center, starting at the tip, over a length of 22 m, and then bonded to the sides. A combination of a particularly tough thermoplastic polyurethane elastomer film and high-quality acrylic adhesive provides excellent protection from erosion.

In the manufacture of rotor blades, it is important that the protective foils can conform to slightly deformed surfaces. These pressure-sensitive polyurethane films have been used for some years in the aircraft and automobile industries, where they have been coated with high-quality, low-bake or polyurethane paints at later stages. The protective material used for rotor blades is 0.36 mm thick, highly transparent, and resistant to UV light and, if staining does ever occur after long-term exposure to adverse ambient conditions, it will be hardly perceptible.

8.17.1.2 Lightning Protection and Performance Enhancement

In order to protect wind turbine blades against lightning, a new foil has been developed that evenly distributes the energy potential of the lightning current over the entire length of the blade. It does this by catching the current and leading it either to lightning receptors or directly to the tower. The underlying principle is a resin-embedded copper mesh layer applied between a polyurethane or fluoropolymer film and an adhesive.

Likewise, to enhance the performance of a wind turbine and to reduce its running noise, a self-adhesive foil which is designed to mimic sharkskin and applied in parallel to the direction of the air flow reduces the air resistance by about 8%. Since, in order to operate economically, wind turbines normally require a wind speed of about 4 m s^{-1} at a height of 10 m, this aerodynamic foil allows them to produce electricity at lower wind speeds. 'Sharkskin foil', which is made from adhesive-coated polyurethane film provided with a fluorpolymer 'sharkskin' microstructure, has been successfully applied to aircraft fuselages to aid airflow since the 1990s.

8.17.2
Adhesive Elastomer Composites

The wide range of structural elements made from elastomer composite materials has become an important and indispensable part of everyday life, due to their resilient or elastic deformation behavior when subjected to mechanical stress. The mechanical joining techniques used for metals are often not suitable for rubber materials, and indeed the adhesive bonding of rubber with latex is one of the most ancient adhesive applications used by mankind. Latex or rubber milk produced from India rubber trees (a dispersion of natural latex in water) was first used about 8000 years ago to produce elastic composites for tool manufacture, and served later to join textile materials or to make them water-repellent.

The use of rubber on a technical level dates back to 1840, when Charles Goodyear discovered that natural rubber could be crosslinked ('vulcanized') by adding sulfur prior to any thermal treatment. Shortly afterwards, Horace H. Day, a physician, discovered that a mixture of natural rubber and resin, when dissolved at a defined ratio and brushed out onto fabric, was very sticky after evaporation of the solvent. Thus, the first PSA tape was born and patented in 1845 (US Patent # 3965; see Chapter 2).

Today, the technical importance of elastomer composites is due mainly to Goodyear's discovery of hot vulcanization, which increases the strength and durability of elastomers. Another major aspect of this technology is that sulfur accounts for the

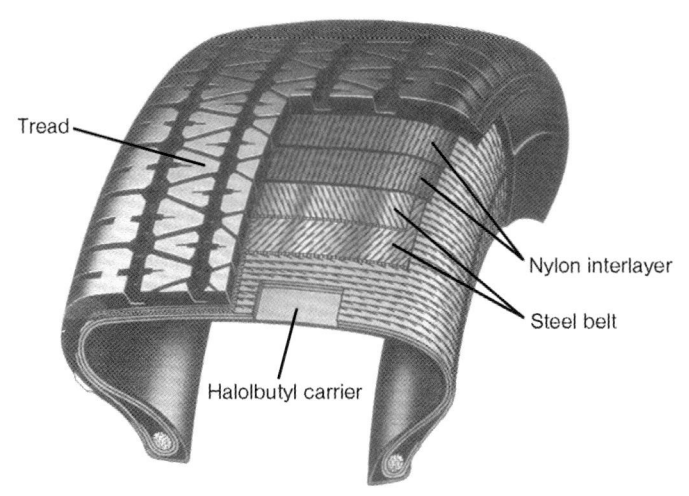

Figure 8.90 Structure of a car tire ('Firestone Tires Report').

build up of a high bonding strength of rubber to brass or brass-coated steel during the vulcanization process, which today remains a vital element of tire manufacture. Adhesive bonds between elastomers crosslinked by sulfur and brass-coated steel fibers must cope with highest mechanical and thermal stress during the service life of tires used for automobiles, trucks and aircrafts (Figure 8.90).

Brass coating not only promotes the drawing capability of steel fibers, but also plays a crucial role in the adhesion mechanisms building up between rubber and metal.

During sulfur vulcanization, a complex layer of CuS dendrites is formed on the brass surface, which protrudes into the elastomer network and is crosslinked to it by sulfur bridges. It has been found that a low content of cobalt in the brass coating stabilizes this layer and/or reduces the formation of zinc compounds that may impair adhesion [119].

To date, sulfur is still an unmatched crosslinking agent that provides a very high degree of adhesion within the composite material obtained. However, the properties of elastomers vulcanized with sulfur can only be modified to a limited extent.

In the production of other rubber–metal composites (Figure 8.91), it has been possible to obtain adequate bonding strength and durability without relying on sulfur by using special adhesives combined with adhesion promoters such as silanes. In the manufacture of composite tires, however, vulcanization remains indispensable.

8.17.3
Light Bulbs

The mounting of halogen light bulbs into glass reflectors is an example of applications where organic adhesives fail. Long-term resistance to elevated temperatures from 300 °C to 1500 °C can only be achieved with inorganic systems (Figure 8.92).

Inorganic, air-drying adhesives are usually water glass formulations with oxidic fillers (aluminum oxide, silicon oxide, zirconium oxide, magnesium oxide, mica, etc.).

Figure 8.91 A rubber–steel damping element (Schwingmetall®).

Water glass is produced from silica by dissolving quartz in molten alkaline carbonate. Water glass is obtained in aqueous solution (2–4 moles of SiO_2 per mole sodium or potassium oxide) [120].

A filled water glass sets by the evaporation of water, a process which is similar to the physical setting of organic adhesives. Setting is enhanced by heating to up to $+60\,°C$ and by circulating air. Higher temperatures at the start of the setting process may impair the bond-line due to higher water-vapor pressure causing pores or cracks in the adhesive layer. Inorganic adhesives are not suited to large-area bonding. Products which contain components that are more soluble in water set by crystallization, with the crystals growing during the course of the setting process. At room temperature, however, this process may continue for weeks.

Colloid oxidic fillers set by the formation of hydrated gel, a process that requires less time than does crystalline setting.

Chemically setting inorganic adhesives usually contain phosphates and alkaline silicates or alumosilicates. Depending on the type of adhesive used, the setting

Figure 8.92 A halogen light bulb (20 W) mounted in a glass reflector using inorganic adhesive.

process is enhanced by heating to 150 °C [121]. In an acid medium, water-soluble, low-molecular-weight silica is transformed into amorphous SiO_2 with a three-dimensional network structure, which is insoluble in water, by a condensation reaction that splits off water.

The filler, purity and particle-size distribution mainly determine the properties of an inorganic adhesive. Contaminations in the filler may impair, for example, the electrical properties, whereas a nonuniform grain-size distribution may disturb the three-dimensional network structure, and hence the mechanical properties.

Today, only a few knowledgeable manufacturers are capable of producing high-temperature-resistant adhesives.

8.17.4
Bonding in Art, Jewelry Making and Archeology

8.17.4.1 Art
When admiring the 'Golden Horseman' – an equestrian statue of King August the Strong of Saxonia, which was erected in 1736 on the New City Market ('Neustädter Markt') in Dresden, Germany – scarcely anybody realizes that the monument is a miracle of adhesive bonding technology. It was built by layering heated copper over an iron frame, then covered with 0.1 µm-thick gold leaf affixed by so-called 'gold size'. The statue was refurbished with new gold plate in 1965 which, until 2002, successfully fought against the adverse environmental conditions for almost 40 years, when it was finally regilded (Figure 8.93).

Figure 8.93 The monument of 'August the Strong' (August der Starke) in Dresden.

Figure 8.94 A gilded statuette from Thailand.

The Ancient Egyptians were master gilders, who used wax as a binder. 'Gold size', which is still used today by restorers of gilded monuments, is a fine coating which consists of polyunsaturated monocarboxylic acids that set by radical polymerization under the effect of oxygen from the air and subsequent auto-oxidation, to produce solid, high-molecular-weight resins. The addition of 1–5% siccatives to the composition (e.g. cobalt, manganese or lead salts of linoleic acid or resin acid) speeds up the slow reaction.

Overlays of gold are found on a wide range of objects, from the golden rims of drinking glasses to the inscriptions on memorial stones. They are also applied, by using a different procedure, to the covers of books, or by the classical gilding method, on statuettes brought home from Asia as souvenirs (Figure 8.94).

When *gilding*, it is not easy to smoothly lay down the sensitive material over the entire surface, as can be seen in Figure 8.94. Folds that can be seen on the tip of the statuette must not appear on a larger surface. Gilding, as restorers also perform it, is covering a clean and smooth surface with a uniform layer of gold size, and patiently allowing it to 'dry' for hours; the gold leaf, which is delivered on tissue paper, is then laid on the gold size without removing the tissue, and gently pressed down. The tissue is then removed from the gold leaf, which is then smoothed out on the gold size with a pad of cotton wool or a very soft, fine brush. It is easily understood that in the majority of cases, the cost of labor is much higher than the value of the gold. For example, a total of 38 g gold (23 carat, corresponding to a purity rate of 99.5%) was laid on August the Strong (1 g of gold is worth €10; gold leaf is worth slightly more).

For adhesion research, gilding has proved to be a phenomenon (see Chapter 3) since, even under adverse conditions, it has a very long life span and provides excellent protection against pitting corrosion on other metals (electroplated gold can only achieve this when a layer 10–20 μm thick is applied). The reason for this is probably that gold size electrochemically 'decouples' gold as a noble metal from the common base metal. The shortcomings of gold leaf become evident when it is exposed to mechanical stress because, due to its relatively high purity, it is very soft and has only a low resistance to mechanical wear.

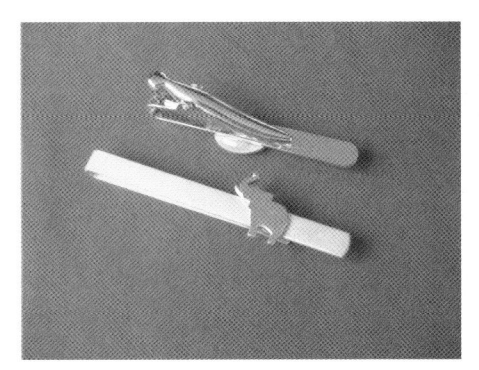

Figure 8.95 Bonded tie-pins.

8.17.4.2 Jewelry Making

Noble metals are also processed by the jewelry industry. In former times, these were soldered with each other, or one with the other, but entailing labor-intensive reworking to remove the flux residues or staining caused by elevated temperatures affecting the surface. Bonding is an attractive option that helps to save such costs (Figure 8.95).

On the reverse side of the tie-pin made from gilded brass (see Figure 8.95) the adhesive has escaped from the bond-line, whereas the bonding of a small elephant made from solid gold but affixed to a tie-pin made from sterling silver, is invisible, until it fails, for example if the tie-pin were to be dropped onto a brittle material such as glass or stone. The owner would be disappointed, not because 'bonding is a cheap imitation, after all', but because the jeweler used an adhesive that was not tolerated by the sensitive silver surface (tarnishing effect), or was too brittle.

It is advisable, however, not to bond decorative objects in the way that the medals shown in Figure 8.96 were bonded. The adhesive residue between the golden center plate and edelweiss and between the edelweiss and base plate, respectively, has spoiled 25 years of membership (left) compared to the reward for 40 years of paid membership (right).

Figure 8.96 Bonded medals.

8.17.4.3 Archeology

Today, both restorers and archaeologists alike would struggle to work without the aid of bonding, not only because this joining technique does not harm sensitive substrates but it also allows dissimilar materials to be joined to each other. Yet, it is often not easy to decide whether to use a modern or an ancient, original adhesive formulation. It is clear that the ancient binders have 'got along' with the adherents over the years (hundreds or thousands of years, to be precise), whereas the long-term behavior of modern formulations, when 'imported' into restored objects, can only be determined using costly analyses.

An example of the use of bonding in restoration is that of Abu Simbel, an archeological site comprising two massive rock temples in southern Egypt. This was relocated during the creation of Lake Nasser, an artificial water reservoir formed after the building of the Aswan dam on the Nile River. The entire site was cut into blocks and reassembled in a new location using modern adhesives, which are not visible at all (and will hopefully remain invisible for the next 2000 years).

Finally, a somewhat strange example is the Palastaula in Trier, Germany, known today as the 'Constantine Basilica', which was erected by the Roman Emperor Constantine the Great in 310 AD (Figure 8.97). The original classical brick building was plastered and painted, but in the 19th century it was rebuilt on a major scale using the Roman fragments of the walls, at the request and expense of King Frederick William IV of Prussia. For this, sandstone and modern mortar were used instead of bricks, and the building was used as a church in later times. During the Second World War, the building was largely destroyed by bombardment but then rebuilt again, this time in brick construction. During the 1950s, an archeologist – who obviously had participated in this second restoration – reported that the original Roman wall had survived the attacks virtually unharmed, whereas the main damages had been found in that part of the building which had been restored in the 19th century. Remembering that the Ancient Romans used to reinforce their mortar with bovine milk and blood (casein and blood albumin, see Chapter 2), and that this was still standard practice in medieval times but no longer in the 19th century, provided the detail that makes the story believable. Even in unprotected condition and without plaster, the

Figure 8.97 The Roman Constantine Basilica in Trier.

Roman mortar could not be scraped off, even with a sharp pocket-knife, as long as 2000 years after its creation, whereas the modern mortar readily crumbled after only 50 years. If the 19th century restoration had given consideration to the ancient materials, the joints would have been 'bombproof', saving the 20th century restorers much work (see Section 8.3).

8.17.5
Removable Adhesive Joints

Bonding as a joining technique is usually classified as 'nondetachable'. To date, this has applied to many applications, and particularly to structural adhesive-bonded joints. However, this classification is incorrect whenever 'nondetachability' is defined as an inherent property of a joint *not* to allow separation (and reconstitution) of the joint without irreversibly changing its elements (or adding new elements). This is illustrated by the following two examples.

First, in olden days, knife blades were mounted into the handles made from precious materials (e.g. silver) by bonding the shank and using rosin as the adhesive. When the blade needed to be replaced due to corrosion or damage, the handle was gently heated with a flame to melt the rosin, the shank was carefully removed from the handle, and a new blade immediately mounted. Dexterous fingers achieved this without applying any new adhesive.

Second, all of us know the Post-it® notes that are manufactured by the 3M company. These notes are easily removed from the block without leaving any residues (at least on visual inspection), and can be reattached to the block or to other surfaces, again and again (see Section 8.17.5.1).

No further attempt to clarify the definition of 'detachability' will be made here, because tricky discussions of terminology hinder rather than benefit the development of a technique. In the following sections the detachability of a bond will be confined to the possibility of separating a bond on demand, easily and without leaving major residues. This is particularly interesting with regards to repair and recycling aspects, which are reflected by a variety of applications that have been filed for patents during the past few years. A choice of investigations is presented in Refs [122–124].

8.17.5.1 Commercially Available Removable Adhesive Systems
The detachable adhesive systems that are currently available on the market are based on PSAs. A variety of detachable notes, similar to Post-it® notes, are produced by other manufacturers: in Germany, double-sided stretchable adhesive tape pads with a pull tab for removal (stretch release tapes) are sold as Power Strips ™ by Beiersdorf, or more recently, as 'Scotch Commands' by 3M.

In principle, PSAs can be formulated in such a way that they can be peeled off a solid surface without leaving any trace. This applies to the PSA tape 'Tesafilm' (Beiersdorf) when bonded to a glass surface. However, when bonded to paper its removal causes paper damage because the inherent strength of the paper, perpendicular to the surface, is rather low. One way of solving this problem could be to reduce the adhesive capacity of the PSA film, but this would result in a nonloadable,

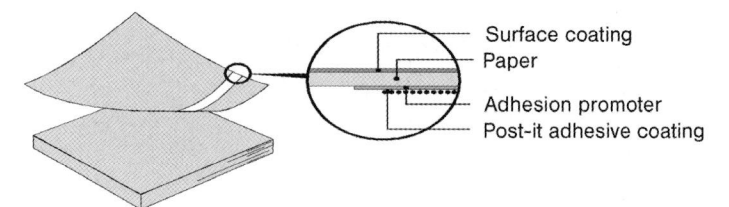

Figure 8.98 The structure of the 'Post it®' bonding system. (Illustration courtesy of 3M).

and hence nonreliable, bonding. To make the best of this bad job, a multilayer coating (Figure 8.98) has been conceived that has been carefully adjusted to the requirements of the paper. Here, a special coating is applied to the writable side of the note, while on the reverse side an adhesion promoter is applied to ensure that the adhesive adheres well to this side of the paper. Finally, a pressure-sensitive microsphere-type adhesive is applied (i.e. microscopic adhesive 'balls').

Figure 8.99 shows in detail the multilayer coating that must be applied very carefully to ensure that the extent of surface area contact between the adhesive and substrate is very low, and that the microspheres are subject to maximum peel stress when peeled off, which enhances removal of the Post-it note. The adhesive must be formulated in such a way as to allow repeated repositioning without consuming the microsphere coating. A scanning electron microscopy image of the sticky surface of a Post-it note is shown in Figure 8.100.

It has become clear that this inconspicuous office supply staple, which was introduced in Germany as early as in 1981, is not based on a 'weak adhesive' but rather on sophisticated technology. However, the principle cannot be transferred arbitrarily to bonds that require higher strength values, and consequently another system was developed based on the effect of stretched adhesive tapes. This had been described in an US patent in 1974 and dealt with the mounting of lenses into grinding machines (see Section 8.8.2), but did not attract much interest. This effect is easily reproduced by bonding a transparent, single-sided adhesive tape (available in every household) to a plastic or glass surface, for example, and allowing a 'handle' of adhesive tape to overlap on each side of the surface. When the free ends of the tape are stretched in parallel to the adhesive layer, the adhesive will undergo debonding from the end of the surface and along the tape borders. The tape then frequently tears up outside the bonded area. When applying adhesive to a highly stretchable carrier material (the 3M system), or when using thicker adhesive layers made from a highly stretchable PSA without a carrier (the Beiersdorf system), a double-sided adhesive

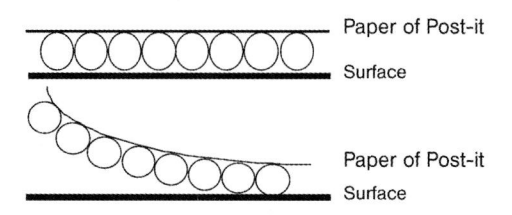

Figure 8.99 Structure of the Post-it® adhesive coating.

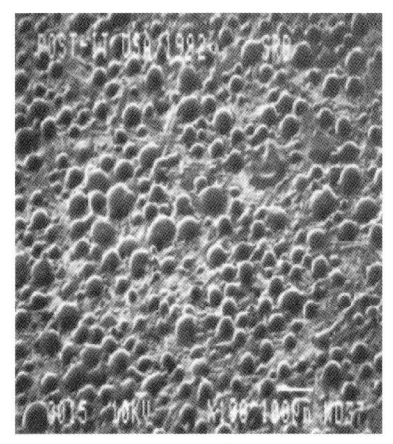

Figure 8.100 Scanning electron microscopy image of the sticky surface of a Post-it® note (original magnification × 60). (Photo courtesy of 3M).

tape that joins two rigid substrates can be released from the surfaces by stretching it in parallel to the bonding area. This process occurs due to the transverse contraction of the adhesive tape at elongation, which causes peeling effects in the adhesive layer perpendicular to the tensile force. This striking removal process has not yet been completely clarified, however, because another explanation suggests that extreme stretching impairs the rotation and vibration of the adhesive molecules, which are important requisites for 'dynamic adhesion' (see Chapter 3).

Stretch-release adhesive tapes have a high potential for new developments in technical applications. For example, they could allow the sound emission of piston engines to be reduced in cars if the valve cap were to be affixed to the piston engine by using stretch-release tapes instead of screws; this would significantly dampen the noise of the engine, more than 20% of which is emitted through the valve cap. Since, today, valves no longer require permanent maintenance, the greater challenge of reassembly of the valve cap would be outweighed by the noise-dampening effect of the bonded joint.

8.17.5.2 High-Strength Removable Adhesives

The above-described removable systems are based upon PSAs, the load capacity and creep resistance of which are limited (see Section 5.1). For the development of removable, high-strength adhesives, other solutions must be found. There are three main mechanisms that lend themselves to the removal of an adhesive bond.

First, severe subsequent embrittlement of the adhesive within its layer has been known to weaken the joint (or to cause overall delamination) since the 1960s, when small amounts of low-molecular-weight phenolic resin were added to solvent-containing neoprene-based contact adhesives to increase their long-term durability. Unfortunately, this well-intended increase of the phenolic resin content induced delamination of the bond (e.g. in doors) after about 10 years; this effect was explained by the continuous subsequent crosslinking of the then-dominant phenolic resins, which increasingly embrittled the adhesive. Attempts to initiate subsequent embrittlement on demand – that is to say to 'incorporate' a subsequent crosslinking option into the adhesive – have failed to date [125], although the present investigations seem to be on the right path.

Second, adhesion promoters that are destroyed on demand would represent an optimum possibility also to remove a high-strength bond, but these have not yet been developed either.

The third intelligent possibility (besides thermal weakening or cold embrittlement) would be to mechanically destroy the bond-line on demand by using additives incorporated in the adhesive; these would expand greatly under external influences that would not occur in ordinary use, thereby breaking the adhesive joint. The first investigations with toluene sulfonic acid hydrazine, which splits off low-molecular-weight compounds at elevated temperatures, failed. However, when encapsulated foaming agents – which expand by physical change of the filler (e.g. evaporation at defined temperatures) – were used, the bond-lines created with dispersion adhesives (like those used to lay PVC floor coverings) were successfully weakened. The expansion was initiated by microwave irradiation; some test results are illustrated in Figure 8.101 [126].

The mere addition of different amounts of Micropearl F-30 (an expanding filler) (Follmann) to dispersion adhesives reduced the peel strength of the bond due to fracture mechanics. If, in addition, the bond was exposed to microwave irradiation, there was mainly a drastic fall in peel strength, with the adhesive being found (in large part) on the surface of the wooden substrate, which was a desired effect. Storage of the filled adhesive for several months did not impair the foaming effect. However, the question remained as to whether this would still be the case if the adhesive bond-line were to be exposed for long periods of time to the conditions prevailing in actual applications of the building industry.

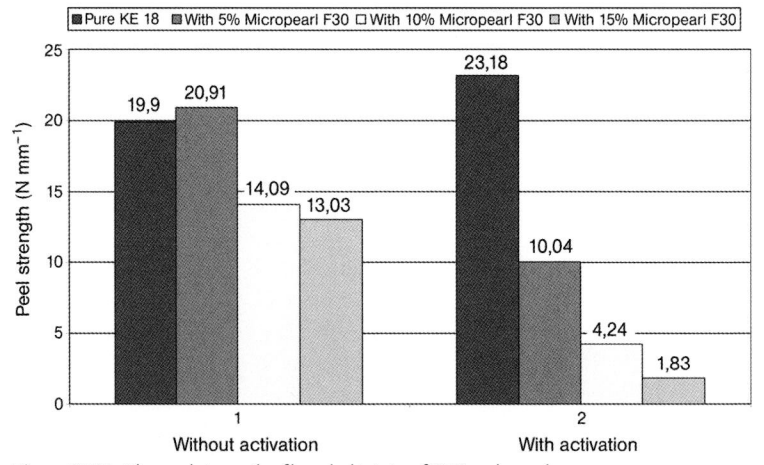

Figure 8.101 The peel strength of bonded joints of PVC and wood. Adhesive: Dispersion adhesive with different amounts of filler, with and without microwave activation.

Attempts to 'burst open' crosslinked adhesives with the aid of these fillers have failed [127]. Whilst it has not yet been decided which of the above-mentioned (or other) methods should be investigated further, one thing is certain – removable adhesive bonds represent a significant step forward, even if they are rarely created by simple means.

9
Perspectives

9.1
Economic Trends

The global market for adhesives is forecast to grow further over the next years. According to the German Adhesives Association (Industrieverband Klebstoffe e.V., Düsseldorf), the worldwide sales of adhesives and sealants were €39.4 billion in 2007. Geographically, Asia, Europe and North America together form the largest group of consumers, each with 30% of the global market. Between 2004 and 2007, adhesives consumption in Europe increased by 2.6%, and sales volume by 5.2% per year.

On a quantity basis, the paper and packaging industries are the major users of adhesives, followed by the construction and building industry (Figure 9.1).

The German Adhesives Association considers Germany to be the world leader in adhesives raw materials technology, with the leading machine manufacturers playing an important role in the continued growth of adhesives applications. In the search for innovative approaches to the processing of adhesives, the German companies contribute to the advance of adhesive bonding technology.

The major tasks for the future are to:

- foster and advance the existing knowledge base
- provide for a practice-oriented transfer of technology
- promote staff exchange between science and industry

The task of research policy will be to create a performing infrastructure in research and training, and also to foster and promote projects in innovation areas with high sustainability.

9.2
Technical Trends

Today, in adhesive bonding technology there is still much empiricism. Whereas, in the design, calculation, creation and quality assurance of welded joints there exist fully developed standard sets of rules, guidelines and specifications that have been

Adhesive Bonding: Materials, Applications and Technology
Walter Brockmann, Paul Ludwig Geiß, Jürgen Klingen, and Bernhard Schröder
Copyright © 2009 WILEY-VCH Verlag GmbH & Co. KGaA, Weinheim
ISBN: 978-3-527-31898-8

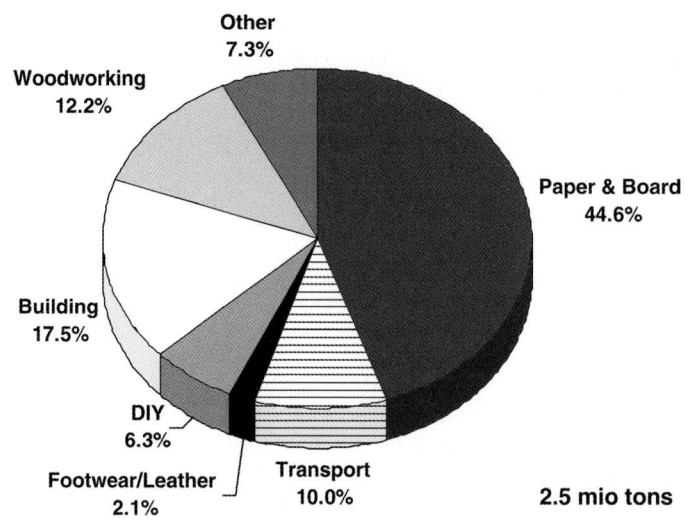

Figure 9.1 The Western European adhesives market, 2007. The key market segments are indicated; values shown are in tons. (Illustration courtesy of Industrieverband Klebstoffe e.V.).

- Paper & Board 44.6%
- Building 17.5%
- Woodworking 12.2%
- Transport 10.0%
- Other 7.3%
- DIY 6.3%
- Footwear/Leather 2.1%
- 2.5 mio tons

accepted and approved, this is not the case for bonded joints. Neither are there any well-defined criteria for the selection of adhesives and the determination of specific process parameters. For example, although the surface preparation of substrates is virtually always demanded, it is definitely not always needed.

With few exceptions, the trend in adhesives development will not be towards further increases in bonding strength, as this is not necessary in most cases. Although long-term durability often seems to have room for improvement, it is aspects of easy processability, high safety of the resultant product, ecological aspects and physiological compatibility that will rather shape further developments.

These requirements can be easily met if the chemical reactions involved when creating an adhesive bonded joint take place, as far as possible, at the adhesive's manufacturing stage. To this end, adhesives should be delivered with as high a molecular weight as possible, the potential candidates for this approach being hot melts and dispersion adhesives. Pressure-sensitive adhesives will also play an important role, being perfectly suited to semi-structural or even structural bonding applications. For example, bus body panels can be fixed very effectively to frames, stiffening profiles into truck cabs, and sectioned headlinings into car roofs, in a single operational sequence using PSAs with very high long-term durability. Nonetheless, the mechanical characteristics of these materials still need to be improved.

One interesting aspect in the development of adhesive bonding technology is that of 'bonding on demand' adhesives; these already exist in the medical sector, and everybody has heard of Post-it® notes in the office. Stretch-release tapes are also available; these can be removed from a joint by pulling a tab and stretching it in parallel to the bonding area. Yet further developments are currently under investigation and will surely open up new markets.

Engineers often underestimate the structural performance of adhesives, and use them only for nonstructural applications, tending to select products with a strength-oriented approach. Two clearly provocative questions illustrate this way of thinking. How do you define the strength that is actually required from an adhesive in the lap joint doublers of commercial aircraft such that it can prevent the formation of uncontrolled cracks – which it clearly does? Also, how do you define the strength required from a car body adhesive so that it can drastically increase the fatigue strength of a spot-welded adhesive-bonded joint – as it often does?

It is clear that adhesive bonding technology offers far more than today's applications, and has enormous future potential, despite it rarely being possible to implement adhesive-bonded joints straight from the drawing board on the basis of sets of rules and specifications. In the future, manufacturing technology will undoubtedly change to allow the performance and risk assessment of adhesive bonded joints to be more accurate and reliable. When compared to former times, today's products are designed and optimized under completely different rules, with the main focus being on vital questions such as shortages of raw materials, total energy content and the environment-friendly disposal or reuse of a product. Although we must accept these facts, we must at the same time try to change our habits without impairing our quality of life. A car body does not necessarily require many hundreds of welding spots in order to be solid and durable – at the end of the car's life each welding spot will need to be crushed and shredded. Likewise, bonded windscreens and rear windows that cannot be removed cause major problems in recycling. Both of these items could just as easily be bonded on demand – a subject which must be worthy of great consideration.

Appendix

Table A.1 Adhesive Bonding: Participating Institutions and Companies, and their Current Research and Development Topics.

Address	Contact	Topic(s) of Research
Academia Sinica Institute of Chemistry P.O. Box 2709 Beijing P.R. China	Prof. Dr. Yunzhao Yu Ph.D., M.S. yuyuzh@bj.col.com.cn	• Epoxide-polycarbonate copolymer networks • Silicone resins
Adhesive Research Institute Shanghai Kangda Chemical Co., Ltd No. 655 Qingda Road Pudong Shanghai P.R. China	Prof. Yibin Hou Ph.D., M.S. shhouyb@yahoo.com.cn http://www.kangda-sh.com	• Acrylate, epoxy adhesive • Silicone materials • Polyurethane sealant
Beijing University of Chemical Technology School of Materials and Engineering 15 Beisanhuan East Road Chaoyang District 100029 Beijing P.R. China	Prof. Junying Zhang Ph.D., M.S. zhangjy@mail.buct.edu.cn	• Design and synthesis of resins and prepolymers for adhesives, including epoxy, silicon and polyurethane resin • Studies of the relationships between performances and structures of adhesives, especial the curing properties and mechanical performance • Development of novel adhesives, such as high-temperature-resistant adhesive, chemical-resistant adhesive, thermal and electrical conductive adhesive, sealants

Table A.1 (*Continued*)

Address	Contact	Topic(s) of Research
Case Western Reserve University Department of Macromolecular Science 10900 Euclid Avenue Cleveland, OH 44106-7202 USA	Prof. Hatsuo Ishida Ph.D., M.S. hxi3@po.cwru.edu	• Surface and interface characterization of composites and coatings • Nanocomposites • Silane coupling agents • Synthesis, characterization, processing and application of polybenzoxazines • Molecular and thermal characterization of polymers (FTIR, Raman, NMR, MDSC, TGA, GC-MS, AFM/microthermal imaging, laser flash photometry) • Non-linear Fourier rheology, linear rheology, dynamic mechanical spectrometry
CNRS & Université Paris-Sud 11 Laboratoire de physique des solides Bât. 510 – Campus d'Orsay 91405 Orsay Cedex France	Prof. Dr. Liliane Léger leger@lps.u-psud.fr	• Entangled polymers, polymer dynamics, reptation • Polymers at interfaces • Adhesion • Lubrication • Flow with slip at the wall • Wetting • Surface modification, characterization of interfaces • Polymer thin films, confinement and dynamics
Columbia University Chemical Engineering Department 500 West 120th Street New York, NY 10027 USA	Prof. Jeffrey T. Koberstein B.S., Ph.D. jk1191@columbia.edu	• Fundamental relationships between molecular structure and properties of polymers and other soft matter • The design of polymer surfaces and interfaces from a molecular perspective
Cornell University 322 Thurston Hall Ithaca, NY 14853 USA	Prof. Chung-Yuen Herbert Hui B.Sc., M.S., Ph.D. ch45@cornell.edu	• Fracture mechanics of aging aircraft • Fracture mechanisms of polymer/polymer interfaces • Fracture mechanics and statistical theory of failure of composite materials

Table A.1 (*Continued*)

Address	Contact	Topic(s) of Research
		• High-temperature creep and crack growth in metals • Swelling kinetics and diffusion of organic molecules in polymer glasses • Hypersingular integrals and boundary element method • Sintering of ceramics • Fluid mechanics of aircraft • Electronic packaging and micro-electromechanics systems
3M Company 3M Corporate Research Materials Laboratory 3M Center-Bldg. 201-3E-03 St. Paul, MN 55144 USA	Dr. Alphonsus V. Pocius B.A., Ph.D. www.mmm.com	• Mechanical testing, surface science and surface analysis [ESCA (electron spectroscopy for chemical analysis) and SSIMS (static secondary ion mass spectrometry)] • Development of water-based primers
Department of Chemical Engineering Room 409, Chemistry Building Tsinghua University Beijing 100084 P.R. China	Prof. Chengyou Kan Ph.D., M.S. kancy@tsinghua.edu.cn http://www.chemeng.tsinghua.edu.cn/scholars/kanchy/index.htm	• Emulsion polymerization • Water-borne polymer products (coatings, paints, adhesives, etc.) • Functional polymeric microsphere • Silicone materials
Fachhochschule Gelsenkirchen Abt. Recklinghausen Fachbereich Angewandte Naturwissenschaft Labor für Organische Chemie und Polymere August-Schmidt-Ring 10 45665 Recklinghausen Germany	Prof. Dr. Klaus-Uwe Koch klaus-uwe.koch@fh-gelsenkichen.de	• Adhesion promoters • Formulation of adhesives • Testing of adhesives • Chemical setting resins • Synthesis of polymers
Fachhochschule München Verfahrenstechnik Papier und Verpackung Lothstrasse 34 80335 München Germany	Prof. Dr. Dirk Burth burth@fhm.edu	• Adhesive bonding with radiation curing systems • Hot-melt pressure-sensitive adhesives for labels and tapes • Coating of paper • UV- and electron beam curing adhesives

Table A.1 (*Continued*)

Address	Contact	Topic(s) of Research
Fachhochschule Münster Fachbereich Maschinenbau Stegerwaldstrasse 39 48565 Steinfurt Germany	Prof. Dr.-Ing. Gerhard Kötting www.fh-muenster.de/fb3/ personen/lehrende/ koetting/koetting.php	• Process-dependent eigenstresses in bonded joints • Surface treatments of steel- and aluminum-alloys • Durability testing methods for bonded joints • Processing of spot-weld and adhesively bonded joints
Fraunhofer Institut für Fertigungstechnik und Angewandte Materialforschung – IFAM Bereich Klebtechnik und Oberflächen Wiener Strasse 12 28359 Bremen Germany	Dr.-Ing. Helmut Schäfer www.ifam.fhg.de	• Development and characterization of polymers • Network polymers • Formulation of adhesives • Chemical and physical analysis • Fundamentals of adhesion • Development of special polymers • Structural and functional polymers • Adhesives, sealants and coating materials • Electrical and optical conducting contact systems • Polymer properties in thin layers
Fraunhofer-Institut für Holzforschung Wilhelm-Klauditz-Institut Bienroder Weg 54E 38108 Braunschweig Germany	Prof. Dr. Rainer Marutzky reiner.marutzky@wki. fraunhofer.de	• IR- characterization of adhesives for wood • Gluing with tannin-formaldehyde adhesives
Fraunhofer-Institut für Werkstoff- und Strahltechnik – IWS Klebtechnikum an der TU Dresden Institut für Oberflächen- und Fertigungstechnik Winterbergstrasse 28 01277 Dresden Germany	Dr. Irene Jansen irene.jansen@iws. fraunhofer.de	• Surface treatment by plasma- and laser techniques • Structural bonding of metals, polymers, glass and wood • Conducting bonded joints • Characterization of surfaces and bonded systems • Strength and durability of bonded joints
Fraunhofer-Institut für Zerstörungsfreie Prüfverfahren IZFP Im Stadtwald, Geb. 37 66123 Saarbrücken Germany	Dr. Eckhardt Schneider eckhardt.schneider @izfp.fraunhofer.de Dr. Bernd Wolter bernd.wolter@izfp. fraunhofer.de	• Testing of thick elastomer adhesive layers in direct glazing systems • Characterization of bonded systems by NMR-analysis

Table A.1 (*Continued*)

Address	Contact	Topic(s) of Research
IFF-Institut für Fügetechnische Fertigungsverfahren GmbH Krausstrasse 22a 85737 Ismaning Germany	Prof. Dr. Christian Lammel christian.lammel @iff-gmbh.de	• Bonding techniques, consulting, testing of bonded joints, induction techniques
IFT-Institut für Fenstertechnik e.V Theodor-Gietl-Strasse 7–9 83026 Rosenheim Germany	Dr. Helmut Hohenstein www.ift-rosenheim.de	• Design methods for structural glass-wood systems • Bonded glass elements in structural wood systems
Imperial College London Department of Mechanical Engineering South Kensington Campus London, SW7 2AZ United Kingdom	Prof. Anthony J. Kinloch PhD, DSc(Eng), FREng, FCGI, FIM, FRSC, CEng, CChem a.kinloch@imperial.ac.uk	• Engineering adhesives, composites, structure-properties relationships in engineering polymers • Nanomodified adhesives • Alignment of carbon nanotubes • Structure-property relationships of nanocomposite materials • Mechanical properties and fracture performance of nanocomposites • Toughening polymers using nano- and microparticles • Failure analysis and mechanisms; finite element modeling of nanocomposites
Institut für Fügetechnik und Werkstoffprüfung gGmbH Bereich Sonderfügeverfahren Otto-Schott-Strasse 13 07745 Jena Germany	Dr. rer. nat. Ursula Basler ubasler@ifw-jena.de	• Adhesive bonding of glass, ceramics, glass ceramics and metals • High-temperature-resistant bonded joints with inorganic adhesives • Characterization of adhesives
Institut für Polymerforschung Dresden e.V. Hohe Strasse 6 01069 Dresden Germany	Dr. rer. nat. Rüdiger Häßler rhaesz@ipfdd.de	• Measuring of glass transition temperature in adhesive layers • Characterization of boundary layers by AFM techniques • Measuring of polymerization reactions by thermal analysis • Influence of surface treatments on polymerization of adhesives • Curing reactions • Temperature-dependent properties of bonded joints

Table A.1 (Continued)

Address	Contact	Topic(s) of Research
Institute of Petrochemistry Heilongjiang Academy of Science No. 164 Zhongshan Road Harbin P.R. China	Prof. Bin Zhang Ph.D., M.S. Harbinzb@yahoo.com	• Epoxy structural adhesives, including room temperature, intermediate and high-temperature curing epoxy resin structural adhesives • Synthesis of special epoxy resin and types of oligomer used in the toughening of epoxy resin (e.g. the preparation of acrylate oligomer PES and PU oligomers) • Polyimide high-temperature-resistant adhesives, including synthesis of thermal curing polyimides terminated with various functional groups and toughening methods (e.g. preparation of polyimides end-capped with alkynyl and nadic groups) • Investigation of high-temperature-resistant adhesives. • Phenolic resin and polyurethane adhesives; special sealing materials including polysulfide-rubber polyurethane and polyacrylate
The Johns Hopkins University Dept. of Physics & Astronomy 3400 North Charles Street Baltimore, MD 21218-2686 USA	Prof. Mark O. Robbins B.A., M.A., Ph.D mr@pha.jhu.edu	• Non-equilibrium processes such as friction adhesion, spreading, fluid invasion, and shear-induced phase transitions
Lehigh University Center for Polymer Science and Engineering (CPSE) P.C. Rossin College of Engineering and Applied Science Packard Laboratory 9 Memorial Drive West Bethlehem, PA 18015 USA	Prof. Anand Jagota Ph.D., M.S. anj6@lehigh.edu.	• Adhesion • DNA/carbon nanotube hybrids • Biomimetics of metal adhesion • Mechanics and computational methods

Table A.1 (*Continued*)

Address	Contact	Topic(s) of Research
Lehigh University Center for Polymer Science and Engineering (CPSE) P.C. Rossin College of Engineering and Applied Science Packard Laboratory 9 Memorial Drive West Bethlehem, PA 18015 USA	Prof. R.A. Pearson B.A., M.S., Ph.D. rp02@lehigh.edu	• Fatigue, fracture and yielding mechanisms in homopolymers, copolymers and polymer blends • Processing of engineering thermoplastics and high-performance thermosets
Lehigh University Department of Chemical Engineering 111 Research Drive Iacocca Hall Bethlehem, PA 18015 USA	Prof. Manoj K. Chaudhury B.Sc., M.S., Ph.D. mkc4@lehigh.edu	• Surface modification of organic and inorganic materials • Adhesion and friction of soft polymers • Drop fluidics • Metal-polymer adhesion • Adhesion to skin and mucous membrane • Instabilities in thin elastomeric films
Michigan State University Composite Materials and Structures Center 2527 Engineering Building East Lansing, MI 48824-1226 USA	Prof. Lawrence T. Drzal B.A., M.S., Ph.D. drzal@egr.msu.edu	• Surface and interfacial phenomena • Adhesion • Fiber-matrix bonding • Surface modification of polymers
NMI Naturwissenschaftliches und Medizinisches Institut an der Universität Tübingen Markwiesenstrasse 55 72770 Reutlingen Germany	Dr. rer. nat. Bernhard Schröder schroeder@nmi.de	• Surface characterization XPS (X-ray photoelectron spectroscopy) and SIMS (secondary ion mass spectrometry)] of bonded parts • Strength and long-term durability of adhesive joints • Adhesive bonding in medical technology
NYU College of Dentistry 806S Schwartz 345 East 24 Street New York, NY 10010 USA	Prof. P. van Thompson D.D.S, Ph.D., B.S. van.thompson@nyu.edu	• Performance and properties of brittle materials • Tissue response to scaffolds • Adhesion to hard tissues • Diagnosis and treatment of hidden caries

Table A.1 (Continued)

Address	Contact	Topic(s) of Research
ofi Institut für Klebetechnik Österreichisches Forschungsinstitut für Chemie und Technik Arsenal, Objekt 213 Franz-Grill-Strasse 5 1030 Wien Austria	Dr. Werner Preusser werner.preusser@ofi.co.at	• Adhesive bonding onto lacquer • Quality aspects of adhesive-bonded rail vehicle • Adhesive bonding of trains • Adhesive bonding of metals
Oxford Brookes University Gipsy Lane Headington Oxford OX3 0BP United Kingdom	Prof. Allan Hutchinson M.Sc., Ph.D arhutchinson@brookes.ac.uk	• Joining technology • Adhesion, adhesives, coatings and sealants: science and technology • Surface characterization and surface treatments • Polymer composite materials • Aerospace applications of composites and sealants • Automotive: composites, joining strategies and vehicle recycling • Construction: composite materials systems, externally bonded FRP reinforcement, bonded connections in timber structures and sealing of curtain walling in building
Prüf- und Forschungsinstitut Pirmasens e.V. Marie-Curie-Strasse 19 66953 Pirmasens Germany	Dr. rer. nat. Gerhard Nickolaus info@pfi-ps.de	• Adhesive bonding for automatic joining of shoes • Dispersion adhesives in footwear
RWTH Aaachen ISF – Institut für Schweisstechnik und Fügetechnik Pontstrasse 49 52062 Aachen Germany	Prof. Dr.-Ing. U. Reisgen office@isf.rwth-aachen.de www.isf.rwth-aachen.de	• Microbonding • Development and optimization of micrometering techniques • Rheological characterization of adhesives. • Production and characterization of microbonds • Processing of microbonding
SINTEF Synthesis and Properties R. Birkelands vei 2B 7465 Trondheim Norway	Otto Reidar Lunder Ph.D., M.Sc. Otto.R.Lunder@sintef.no	• Pretreatment for adhesive-bonded automobile parts • Micro-featured polymer components • Adhesion and surface modification

Table A.1 (*Continued*)

Address	Contact	Topic(s) of Research
Studiengesellschaft Stahlanwendung e.V. Sohnstrasse 65 40237 Düsseldorf Germany	Dr.-Ing. Hans-Joachim Wieland stud.ges@stahlforschung.de	• Adhesive bonding of steel for car manufacturing • Adhesive bonding for the production of oooooooo? • Adhesive bonding for the building industry • Standardization, quality control and test engineering
Szczecin University of Technology Institute of Chemical Organic Technology ul. Pulaskiego 10 70 322 Szczecin Poland	Prof. Dr. Zbigniew Czech psa_czech@wp.pl	• Synthesis and modification of solvent-borne acrylic pressure-sensitive adhesive (PSA) • UV crosslinkable PSA systems • Water-soluble PSA • Silicone PSA and polyurethane PSA • Crosslinking agents and photoinitiators
Technical University Delft The Adhesion Institute Kluyverweg 1 2629 HS Delft The Netherlands	Dr. J. A. Poulis J.A.Poulis@tudelft.nl www.hechtingsinstituut.nl	• Constructive adhesive bonds and coatings in a wide range of applications, but mainly for aerospace and civil engineering • Surface treatment (both metals and polymers) • Adhesive selection • Tests: initial strength and durability tests on coatings and adhesive bonds
Technische Universität Braunschweig Institut für Füge- und Schweißtechnik Langer Kamp 8 38106 Braunschweig Germany	Prof. Dr.-Ing. Prof. h.c. Klaus Dilger Prof. Dr.-Ing. Prof. h.c. Stefan Böhm ifs-bs@tu-bs.de	• Adhesive bonding of technical textiles, adhesive bonding for the building industry, adhesive bonding in medicine (teeth) • Close to practice and nondestructive controls • Automation of adhesive bonding • Quick-time curing, aging, surface conditioning • Polymer foams, pressure-sensitive adhesives, computer-aided methods • Micro-adhesive bonding, use of adhesives with reduced viscosity, application techniques • Micro-dose methods, adhesive bonding with batch technology

Table A.1 (*Continued*)

Address	Contact	Topic(s) of Research
Technische Universität Dresden Fügetechnik und Montage George-Bähr-Strasse 3c 01069 Dresden Germany	Prof. Dr.-Ing. habil. Uwe Füssel ftm@mciron.mw.tu-dresden.de	• Development of operations and tools for the category groups of thermal joining, adhesive bonding, hybrid joining and mechanical joining (joining by screws, pressing) • Integral strategy of processes for mounting, handling and joining
Technische Universität Berlin Fügetechnik und Beschichtungstechnik im Institut für Werkzeugmaschinen und Fabrikbetrieb Pascalstrasse 13 10587 Berlin Germany	Prof. Dr.-Ing. Dr. h.c. Lutz Dorn Lutz.Dorn@tu-berlin.de	• Environmentally friendly surface treatments for adhesive bonding of plastic material (low-pressure plasma, corona, flame impingement) • Investigation of adhesive bondings of plastic materials with metals (strain status, static and dynamic stability, creep behavior in tension and aging, design and dimensioning) • Combination of joining techniques for metals, such as spot-welding with adhesive bonding
Technische Universität Kaiserslautern Arbeitsgruppe Werkstoff- und Oberflächentechnik Erwin-Schrödinger-Strasse 67653 Kaiserslautern Germany	Univ. Prof. Dr.-Ing. Paul L. Geiß geiss@mv.uni-kl.de	• Adhesive bonding for light construction • Adhesive bonding of glass • Adhesive bonding with structural and elastic adhesives • Pressure-sensitive adhesives • Calculation and simulation of adhesive bondings • Adhesive bonding of aluminum with long-term stability • Analysis of adhesives and polymers
Technische Universität München Institut für Werkzeugmaschinen und Betriebswissenschaften Boltzmannstrasse 15 85747 Garching Germany	Prof. Dr.-Ing. Michael Zäh michael.zaeh@iwb.tum.de	• Automated nondestructive control of adhesive bondings • Simulation of adhesive bonding • Investigation of the behavior of the adhesive bonding process • Further developments of installation engineering

Table A.1 (Continued)

Address	Contact	Topic(s) of Research
Tel Aviv University School of Mechanical Engineering The Iby and Aladar Fleischman Faculty of Engineering Ramat Aviv 69978 Israel	Prof. Chai Herzl Ph.D., M.S. herzl@eng.tau.ac.il	• Experimental and analytical aspects of solid mechanics, with special emphasis on the deformation and fracture of composite materials, adhesively bonded joints and thin-film interlayers • Contact buckling and postbuckling of beams and plates • Evaluation of the deformation and fracture characteristics of thin-bondline adhesive joints under mode I, mode II and mode III loading • Application of Moire interferometry for studying in-plane deformation around stress risers • Mechanical evaluation of metal-based multilayer thin films • Indentation strength of ceramics bilayers
Tampere University of Technology Department of Materials Science Korkeakoulunkatu 33720 Tampere Finland	K. Hakala M.Sc., Ph.D Kati.hakala@tut.fi	• DSC measurements for the PU system • Fluorescence sensing for monitoring the curing and aging of adhesives and polymer matrix composites • Durability prediction of adhesively bonded stainless steel
Universität des Saarlandes Arbeitsgruppe Strukturforschung, Polymere, Grenzschichten (ASPG) Postfach 151150 66041 Saarbrücken Germany	Prof. Dr. Wulff Possart w.possart@mx.uni-saarland.de	• Basic mechanisms of adhesion • Hypermolecular structure and mobility of macromolecules at interfaces • Chemical reactions and aging in interfaces, in metal joinings with polymers, on the surfaces of polymers, and on films of polymers

Table A.1 (*Continued*)

Address	Contact	Topic(s) of Research
Universität Göttingen Institut für Holzbiologie und Holztechnologie Lehrbereich Holzchemie und Holztechnologie Büsgenweg 4 37077 Göttingen Germany	Prof. Dr. Holger Militz hmilitz@gwdg.de	• Technology of wood • Protection of wood • Chemical modification of wood materials • Chemistry of wood • Adhesive gluing of wood • Coating technology for wood • Wood products for structural engineering • Panel materials from cellulose
Universität Kiel Lehrstuhl für Materialverbunde Kaiserstrasse 2 24143 Kiel Germany	Prof. Dr. Franz Faupel ff@tf.uni-kiel.de	• Nucleation and growth of metallic films on polymer surfaces • The interplay of metal diffusion and aggregation at polymer surfaces and inside polymers • Glass transition on polymer surfaces • Metal-polymer interface and interfacial chemistry • Condensation (sticking) coefficients of metals on polymers • Correlation between microscopic parameters of the interface and adhesion • Ion beam modification of polymer surfaces • Tailoring of the interface to optimize adhesion and other properties
Universität Paderborn Labor für Werkstoff- und Fügetechnik LWF Pohlweg 47–49 33098 Paderborn Germany	Prof. Dr.-Ing. Ortwin Hahn ortwin.hahn@lwf.upb.de	• Application of alternating field with high frequency for curing of reaction polymers for adhesive bonding • Shaft to collar connection by use of adhesive bonding • Structural damages in the elements of adhesive bondings by shrinkage • Adhesive bonding of knop elements for spaceframe constructions

Table A.1 (*Continued*)

Address	Contact	Topic(s) of Research
Universität Stuttgart Materialprüfungsanstalt Otto-Graf-Institut (FMPA) Pfaffenwaldring 40 70569 Stuttgart (Vaihingen) Germany	Prof. Dr.-Ing. H.W. Reinhardt http://www.fmpa.de	• Structural sealant glazing • Investigations of load capacity and long-term durability of adhesive joints • Monitoring and performance of verification of suitability • Investigation of case of damage
Universität Ulm IdM-Institut für dynamische Materialprüfung Helmholtzstrasse 20 89081 Ulm Germany	Prof. Dr. W. Pechhold wolfgang.pechhold@physik.uni-ulm.de www.uni-ulm.de/institute/idm	• Viscoelastic real-time spectroscopy in the frequency area of 10^{-3} to 10^{7} Hz
Université Bordeaux 1 Laboratoire de Mécanique Physique (LMP) UMR CNRS 5469 351 Cours de la Libération 33405 Talence Cedex France	Prof. Dr. Martin E.R. Shanahan m.shanahan@lmp.u-bordeaux1.fr www.u-bordeaux1.fr/	• Mechanics of adhesive bonds • Physicochemistry and electrochemistry of adhesion • Aging and degradation phenomena • Capillary and wetting phenomena
University of Akron Department of Polymer Science Akron, OH 44325-3909 USA	Prof. Mark D. Foster B.S., Ph.D. foster@polymer.uakron.edu	• Thermodynamics and dynamics in blends of polymers that differ in molecular architecture (e.g. star and linear chains) • Properties of the surfaces of pressure-sensitive adhesives • Adsorption of proteins to well-defined surfaces • Surface behavior of polymer brushes
University of Alicante Adhesion and Adhesives Laboratory 03080 Alicante Spain	Prof. Dr. José Miguel Martín-Martínez jm.martin@ua.es www.ua.es/grupo/laa	• Development of new environmentally friendly surface treatments to enhance adhesion in polymers, plastics, rubbers and composites • Synthesis, formulation and characterization of polyurethane adhesives • Structure-adhesion properties relationships • Development of thermoset coatings and acrylic polymers for protection of cultural heritage

Table A.1 (*Continued*)

Address	Contact	Topic(s) of Research
		• New bioadhesives for medical applications • Bonding optimization in materials, adhesives and joints in footwear • Organic-inorganic hybrid adhesives: optimization of interfacial interactions
University of Bristol Queen's Building University Walk Clifton Bristol BS8 1TR United Kingdom	Prof. Robert Adams BSc, DSc, PhD, CEng, ACGI, FIMechE, FInstP, FIM r.d.adams@bristol.ac.uk	• Recycling of structural composites • Adhesive joints; composite materials • Vibration of composites and bonded joints • Material and structural damping • Development of CFRP string instruments
University of California at Santa Barbara Department of Chemical Engineering 3357 Engrg II Santa Barbara, CA 93106-5080 USA	Prof. Jacob Israelachvili Ph.D., MS, Honorary Degree of Doctor of Engineering University of South Florida Honorary Degree of Doctor sc. h.c. ETH Zurich jacob@engineering.ucsb.edu	• Directly measuring the forces between surfaces in vapors and liquids • Interactions at the molecular level • Solid-liquid interfacial phenomena • Surface forces and the physical properties of very thin films • Rheology and tribology of surfaces and thin films at the molecular level
University of California at Santa Barbara Materials Department Santa Barbara, CA 93106-5050 USA	Prof. Edward J. Kramer B.Ch.E, Ph.D. Docteur honoris causa Ecole Polytechnique Federale de Lausanne edkramer@mrl.ucsb.edu	• Fracture of polymers • Diffusion of block copolymers • Polymer interfaces and surfaces • Polymer blends • Self-assembly • Phase behavior in thin films
University of Maine Department of Physics and Astronomy Orono, ME 04469-5708 USA	Prof. William N. Unertl Ph.D., M.S., B.S., B.S unertl@maine.edu	• Thermodynamics of contact angles, wetting and adhesion • Generalization of the classic theory of capillarity to high-curvature situations • Applications of digital image analysis and processing to interfacial tension and other surface science problems

Table A.1 (*Continued*)

Address	Contact	Topic(s) of Research
		• Modelling of cell and protein adhesion to synthetic and natural biomaterials • Applications of surface thermodynamics to biotechnological problems
University of Massachusetts Department of Polymer Science and Engineering Conte Research Center 120 Governors Drive Amherst, MA 01003 USA	Prof. Thomas J. McCarthy B.S., Ph.D. tmccarthy@polysci.umass.edu	• Nanoscopic polymer objects • Fundamental wettability studies using surfaces of defined chemistry, topography and pattern • Ultrahydrophobic and ultrahydrophilic surfaces • Nonfouling surfaces • Chemistry in supercritical fluids – swollen polymers • Polymer adsorption • Covalently attached monolayers • Polymer and inorganic surface modification • Surface acoustic wave sensor design • Green conformal coatings technologies • Chemical/topographic lithography • Wound dressings
University of Michigan Department of Materials Science and Engineering 2282 G. G. Brown Building Ann Arbor, MI 48109-2125 USA	Prof. Michael Thouless B.A., M.S., Ph.D. thouless@umich.edu	• Micromechanical modeling of materials • Interfacial fracture mechanics • Adhesion • Mechanical properties of adhesive joints • Mechanical properties of thin films and coatings • Mechanics of nano- and bio-systems
University of Rochester Department of Mechanical Engineering Hopeman 218 Rochester, NY 14627 USA	Prof. David J. Quesnel Ph.D., M.S., B.E. dque@me.rochester.edu	• Mechanics and materials science of joining • Mechanics of adhesive bonding • Molecular mechanics, mechanics of adhesion and micromechanics of fracture

Table A.1 (Continued)

Address	Contact	Topic(s) of Research
University of Surrey Faculty of Engineering and Physical Sciences Guildford Surrey GU2 7XH United Kingdom	Prof. J. F. Watts BSc, PhD, DSc, CEng, FIM, FInstP, CPhys J.Watts@surrey.ac.uk	• Application of X-ray photoelectron spectroscopy (XPS) to materials science • Application of time-of-flight secondary ion mass spectrometry (ToF-SIMS) to materials science • Cathodic delamination of organic coatings and in adhesive joints • Organic coatings for corrosion prevention • Adhesive bonding for aerospace and automotive applications • Surface modification of polymers • Environmentally friendly pretreatments for adhesive bonding • High-energy XPS • Acid-base interactions in adhesion
University of Surrey Faculty of Engineering and Physical Sciences Guildford Surrey GU2 7XH United Kingdom	Prof. A. D. Crocombe B.Sc.(Eng), Ph.D., FIMechE, CEng A.Crocombe@surrey.ac.uk	• Adhesive bonding (fatigue damage and environmental degradation) • Composite materials • Biomechanics • Damage modeling (predictive modeling techniques developed for quasi-statically loaded structures to embrace loading conditions such as static and dynamic fatigue and environmental degradation)
University of Texas at Austin Center for Mechanics of Solids, Structures and Materials W.R. Woolrich Laboratories (WRW) University Station C2100 Austin, TX 78712-0284 USA	Prof. Kenneth M. Liechti B.Sc., M.S., Ph.D. kml@mail.utexas.edu	• Interfacial force microscopy • Nano indentation of ultra-thin films and self-assembled monolayers • Mechanics of adhesion and friction • Interfacial fracture mechanics • Experimental mechanics • Viscoelasticity • Composite materials • Microelectronics packaging

Table A.1 (*Continued*)

Address	Contact	Topic(s) of Research
University of Tokyo Graduate School of Agricultural and Life Sciences Yayoi, Bunkyo-ku, 113-8657 Tokyo Japan	Prof. Hirokuni Ono Ph.D., M.S. ahono@mail.ecc.u-tkyo.ac.jp	• Acrylic pressure-sensitive adhesives • Adhesives from lignocellulosic waste
University of Toronto Mechanical & Industrial Engineering 5 King's College Road Toronto, Ontario, M5S 3G8 Canada	Prof. Dr. A. Wilhelm Neumann neumann@mie.utoronto.ca	• Thermodynamics of contact angles • Wetting and adhesion generalization of the classical theory of capillarity to high-curvature situations • Applications of digital image analysis and processing to interfacial tension and other surface science problems • Modeling of cell and protein adhesion to synthetic and natural biomaterials • Applications of surface thermodynamics to biotechnological problems
University of Utah Department of Mechanical Engineering 50 s. Central Campus Drive Salt Lake City, Utah 84112 USA	Prof. K. L. DeVries Ph.D., B.S., A.S. kldevries@eng.utah.edu	• Failure in materials • Material selection and substitution • Mechanical properties of polymers • Behavior of adhesive joints • Mechanical testing of materials and biomaterials.
University of Wollongong Polymer Properties Group Faculty of Engineering Wollongong, NSW 2522 Australia	Prof. Hugh Brown B.Sc., M.S., Ph.D. hbrown@uow.edu.au	• Adhesion of polymers, coatings and paints • Characterization of thin polymer films • Mechanical properties and failure of polymers
Virginia Polytechnic Institute and State University Department of Chemistry 2107 Hahn Hall Blacksburg, VA 24061 USA	Prof. Thomas C. Ward B.Sc., M.S., Ph.D. tward@vt.edu	• Adhesion between a fiber and a matrix in a composite, or between an adhesive and a metal • Modeling of long-term durability of adhesive and sealant joints

Table A.1 (*Continued*)

Address	Contact	Topic(s) of Research
Virginia Polytechnic Institute and State University Department of Chemistry 2107 Hahn Hall Blacksburg, VA 24061 USA	Prof. Timothy E. Long B.Sc., M.S., Ph.D. telong@vt.edu	• Mechanics of hierarchical structures • Polymer adhesion • Biomimetic materials design • Responsive surfaces and materials • Elastic instabilities • Deformation and fracture of thin films • Polymer patterning • Nanocomposites • Combinatorial methods
Virginia Polytechnic Institute and State University Department of Chemistry 2111 Hahn Hall Blacksburg, VA 24061 USA	Prof. James E. McGrath B.S., M.S., Ph.D jmcgrath@vt.edu	• Synthesis and characterization of high-performance matrix polymers and structural adhesives • High-temperature polymer dielectrics for computers • Fire-resistant polymers and composites • New, directly copolymerized sulfonated aromatic copolymers for proton-exchange membranes (fuel cells)
Virginia Polytechnic Institute and State University Science & Mechanics 120E Patton Hall Blacksburg, VA 24061 USA	Prof. David A. Dillard Ph.D., M.S., B.S. dillard@vt.edu	• Design and characterization of adhesive joints • Fracture and fatigue of adhesive bonds • Mixed mode fracture studies of bonded joints • Performance and durability of adhesives, coatings and sealants • Viscoelastic characterization of adhesives, sealants and membranes
Wehrwissenschaftliches Institut für Werk-, Explosiv-und Betriebsstoffe Institutsweg 1 85435 Erding Germany	Prof. Dr. Jürgen von Czarnecki www.bwb.org/wiweb WIWEBpostein-gang@bwb.org	• Early detection of cases of damage, monitoring of bonding status • Analysis of cases of damage to adhesive bondings • Repairing procedures for adhesive bondings • Prediction of durability of adhesive bondings

References

Chapter 2

1 G.S. Plinius, Senior (1857) *Historia Naturalis*, Vol. 23, Chap. 43, Teubner Verlag, Leipzig.
2 Johnson & Johnson (1972) *Professional Uses of Adhesive Tapes*, 3rd edition, New Brunswick, New Jersey.
3 W.H. Shecut and H.H. Day (1845) U.S. Patent 3, 965.
4 R.G. Drew (1930) U.S. Patent 1, 760, 820.
5 O.J. Hendricks and C.A. Dahlquist (1967) in *Pressure Sensitive Tapes Adhesion and Adhesives*, Vol. 2 (eds R. Houwing and G. Salomon), Elsevier, Amsterdam, pp. 387–408.
6 A. Matting (ed.) (1969) *Metallkleben*, Springer-Verlag, Berlin, Heidelberg, New York.
7 J.A. Bishopp (1990) The history of redux and the redux bonding. *International Journal of Adhesion and Adhesives*, **17** (49), 287–301.
8 (a) E. Preiswerk (1965) Die Bindefunktion der Äthoxylin-(Epoxi-) Harze. *Technica*, **4**, 247; (b) E. Preiswerk (1965) Die Bindefunktion der Äthoxylin-(Epoxi-)Harze. *Technica*, **5**, 355.
9 O. Gerngross and E. Goebel (eds) (1933) *Chemie und Technologie der Leim- und Gelatinefabrikation*, Verlag Th. Steinkopf, Dresden, Leipzig.

Chapter 3

1 T. Young (1805) Cohesion of fluids. *Transactions, Royal Society of London*, 95.
2 A. Dupré (1896) *Théorie méchanique de la chaleur*, Gauthier-Villars, Paris.
3 D.D. Eley (1961) *Adhesion*, Oxford University Press, London.
4 R.N. Wenzel (1949) Surface roughness and contact angle. *Journal of Physical and Colloid Chemistry*, **53** (1949) 1466.
5 K. Tingey (1996) *Surface restructuring of polyurethanes and its control by plasma treatment*. Proceedings of the 19th Annual Meeting of the Adhesion Society (ed. Th. Ward), Myrtle Beach, USA, pp. 436–437.
6 N.A. de Bruyne (1957) *Klebtechnik – Die Adhäsion in Theorie und Praxis*, Berliner Union, Stuttgart.
7 P.W. Atkins and J.A. Beran (1996) *Chemie – einfach alles*, VCH-Verlagsgesellschaft, Weinheim.
8 W. Possart and A. Röder (1984) Measurement of electrical potential distribution in a polymer near the contact of a metal by means of SEM. *Physica Status Solidi A*, **84**, 319–325.
9 W. Brockmann and P. Geiß (1994) Neue Erkenntnisse bezüglich der Fraktoemission bei Haftklebstoffen durch Kopplung von Autografie und Ladungsmessung beim Schälversuch, in *Kleben –*

SWISS BONDING '94 (ed. E. Schindel-Bidinelli), Print Service, Mülheim.
10 W. Brockmann and R. Hüther (1992) Natur der Adhäsionsbindungen von Haftklebstoffen, in *EURADH '92 – Tagungsband*, Dechema, Frankfurt, pp. 444–451.
11 S.S. Voyutskii (1963) *Autohesion and Adhesion of High Polymers*, Interscience John Wiley & Sons, New York.
12 J.W. McBain and W.B. Lee (1927) Adhesives and adhesion: gums, resins and waxes between polished metal surfaces. *Journal of Physical Chemistry*, **31**, 1644–1680.
13 G. Sandstede, E. Robens and G. Walter (1960) *Über die Bindung von aliphatischen Carbonsäuren an Metalloberflächen, untersucht durch gravimetrische Gassorptionsmessung*, International Kongress für grenzflächenaktive Stoffe, Vol. III, Universitätsdruckerei Mainz, pp. 409–414.
14 G.J. Kautsky and M.R. Barusch (1965) Adsorption and desorption of a surface active aminoamide on an oxidized iron surface. *Industrial and Engineering Chemistry Production Research*, **4** (4), 233–236.
15 H. Dunken (1962) Über physikalische und chemische Adhäsion. *Plaste und Kautschuk*, **9** (H7), S314–S317.
16 L. Dimter and K. Thinius (1964) Zur Kenntnis der duroplastischen Komponente in Kombinationsklebstoffe. *Plaste und Kautschuk*, **11** (6), 328–331.
17 A.F. Lewis and L.J. Forrestal (1964) Chemical nature of polymer to metal adhesion, in *Adhesion*, ASTM Special Technical Publication No. 360, ASTM, Philadelphia, pp. 59–75.
18 I. Langmuir (1916) The constitution and fundamental properties of solids and liquids. *Journal of the American Chemical Society*, **38**, 2221–2295.
19 (a) W. Brockmann (1969) Über Haftvorgänge beim Metallkleben. *Adhesion*, **13** (H9), S355; (b) W. Brockmann (1969) Über Haftvorgänge beim Metallkleben. *Adhesion*, **13** (H11), S448–S460; (c) W. Brockmann (1970) Über Haftvorgänge beim Metallkleben. *Adhesion*, **14** (2), 52–56; (d) W. Brockmann (1970) Über Haftvorgänge beim Metallkleben. *Adhesion*, **14** (7), 750–752.
20 W. Brockmann (1977) Untersuchung zu Adhäsionsvorgängen zwischen Kunststoffen und Metallen. *Adhäsion*, **19** (1975) No. 1, pp. 4–14, No. 2, pp. 34–39.
21 W. Brockmann (1977) Interface Reactions and their Influence on the Long-Term Properties of Metal Bonds. *Adhesives Age*, **20** (6), 30–34.
22 S. Lotz (1995) *Untersuchungen zur Festigkeit und Langzeitbeständigkeit adhäsiver Verbindungen zwischen Fügepartner aus Floatglas*, Dissertation, Universität Kaiserslautern.
23 M. Brémont (1994) *Verbesserung der Beständigkeit von Klebverbindungen aus verzinktem Stahl und einem Epoxidharzklebstoff*, Dissertation, Universität Kaiserslautern; Hinterwaldner-Verlag, München.
24 R. Hüther (1995) *Ein Beitrag zur Klärung der Adhäsionsmechanismen von Haftklebstoffen*, Dissertation, Universität Kaiserslautern; Verlag Shaker, Aachen.
25 T. Neeb (1999) *Adhäsionsmechanismen an mechanisch vorbehandelten Oberflächen*, Dissertation, Universität Kaiserslautern.
26 W. Possart (2005) *Adhesion – Current Research and Applications*, Wiley-VCH Verlag, Weinheim.
27 H. Jopp (1995) *Ein Beitrag zum besseren Verständnis der Wirkungsweise des Strahlens auf die Festigkeit und Beständigkeit von Metallklebungen*, Dissertation, Universität Kaiserslautern; Verlag Shaker, Aachen.
28 S. Emrich (2003) *Untersuchungen zum Einfluss von Oberflächenchemie und -morphologie auf die Langzeitbeständigkeit geklebter Aluminiumverbunde*, Dissertation, Universität Kaiserslautern.
29 W. Brockmann, P.L. Geiß and A. Wagner (2004) *Selbstheilungseffekte der Adhäsion in Klebverbindungen*. Proceedings of

the 18th International Symposium on Swiss-bonding (ed. E. Schindel-Bidinelli), Rapperswil, pp. 17–25.

Chapter 4

1 D. Satas (1999) *Handbook of Pressure-Sensitive Adhesive Technology*, Satas and Associates, Warwick, R.I.
2 W. Brockmann and R. Hüther (1996) Adhesion mechanisms of pressure sensitive adhesives. *International Journal of Adhesion and Adhesives*, **16**, 81–86.
3 G. Habenicht (2002) *Kleben – Grundlagen, Technologie, Anwendungen*. Springer-Verlag, Berlin.

Chapter 5

1 Pressure-Sensitive Tape Council (1994) *Test Methods for Pressure-Sensitive Adhesive Tapes*, Chicago.
2 A.V. Pocius (2002) *Adhesion and Adhesive Technology*, Carl Hanser Verlag, München.
3 W. Bauer (1934) U.S. Patent 1 982 946.
4 E.W. Ulrich (1959) U.S. Patent 2 884 126.
5 J.A. Schlademann (1989) in *The Handbook of Pressure-Sensitive Adhesive Technology* (ed. D. Satas), Van Nostrand Reinhold, New York.
6 Sung Gun Chu (1989) in *The Handbook of Pressure-Sensitive Adhesive Technology* (ed. D. Satas), Van Nostrand Reinhold, New York.
7 C.A. Dahlquist (1966) *Tack, adhesion, fundamentals and practice*, McLaren and Sons Ltd, London.
8 H. Baumann (1967) *Leime und Kontaktkleber*, Springer-Verlag, Berlin.
9 H. Elsner (2002) Schmelzkleben. *Adhäsion Kleben und Dichten*, **3**, 34–35.
10 H.F. Huber and N. Vollkommer (1992) in *Handbuch Fertigungstechnologie Kleben* (eds O.-D. Hennemann, W. Brockmann and H. Kollek), Hanser Verlag, München–Wien, pp. 67–68.
11 Editorial (1998) Die Schmelzklebstoffproduktion wächst weiter. *Adhäsion Kleben und Dichten*, **1–2**, 20–24.
12 R. Jordan (1985/1986) *Schmelzklebstoffe, Band 4a, Rohstoffe und Herstellung*, Hinterwaldner Verlag, München.
13 G. Habenincht (2002) *Kleben: Grundlagen, Technologie, Anwendungen*, Springer-Verlag, Berlin, p. 199.
14 H.F. Huber (1995) *Dauerhaft kleben: Eine Einführung für den Praktiker*, Vincentz Verlag, Hannover.
15 P. Hoessel, H. Schupp, K. Lienert and H. Lehmann (1990) Schmelzklebelack lösung für hitzebeständige Beschichtungen. Deutsche Patentoffenlegung, DE 39 17 197 A1.
16 H.F. Huber and N. Vollkommer (1992) in *Handbuch Fertigungstechnologie Kleben* (eds O.-D. Hennemann, W. Brockmann and H. Kollek), Hanser Verlag, München–Wien, pp. 69–71.
17 G. Habenicht (2002) *Kleben: Grundlagen, Technologie, Anwendungen*, Springer-Verlag, Berlin, p. 88.
18 A. Groß, P. Theuerkauff, H. Kollek and H. Brockmann (1988) Nachvernetzende Epoxidharz-Schmelzklebstoffe. *Adhäsion Kleben und Dichten*, **11**, 6–24.
19 H. Onusseit (2002) Trends in der Hot Melt Entwicklung. *Adhäsion Kleben und Dichten*, **4**, 27–30.
20 B. Schröder and P. Pfizenmaier (1998) *Verschluss zum Verschließen eines Beutels sowie Verwendung eines derartigen Verschlusses*. Deutsche Patentoffenlegung PE 44 18 877 C2.
21 F. Nerdel (1964) *Organische Chemie*, Walter de Gruyter u. Co., Berlin.
22 W. Brockmann, H. Brockmann, Jr and H. Buddziekiewicz (1968) Die Konstitutionsermittlung eines Phenol-Formaldehydharzes. *Kautschuk Gummi Kunststoffe*, **12**, 679–683.
23 N.A. de Bruyne (1996) *My Life*, Midsummer Books, Cambridge.
24 (a) W. Brockmann, O.-D. Hennemann, H. B. Kollek and C. Matz (1986) Adhäsion in Aluminiumklebungen des

Fahrzeugbaus. *Adhesion,* **7**–**8**, 31–38; (b) W. Brockmann, O.-D. Hennemann, H. B. Kollek and C. Matz (1986) Adhäsion in Aluminiumklebungen des Fahrzeugbaus. *Adhesion,* **9**, 24–30; (c) W. Brockmann, O.-D. Hennemann, H. B. Kollek and C. Matz (1986) Adhäsion in Aluminiumklebungen des Fahrzeugbaus. *Adhesion,* **10**, 20–35.

25 A. Ravve (1995) *Principles of Polymer Chemistry,* Plenum Press, New York.

26 W.F. Gum (ed.) (1992) *Reaction Polymers: Polyurethanes, Epoxies, Unsaturated Polyesters, Phenolics, Special Monomers and Additives; Chemistry, Applications, Markets,* Hanser, Munich and Oxford University Press, Oxford.

27 J.D. Barwich, D. Guse and H. Brockmann (1989) Härtung von Epoxidharzen mit Dicyandiamid und Monuron. *Adhesion,* **5**, 27–30.

28 A. Groß (1987) *Modellreaktionen zum Aushärtungsverhalten von Epoxidharzklebstoffen.* Dissertation, Universität Bielefeld.

29 K.C. Frisch and L.P. Rumao (1970) *Journal of Macromolecular Science – Reviews in Macromolecular Chemistry,* C-5, 103.

30 J.W. Baker and I.B. Haldsworth (1974) *Journal of the Chemical Society,* 713.

31 J.W. Baker and J. Gaunt (1949) *Journal of the Chemical Society,* 9, 19, 24, 27.

32 K.N. Edwards(ed.) (1981) *Urethane Chemistry and Applications,* ACS Symposium Series 72, Washington, DC.

33 P.C. Briggs Jr. and L.C. Muschiatti (1975) U.S. Patent 3, 890, 407.

34 G.M. Estes, S.L. Cooper and H.V. Toblsky (1970) *Journal of Macromolecular Chemistry,* C-4, 313.

35 B. Mueller and W. Rath (2004) *Formulierung von Kleb- und Dichtstoffen,* Vinzenz Network.

36 Glasübergangstemperaturen von Monomeren, Übersicht von Sartomer, verfügbar unter www.sartomer.com. (last accessed 20 June 2008).

37 C.W. Boeder (no date) Adhesives, Coatings and Sealers Division, 3M, St. Paul, MN, USA.

38 National Starch & Chemicals (no date) Dymax, Loctite, Lord Product information.

39 Organische Peroxide (no date) Broschüre der Peroxid-Chemie GmbH, Pullach.

40 Produkbeispiele unter www.wellomer.com; www.delo.de; www.panacol.com; www.dymax.com (last accessed 20 June 2008).

41 Firmenbroschüre der Hönle AG, www.hoenle.de (last accessed 20 June 2008).

42 W. Beyer and W. Walter (1984) *Lehrbuch der organischen Chemie,* S. Hirzel Verlag, Stuttgart.

43 H. Baumann (1967) *Leime und Kontaktkleber,* Springer-Verlag, Berlin.

44 H. Onusseit (2003) Klebstoffe auf Basis natürlicher Rohstoffe. *Phänomen Farbe,* **23** (H9-10), pp. 14–19.

45 O. Gerngross and E. Goebel (eds) (1933) *Chemie und Technologie der Leim und Gelatinefabrikation,* Verlag Th. Steinkopf, Dresden, Leipzig.

46 E.P. Plueddemann (1991) *Silane Coupling Agents,* 2nd edition, Plenum, New York.

47 M. Brémont (1994) *Verbesserung der Beständigkeit von Klebverbindungen aus verzinktem Stahl und einem warmhärtenden Epoxidharzklebstoff,* Hinterwaldner-Verlag, München.

48 W. Brockmann (1977) Interface reactions and their influence on the long-term properties of metal bonds. *Adhesives Age,* **20**, 30–34.

49 H. Brockmann (1992) Zusammenhang von Adhäsion-Kohäsion und molekularer Struktur, in Proceedings Tagungband Swissbonding' 92, Rapperswil, pp. 62–70.

50 B. Wixmerten (2001) *Primer für Edelstahlklebungen mit Polyurethan- und Epoxidharzklebstoffen.* Dissertation, Universität Bielefeld.

Chapter 6

1 W.A. Lees (1984) *Adhesives in Engineering Design,* Springer-Verlag, Berlin, pp. 48–55.

2 R.J. Schliekelmann (1971) *Metallkleben – Konstruktion und Fertigung in der Praxis,* DVS Verlag, p. 37.

3 W. Brockmann and G. Fauner (1986) Fügen durch Kleben, in *Handbuch der Fertigungstechnik, Band 5, Fügen, Handhaben, Kleben* (ed. G. Spur), Hanser Verlag, p. 476.
4 H. Degner (1975) Berechnung von Klebverbindungen. *Schweißtechnik*, **25** (3), 117–121.
5 F. Mittrop (1962) Metallklebverbindungen und ihr Festigkeitsverhalten bei verschiedenen Beanspruchungen. *Schweißen und Schneiden*, **14** (9), 394–401.
6 W. Brockmann (1971) *Grundlagen und Stand der Metallklebtechnik*, VDI-Verlag, Düsseldorf.
7 D. Schlottmann (1983) *Konstruktionslehre, Grundlagen*, 2nd edition, Springer, New York.
8 A. Matting (1969) *Metallkleben*, Springer, Berlin.
9 W. Althof, G. Klinger, G. Neuman and J. Schlothauer (1977) *Klimaeinfluß auf die Kennwerte des elasto-plastischen Verhaltens von Klebstoffen in Metallklebungen*. DFVLR Deutsche Forschungs- und Versuchsanstalt für Luft- und Raumfahrt, Institut für Strukturmechanik, Research Report 7763, Braunschweig.
10 H. Kollek (1996) *Reinigen und Vorbehandeln*, Vincentz Verlag.
11 Hugo Kern-Liebers GmbH & Co. (2001) *KernLiebers Dosier-Kompendium*, Schramberg.

Chapter 7

1 A. Pocius (1997) *Adhesion and Adhesives Technology*, Hanser Publishers, Munich.
2 ASTM D 3983 (no date needed) *Standard test method for measuring strength and shear modulus of nonrigid adhesives by the thick-adherend tensile-lap specimen*. American Society for Testing and Materials.
3 B.R.K. Blackman and A. Kinloch (2001) Fracture tests for structural adhesive joints, in *Fracture Mechanics Testing Methods for Polymers, Adhesives and Composites* (eds A. Pavan, D.R. Moore and G.J. Williams), Elsevier Science, Amsterdam.
4 ISO 11343 (no date needed) *Adhesives – Determination of Dynamic Resistance to Cleavage of High Strength Adhesive Bonds under Impact Conditions – Wedge Impact Method*. International Organization for Standardization.
5 Guidance Document 003 (no date) European Organisation for Technical Approvals (EOTA).
6 D. Hasenberg (2006) *Zerstörungsfreie Prüfung geklebter Verbindungen mittels Ultraschall angeregter Thermografie*. Dissertation, Technische Universität Braunschweig.
7 M.F. Zäh, D. Kosteas, C. Lammel, Th. Mosandel, M. Michaloudaki and F. Dirscherl (2003) Zerstörungsfreie Prüfverfahren – Erzielbare Resultate und industrielle Umsetzbarkeit. *Adhäsion Kleben und Dichten*, **47** (7–8), 18–22.
8 D. Hasenberg, et al. (2004) *Charakterisierung von Klebverbindungen mittels frequenzmodulierter Ultraschallthermografie*, Dach-Jahrestagung, ZfP in Forschung, Entwicklung, Anwendung, Salzburg.
9 Th. Neeb (1999) *Adhäsionsmechanismen an mechanisch vorbehandelten Metalloberflächen*. Dissertation, Universität Kaiserslautern.
10 P.L. Geiß (1998) *Verarbeitungskonzepte und Belastungskriterien für Haftklebstoffe*. Dissertation, Universität Kaiserslautern; Hinterwaldner-Verlag, München.
11 W. Brockmann (1971) *Grundlagen und Stand de Metallklebtechnik, VDI Taschenbücher*, Chapter T22, VDI-Verlag GmbH, Düsseldorf.
12 W. Althof and G. Neumann (1974) Verfahren zur Ermittlung von Schubspannungs-Gleitungs-Diagrammen von Konstruktionsklebstoffen. *Mater. Prüf.*, **16** (H10), pp. 387–388.
13 G. Habenicht (2002) *Kleben, Grundlagen, Technologien, Anwendung*, 4. Auflage, Springer-Verlag, Berlin.

14 W. Brockmann, P. Geiß and C. Eicher (2003) *Klebbarkeit von Edelstahl rostfrei im Automobilbau*, FOSTA-Forschungsbericht P 553, Düsseldorf.

15 W. Brockmann, et al. (1986) Adhesion in bonded aluminium joints for aircraft construction. *International Journal of Adhesion and Adhesives*, **6** (3), pp. 115–143.

16 W. Brockmann and H. Kollek (1991) Durability assessment and life prediction for adhesive joints, in *Engineered Materials Handbook*, Volume 3, ASM International, Materials Park, pp. 663–672.

17 (a) W. Brockmann and S. Emrich (2002) Wie lange halten vorbehandelte Aluminiumklebungen Teil 1. *Adhäsion Kleben und Dichten*, **45** (5), 34–39; (b) W. Brockmann and S. Emrich (2002) Wie lange halten vorbehandelte Aluminiumklebungen Teil. *Adhäsion Kleben und Dichten*, **45** (7–8), 36–41.

18 H. Kleiner, J. Gehrke and H. Zschipke (2001) Mit Trockenschmierstoff beschichtete Aluminiumbleche erfolgreich kleben. *Adhäsion Kleben und Dichten*, **44** (11), 32–35.

19 D. Altenpohl (1965) *Aluminium und Aluminiumlegierungen*, Springer-Verlag, Berlin.

20 J. Kinloch and N.R. Smart (1981) Bonding and failure mechanisms in aluminium alloy adhesive joints. *Journal of Adhesion*, **12**, S23–S35.

21 W. Brockmann and S. Emrich (2000) *Model studies of the influence of enlarged surface areas due to increased adherend macroroughness on the adhesive strength and long-term-durability of aluminium-epoxide joints*. Proceedings of EURADH 2000, Société Francaise du Vide, Le Vide Science, Technique et applications Paris, No. 296-2/4-2000, pp. 311–319.

22 W. Brockmann, S. Emrich, C. Eicher and P.L. Geiß (2004) *Dominant effects of nano scale surface structuring onto long-term-durability of bonded metal joints*. Proceedings of EURADH 2004, DECHEMA, Frankfurt, pp. 156–160.

23 M. Graf (2004) *Die Langzeitbeständigkeit von Klebverbindungen aus Floatglas*. Dissertation, Technische Universität Kaiserslautern.

24 Jun Qu (2000) *Untersuchung zur Langzeitbeständigkeit adhäsiver Verbindungen aus Floatglas und Kunststoffen*. Dissertation, Technische Universität Kaiserslautern.

Chapter 8

1 K. Hoffer (1980) *Systemoptimierte Verbindungen im Flugzeugbau*. Proceedings, Systemoptimierte Verbindungen VDI-Berichte Nr. 360, Düsseldorf, pp. 31–43.

2 J.A. Bishop (1997) The history of redux and the redux bonding. *International Journal of Adhesion and Adhesives*, **17** (4), 287–301.

3 P. Allbericci (1983) Aerospace applications, in *Durability of Structural Adhesives* (ed. A.J. Kinloch), Applied Science Publishers, London, Chap. 8.

4 R.J. Schickelmann (1972) *Metallkleben – Konstruktion und Fertigung in der Praxis*, Fachbuchreihe Schweißtechnik, Band 60, Deutscher Verlag für Schweißtechnik, Düsseldorf.

5 Chr. Matz (1985) Alterungs- und Chemikalienbeständigkeit von strukturellen Klebverbindungen im zivilen Flugzeugbau. *Aluminium*, **61** (6), 118–121.

6 W. Brockmann, O.-D. Hennemann, H. Kollek and Chr. Matz (1986) Adhesion in bonded aircraft structures. *International Journal of Adhesion and Adhesives*, **6**, 115–143.

7 E.W. Thrall (1979) PABST-Program, Test Results. *Adhesives Age*, **22** (10), 22–33.

8 K.-H. Brück (ed.) (1990) *Fahrzeugverglasung*, Vieweg.

9 D. Symietz and B. Walterspiel (1995) Moderne Scheibenklebstoffsysteme zur Erhöhung von Fahrkomfort, Sicherheit

und Anwenderfreundlichkeit, in *Fahrzeugverglasung, expert verlag.*
10. D. Symietz (2001) Fügetechnik Kleben bei der Fahrzeugverglasung – Status und Trends, in *Glas im Automobil, expert verlag.*
11. H. Flegel (2002) The future of adhesive bonding as a joining technique. *AutoTechnology,* **5**, 64–67.
12. A. Lutz and D. Symietz (2003) Innovative structural adhesive approach for car manufacturing. *AutoTechnology,* **6**, 42–46.
13. bmbf-Forschungsergebnisse Meerestechnik 1994–1998, 14, Abschnitt 8.3, *Kleben im Bauwesen.*
14. R. Peterle (2000) Technisches Wissen – Klebstoffe. Uzin Utz AG, www.uzin-utz.com (last accessed 20 June 2008).
15. A. Dayton, H. Minge, V. Rydl and D. Voigt (2003) *verputzen, streichen, tapezieren: kunterbunt und gesund,* Hobbytip Nr. 341, WDR.
16. K.R. Wisser, J. Kunz and P.L. Geiß (2000) *Behavior and design of adhesive anchors under tensile load.* Proceedings of the 23rd Annual Meeting of the Adhesion Society, Myrtle Beach.
17. R. Eligehausen, R. Mallée and G. Rehm (1997) *Befestigungstechnik. Beton-Kalender 1997,* Verlag Ernst & Sohn, Berlin.
18. Stahlton, A.G. (1995) *Verstärkung von Konstruktionsbeton mit Klebelamellen aus Stahl und CFK. Stahlton AG.*
19. M. Dunky and P. Niemz (2002) *Holzwerkstoffe und Leime,* Springer-Verlag, Berlin.
20. M. Dunky, T. Pizzi and M. Van Leemput (2002) Proceedings, COST Action E14, Wood Adhesion and Glued Products, Working Group 1 Wood Adhesives.
21. J. Gabriel, F. Stoffel and T. Peng Hong (2004) Proceedings, 7th Pacific Rim Bio-Based Composites, Nanjing, P.R. of China.
22. C.J. Johansson, T. Pizzi and M. Van Leemput (2001) Proceedings, COST ActionE13, Wood Adhesion and Glued Products, Working Group 2: Glued Wood Products.
23. M. Kemmsies (1998) Swedish Testing Institute, report 97B2, 2063-1.
24. B. Radovic and C. Rothkopf (2003) Eignung von 1K-PUR-Klebstoffen für den Holzbau unter Berücksichtigung von 10-jähriger Erfahrung. *Bauen mit Holz,* **6** (3), 24.
25. K. Richter. Feuchtebeständige Verklebung von Brettschichtholz durch Einsatz eines Haftvermittlers. www.empa.ch (last accessed 20 June 2008).
26. H. Onusseit (2004) Trends in der Verpackung – Eine Herausforderung für Verpackungsklebstoffe. *Adhäsion Kleben und Dichten,* **48** (9), 21–25.
27. H. Onusseit (2003) Trends bei Verpackungsklebstoffen. *Adhäsion Kleben und Dichten,* **47** (3), 20–24.
28. P. Roier and H. Mensing (2004) Planlage in der Wellpappe. *Allgemeine Papierrundschau,* **4**, 16–21.
29. G. Henke (1996) Vier Typen, die sich ergänzen. *Pack Aktuell,* **15** (11), 18–20.
30. H. Onusseit (1999) Maßgeschneiderte Klebstoffe für jede Verpackung. *Farbe & Lack,* **105**, 160–167.
31. H. Onusseit (2001) Klebeverschlüsse für Fruchtsaftverpackungen. *Flüssiges OBST,* **10**, 611–614.
32. E. Pürkner (1992) Palettensicherung mit Antislipmitteln, H. Onusseit: Nicht immer, aber immer häufiger. *Packung & Transport,* **10**, 30–32.
33. H. Onusseit (2003) Trends bei der Etikettierung. *Flüssiges OBST,* **8**, 475–477.
34. G. Van Halteren (2004) Vom Knochenleim zum Sekundenkleber. *Naturwissenschaft im Unterricht Chemie,* **80**, 44.
35. H. Onusseit (2002) Etikettierung im Wandel – Etikettierklebstoffe für PET Flaschen. *Getränkeindustrie,* **56**, 20–23.
36. M. Dietz (2004) Klebstoffe und Applikationstechnik. *Bauindustrie,* **2**, 31–33.
37. D. Liebau (2001) *Industrielle Buchbinderei, I,* Heinz Verlag Beruf + Schule.
38. H. Onusseit (1990) Klebstoffsysteme. *Bindereport,* **1**, 26–29.

39 H. Onusseit (1997) Dispersionsklebstoffe für die Klebebindung. *Bindereport*, **6**, 373–381.
40 H. Onusseit (2003) PUR-Schmelzklebstoffe. *Bindereport*, **11**, S36.
41 H. Onusseit (2000) Klebebinden mit mehrschichtigem Klebstoffauftrag. *Bindereport*, **11**, 38–39.
42 P. Steinbach (2003) Proteinklebstoffe – Nach dem Vorbild der Natur. *Bindereport*, **11**, 34–35.
43 CEPI (2003) *CEPI Special Recycling 2002 Statistics*, October.
44 A. Ackermann, H.-J. Putz and L. Göttsching (1996) Herkunft und Gehalt klebender Verunreinigungen im grafischen Altpapier. *Wochenblatt für Papierfabrikation*, **124**, pp. 508–516.
45 H. Onusseit (1999) Klebstoffe für die Papier und Verpackungsindustrie – wie beeinflussen sie das Papierrecycling? *Allgemeine Papierrundschau*, pp. 945–951.
46 Verbundvorhaben 4332.62-N 597.7 (1994) Neue Klebtechniken, Wirtschaftsministerium Baden-Württemberg.
47 W. Gruber (2000) *Hightech-Industrieklebstoffe: Grundlagen und industrielle Anwendungen*, Verlag Moderne Industrie, Landsberg.
48 www.loctiteeurope.com/de/literature.htm (last accessed 20 June 2008).
49 A. Groß, O.-D. Hennemann and J. Bischoff (1992) in *Handbuch Fertigungstechnologie Kleben* (eds O.-D. Hennemann, W. Brockmann and H. Kollek), Hanser Verlag, München-Wien, pp. 332–383.
50 W. Brockmann (1998) W. u. R. Lüschen: Das Kleben von Stahl und Edelstahl rostfreu, Merkblatt 382, Stahlinformationszentrum, Düsseldorf.
51 Research Project IGF No. 14704, AIF, Köln (to be published in 2009).
52 U. Giersch and U. Kubisch (1995) *Gummi – Die Elastische Faszination*, Dr. Gupta Verlag.
53 J. Schnetger (1991) *Lexikon der Kautschuktechnik*, 2. Auflage, Hüthig Verlag.
54 R.W. Thomson (1845) GB Patent 10990.
55 D.B. Wootton (2001) *The Application of Textiles in Rubber*, Rapra Technology Limited.
56 W. Loy (1997) *Die Chemiefasern*, Schiele & Schön, Berlin.
57 Personal measurements in the lab of Mehler Engineered Products GmbH in Fulda (no date).
58 H.M. Wenghoefer (1974) *Rubber Chemistry and Technology*, **47** (5), 1066.
59 M. Fahrig (1998) *Rubber Technology International*, 139.
60 Y. Lyengar (1975) *Journal of Applied Polymer Science*, **5**, 514.
61 V. Morin, (no date) Michelin Company report.
62 C.M. Cepeda-Jiménez, M.M. Pastor-Blas, T.P. Ferrándiz-Gómez and J.M. Martín-Martínez (2000) Surface modifications on thermoplastic styrene-butadiene rubber treated with sulfuric acid, in *Polymer Surface Modifications: Relevance to Adhesion*, Vol. 2, VSP, Zeist, pp. 305–334.
63 C.M. Cepeda-Jiménez, M.M. Pastor-Blas, T.P. Ferrándiz-Gómez and J.M. Martín-Martínez (2001) Influence of the styrene content of thermoplastic styrene-butadiene rubbers in the effectiveness of the treatment with sulfuric acid. *International Journal of Adhesion and Adhesives*, **21**, 161–172.
64 M.D. Romero-Sanchez, M.M. Pastor-Blas, J.M. Martín-Martínez, P.A. Zhdan and J.M. Watts (2001) Surface modifications in a vulcanized rubber using corona discharge and ultraviolet radiation treatments. *Journal of Materials Science*, **36** (24), 5789–5799.
65 M.D. Romero-Sanchez, M.M. Pastor-Blas and J.M. Martín-Martínez (2001) Chlorination of vulcanized SBR rubber by immersion or brushing in TCI solutions. *Journal of Adhesion Science and Technology*, **15** (13), 1601–1620.
66 M.D. Romero-Sanchez, M.M. Pastor-Blas and J.M. Martín-Martínez (2002) Improved peel strength in vulcanized SBR rubber roughened before

chlorination with trichloroisocyanuric acid. *The Journal of Adhesion*, **78**, 15–38.
67 J.M. Martín-Martínez (2002) Rubber base adhesives, in *Adhesion Science and Engineering – 2. Surfaces, Chemistry and Applications*, Elsevier, Amsterdam, Chap. 13, pp. 573–675.
68 S.M. Tavakoli (2002) Adhesive bonding of medical and implantable devices. *Medical Device Technology*, **11**, 32–36.
69 B. Schröder, M. Kopf, R. Reichl, W. Brockmann and J. Jopp (1998) Klebstoffeinsatz bei der Herstellung medizinischer Instrumente. *Adhäsion Kleben und Dichten*, **42**, 16–21.
70 Johnson & Johnson (1972) *Professional Use of Adhesive Tapes*, New Brunswick, New Jersey.
71 N. Ewoldsen and R.S. Demke (2001) A review of orthodontic cements and adhesives. *American Journal of Orthodontics and Dentofacial Orthopedics*, **120**, 45–48.
72 H.F. Huber (1995) *Dauerhaft kleben*, Vincenz Verlag, Hannover.
73 Tissure Adhesives & Sealants (1999) *MedProMonth*, **IX** (3), 75–76.
74 Editorial (2003) (no authors; internet journal) *Biomedical Materials*, **2**, 6; www.intnews.com (last accessed 20 June 2008).
75 W.D. Spotnitz (1996) in *Surgical Adhesives and Sealants* (eds D.H. Sierra and R. Saltz), Technomic Publishing Company, Inc., Lancaster, Pennsylvania, pp. 3–7.
76 Polymeric Sealants (1998) *MedProMonth*, **VIII** (3), 85.
77 Fibrin Glues and Surgical Sealants (1997) *MedProMonth*, **VII** (1), 21.
78 W.D. Spotnitz (2002) Use of surgical glue takes hold at Temple. *Temple Times*, **32** (16) www.temple.edu/temple-times/5-18-00/glue.html (last accessed 20 June 2008).
79 Surgical Glues and Sealants (2000) *MedProMonth*, X (3), 86–87.
80 J. Ennker (1994) *Gewerbeklebstoffe in der Thorax- und Kadiovaskularchirurgie*, Steinkopff Verlag, Darmstadt.
81 Progress in Surgical Sealants (1999) *Biomaterials & Surgery*, **11–12**, 171.
82 K. Rischka, A. Hartwig and M. Wiegemann (2004) *Adhäsion*, **9**, 12–17.
83 Editorial (2003) (no authors; internet journal) *Biomedical Materials*, **3**, 9; http://www.intnews.com (last accessed June 2008).
84 W. Nachtigall (1974) *Biological Mechanisms of Attachment*, Springer-Verlag, Berlin.
85 S.N. Gorb (2001) *Attachment Devices of Insect Cuticle*, Kluwer Academic Publishers, Dordrecht.
86 S. Tyler (1988) *Fortschritte der Zoologie*, **36**, 331–347.
87 L.A. Thomas and C.O. Hermans (1985) *Biological Bulletin*, **169** (3), 675–688.
88 D.P. DeVore and R.J. Gruebel (1978) *Biochemical and Biophysical Research Communications*, **80** (4), 993–999.
89 N. Maruyama, H. Etoh, K. Sakata and K. Ina (1991) *Agricultural and Biological Chemistry*, **55**, 2887–2889.
90 J. Pardo, E. Gutierrez, C. Saez, M. Brito and L.O. Burzio (1990) *Protein Expression and Purification*, **1**, 147–150.
91 J.H. Waite, D. Hansen and K.T. Little (1989) *Journal of Comparative Physiology B*, **159**, 517–525.
92 H. Yamamoto, S. Yamauchi and S. Ohara (1992) *Biomimetics*, **1**, 219–238.
93 M.W. Denny (1980) *Dissertation Abstracts International B Sciences and Engineeering*, **40**, 5166–5167.
94 M. Scherge and S.N. Gorb (2001) *Biological Micro- and Nanotribology*, Springer-Verlag, Berlin.
95 J.H. Waite (1983) *Biological Reviews*, **58** (2), 209–231.
96 J.H. Waite (1991) *Chemistry and Industry*, **17**, 607–611.
97 E.C. Bell and J.M. Gosline (1996) *Journal of Experimental Biology*, **199**, 1005–1017.
98 J.E. Smeathers and J.F.V. Vincent (1979) *Journal of Molluscan Studies*, **45**, 219–230.
99 H.M. Feder (1955) *Ecology*, **36** (4), 764–767.

100 M.E. Lavoie (1956) *Biological Bulletin*, **111**, 114–122.
101 V.L. Paine (1929) *American Naturalist*, **62**, 517–529.
102 P. Flammang (1996) *Echinoderm Studies*, **5**, 1–60.
103 J.D. McKenzie (1987) *Cell and Tissue Research*, **248** (1), 187–199.
104 P. Flammang and M. Jangoux (1993) *Zoomorphology*, **113**, 47–60.
105 A.B. Yule and G. Walker (1987) *Crustacean Issues*, **5**, 389–402.
106 D. Lacombe and V.R. Liguori (1969) *Biological Bulletin*, **137**, 170–180.
107 H. Günzl (1978) *Zoomorphologie*, **90**, 197–204.
108 E. Braum, N. Peters and M. Stolz (1996) *International Reviews Ges. Hydrobiology*, **81**, 101–108.
109 H.M. Peters (1965) *Zoologisches Jahrbuch Physiologie*, **71**, 287–300.
110 H.M. Peters and S. Berns (1983) *Zoologisches Jahrbuch Anatomie*, **109**, 59–80.
111 R. Bennemann and I. Pietzsch-Rohrschneider (1978) *Cell and Tissue Research*, **193**, 491–501.
112 X. Turon (1991) *Cahiers de Biologie Marine*, **32**, 295–309.
113 M. Gianguzza and G. Dolcemascolo (1994) *European Archives of Biology*, **105**, 51–62.
114 J.S. Edwards (1992) *International Journal of Insect Morphology and Embryology*, **21** (4), 369–371.
115 T.P. Cogley, J.R. Anderson and J. Weintraub (1981) *International Journal of Insect Morphology and Embryology*, **10**, 7–18.
116 E.K. Tillinghast (1981) *Naturwissenschaften*, **68** (10) 526–527.
117 T. Strohmenger and W. Nentwig (1987) *Zoologischer Anzeiger*, **218** (1–2), 9–16.
118 A.E. Sorensen (1986) *Annual Reviews of Ecological Systems*, **17**, 443–463.
119 W.J. van Ooij, J. Giridhar and J.H. Ahn (1991) *Kautschuk Gummi Kunststoffe*, **44**, 348.
120 (no date) Polytec GmbH, Polytec-Platz 1–7, 76337 Waldbrunn, Germany.
121 (no date) Panacol Elosal GmbH, Obere Zeil 6–8, 61440 Oberursel, Germany.
122 G. Habenicht (2002) *Kleben – Grundlagen, Technologien, Anwendung*, 4. Auflage, Springer-Verlag, Berlin.
123 A. Hartwig and O.-D. Hennemann (1993) Recycling geklebter Materialverbunde. *VDI-Berichte*, **1072**, 225–235.
124 G. Manivannan and S.P. Saran (1995) Unglueing by supercritical fluids. *Adhesives Age*, **38** (9), 34–36.
125 W. Brockmann and B. Schröder (2004) Research Report Deutsche Bundesstiftung Umwelt, Az. 15481, Osnabrück.
126 B. Schröder (2004) Investigations for the development of a controlled removable dispersion adhesive for the building industry, in *Book of Abstracts, EURADH 2004*, Vol. 2, DECHEMA, Frankfurt, pp. 433–438.
127 Y. Nishiyama and C. Sato (2004) Expansion behaviour of thermally expansive micro-capsules for dismantlable adhesives, in *Book of Abstracts, EURADH 2004*, Vol. 1, DECHEMA, Frankfurt, pp. 319–324.

Index

a

Acacia arabica 90
acetic acid vapors 86
acetone-degreased aluminum bondings 177
acetone-degreased aluminum surfaces 181
acid-base bonds 23
acrylate adhesives 41, 84, 71, 227
– formulations 84
acrylate dispersions 71
acrylate/methacrylate adhesive groups 74
acrylate systems 42
acrylic adhesives 47
acrylic diester 31
acrylonitrile butadiene-styrene (ABS) 85
– copolymer 74, 84, 85
adherents deformation 107
adhesion promoters 24, 94–96, 189, 251, 325
– aramid fibers 328
– chemistry 94
– rubber mixtures 328
– textiles 328
adhesion theories 24
adhesive(s) 29, 39, 108
– anchor 238
– bonds 13, 191, 359
– chemistry 39
– classification 29, 32
– elastic-plastic deformation 108
– peel stress 130
– polymers 137
– properties 39
– protein plaque 345, 350
– residues 201, 363
– secretions 352
– substances 354
– surface 352
– survey 29
– systems 11, 28, 32, 115, 238–240, 258, 278, 324, 326, 365
– tapes 32, 40, 299, 302, 305, 317
– technique 1
adhesive bonded joints 101, 103, 104, 115, 373
– design 101, 103
– factors 104
– production 101
adhesive bonding 5, 197, 294, 317
– advantages 317
– historical development 5
– technology 1, 193, 205, 371, 372
adhesive deformation 107
adhesive film(s) 199, 224, 289, 308, 309
adhesive fracture energy 133
adhesive interactions 12
adhesive joints 2, 3, 224
– advantages 2
– characteristic features 2
– disadvantages 2
adhesive-wood interphase 247
adipic acid 68
aerosol adhesives 297
aircraft graphics technology 339, 340
aircraft industries 61
aircraft manufacture 207, 213, 217–219
– advantages 213
– historical development 209
alcoholic bonds 23
aleuritic acid, see aliphatic organic acid
aliphatic isocyanates 67
aliphatic organic acid 21
alizarine 98
alkali metal ions 315
alkoxysilane condensation 97
alkoxysilane hydrolysis 97
alkyl acrylate ester polymers 41
aluminum alloys 169

Index

aluminum oxide 98
aluminum/plastic film 262
aluminum surfaces 27, 179
American Minnesota Mining and Manufacturing Company 8
amine-curing epoxy resin adhesives 99
anaerobic adhesives 72, 73, 76, 318
animal glues 285
animal skin glue, *see* Se-gin
anisotropic adhesives 309
anodization methods 212
anodized surfaces 217
antennulary sucker 353
anti-aging agents 47
– metal oxides 47
anticorrosive agent 171
anti-graffiti foils 234
antislipping products 268
aromatic isocyanates 67
aromatic resins 46
– coumarone-indene 46
– poly(α-methyl styrene) resins 46
Asteronyx loveni 352
American Society for Testing and Materials (ASTM) 125
– databases 126
– standard practice D2094 129
– standards 125
ASTM standard shear tests 131
– ASTM D 1002 131
– ASTM D 2095 129
– ASTM D 3165 131
– ASTM D 5656 131
– ASTM E 229 131
atactic poly-alpha olefins (APAOs) 250
Atlas-Ago bonding technique 9
attic storeys 255
automotive industry 8, 220–224

b

backgluing adhesive 284
ball grid arrays (BGAs) 310
bending moments 105
bifunctional silicon compounds 96
biodegradable protein, *see* prolamin gel
biological attachment devices 346
bis-phenol-A diglycidyl ether 63, 64
bis-phenol-F diglycidyl ether 63
black/dark adhesives 312, 315
blasting method 110
– SACO 110
blast media 110
block-copolymer-based PSAs 42, 46
block copolymer systems 42

bonded joints 166
bonded rotor blade pockets 193
bonding energies 12
– magnitude 12
bonding-on-demand techniques 1
bonding process 231
bonding technology 219
book binding industries 278
boric-sulfuric acid 217
bottle-cleaning machine 288
bovine serum albumin (BSA) 346
brass-coated steel fibers 359
building industries 61
burning process 256
burning reaction 256
butadiene-acrylonitrile rubber 71
butt tensile tests 129
butyl acrylate 47

c

carbon-carbon bonds 12, 299
carbon-fiber Airbus tail-units 208
carbon-fiber reinforced plastics 207
carbon fiber-reinforced polymers (CFRP) 241
– advantage 241
– surface 204
carboxymethyl cellulose 9
cartridge systems 239
casein-based adhesives 270
casein glues 92, 93
catalytic system 52
cationic photocuring 59
cellophane backings 304
cellulose acetate 304
cellulosic fibers 246, 329
cement glands 348
ceramics-based fillers 308
cerebrospinal fluid 345
chelate bond 98
chelate complexes 95
– agents 99
chemical adhesion 22
chemical interactive forces 12
chemical reactions 21–24
chemical surface preparation techniques 111
chemical techniques 146
chemisorption experiments 22
chloroprene 71
chlorosulfonated rubber systems 74
chromatographic capacity 22
chromic acid 212
chromosulfuric acid 211, 217
cichlid fish, larvae 354
cigarette rods 274

circuit boards   311
civil aircraft   208
cladoceran crustaceans   354
classical adhesion theories   14–17
– schematic presentation   15
cleansing agents   109
– organic solvents   109
– water-miscible detergents   109
cloth adhesive tapes   304
coating system   37
coefficient of thermal expansion (CTE)   125, 128
cold-curing epoxy resins   10
cold-curing systems   9
cold-setting adhesives   317
cold-setting epoxy resin systems   217
collagen gel   345
colloid oxidic fillers   360
condensation crosslinking   86
construction engineers   243
consumer-friendly closure systems   266
contact adhesives   33, 47, 48
– chloroprene rubbers   47
– polyethylene copolymers   47
– polyvinyl acetates   47
corrosion-preventing lacquers   11
corrugated board   258, 259
– industry   258
COST project   257
crack initiation   133
crosslinking prepolymers   47
– phenolic resins   47
crosslinking reaction   87
crystalline polymers   350
cumene hydroperoxide   76
curing reaction   115
CuS dendrites   359
cyanoacrylate adhesives   83–85, 295
– anionic polymerization   83
– formulations   85
– super glues   83
cyanoacrylate monomers   85
– polymerization   85
cyanacrylic acid   85
cyclohexyl methacrylate   84

d
damping/insulation material   222
– styrofoam   222
de-adhesive material   353
de-adhesive secretions   352
De diversis artibus   6
degradation process   167
degradation reactions   213

Deir-el-bahari   5
dental treatment   344
detachable adhesives   91
dextrin adhesives   274
dextrin glues   92, 93
dextrin/synthetic resin-based adhesives   286
die-cut film adhesives   122
die-cut pressure-sensitive adhesive films   123
differential scanning calorimetry (DSC)   127
– ASTM D3418   127
difficult-to-bond materials   26
diffusion bonding   71
– cold welding   71
– solvent welding   71
diffusion theory   20
dihydroxy-boranyl   22
N,N-dimethyl-p-toluidine   76
dipole-dipole bonds   18
direct-current resistivity   225
dispersion adhesives   33, 69, 279, 297
dispersion film   282
dispersion-type tile adhesives   237
disposable static mixers   118
dog-bone-shaped test specimens   126
double cantilever beam (DCB)   133
double-layer acrylic film   305
double-sided adhesive tapes   293
Drosera spatulata   356
Drucker–Prager model   103
drum melter   281
dry blast processes   110
– flywheel-blasting   110
– grit-blasting   110
– vacuum-blasting   110
drylub-coated aluminum substrate   172
dry powders   37
dual-lock fasteners   300
duo-glands   347
Dupre equation   16
dynamic adhesion   183
dynamic mechanical analysis (DMA)   44
– plot   102
dynamic mechanical thermoanalysis (DMTA)   125, 127
– ASTM D4065   127
dynamic scanning calorimetry (DSC)   125

e
elastic adhesives   105
elastic deformation   105
elastic modulus   127
elastic PSA buffers   304
elastomer pads   308

elastomers (rubber) 323
– properties 323
electrical heating 155
electrical techniques 146
electrically conductive adhesives 308
electrochemical corrosion 167
electrochemical oxide/acid 111
electrochemical stability 181
electromagnetic spectrum 77
electronic devices 307
electronics industry 307
electrostatic phenomena 19
emulsion polymer isocyanate (EPI) adhesives 250
epoxide reaction 59
epoxy-amine adducts 64
epoxy-based adhesives 211
epoxy-based reactive hot melts 52
epoxy cements 315
epoxy nitrile adhesives 211
epoxy resin 9, 16, 24, 31, 36, 38, 58, 59, 181, 216, 227, 230, 343
– adhesives 39, 55, 58, 60, 61, 63, 176, 189, 212, 214, 320
– AW106 184, 186
– based reactive hot melts 52
– based Terokal 5026 186
– chemistry 58
– formulations 63, 65
– heat resistance 55
– properties 60
– protein-based adhesives 285
– use 9
epoxy/phenolic primers 211
epoxy systems, *see* plastisol adhesives
erosion-protection foils 357
ethylene propylene diene rubber (EPDM) 295
ethylene vinyl acetate (EVA) copolymers 49–51, 250, 271, 277, 283
– acrylates 263
– based dispersion adhesives 260, 286
– based hot melts 279
– based products 265
– hot glues 332
2-ethylhexyl acrylate 41, 47
European Article Numbering (EAN) technology 293
European Association of Adhesives Manufacturers 333
European Cooperation in Science and Technology 257
European Committee for Standardization 125

European Framework Regulation 1935/2004 292
European VOC solvents directive 144
exothermic adsorption 22

f

fabric-reinforced rubber goods 331
facade engineering 241
failure mode and effects analysis (FMEA) 123
fast-setting adhesives 316
– mortars 237
fatigue tests 139
fatty acids 90
fiber-reinforced cement 322
fiber-reinforced plastics 207, 219
fiber-reinforced polymers (FRP) 241
fiber-reinforced rubber composites 324
filament yarn 324
fillers 65, 99, 259
– aerosil 99
– aluminum oxide 65
– chalk 65, 259
– silica 99
film adhesives 121
– processing 121
finger print sensors 310
finite element simulation 138
flash lamps 154
flexible packages 260
floating roller peel test 136
fluorpolymer film 358
flysheet pockets 286
Fokker aircraft 210
fold-glued sheets 286
fold gluing 286
folding cartons 260
footwear industry 333
formaldehyde condensation products 256
formaldehyde condensation resins 249
Fourier transform infrared spectroscopy (FTIR) 22, 23
fracture resistance 133

g

gas discharge lamps 81
gelatin-resorcinol formaldehyde (GRF) tissue 345
gelatin-resorcinol pentanedial ethanedial (GR-DIAL) adhesive 345
gene technology 94
geotextiles 322
German Adhesives Association 371

German Association of Automotive
    Manufacturers 142
German Federal Institute for Risk
    Assessment 292
German Federal Ministry of Research and
    Technology 318
German Naval Airship Department 88
glandular organs 347
glass bubbles 65
glass-fiber patch 200, 201
glass fiber-reinforced plastic 215, 222,
    226
– bus roof 227
glass industry 299
glass primer products 225
glass reflectors 359
glass surface 26, 314
glass transition temperature (Tg) 125
glue boiler, Kollepsos 6
glue factory infrastructure 8
glue spreader 249
glue tablet 5
glutine glue 92
glycerol polyglycidyl ether 328
Goland equations 131
gold 25
graffiti-resistant films 337
graphic film 339
graphic industry 278, 279
graphic products 278
graphics sector 293
gray cast iron 318
grit-blast corundum (GBK) 157, 185
grit-blasted aluminum surfaces 181
grit-blast glass (GBG) 157, 185
grit blasting method 109
grit-blast steel (GBS) 157, 185
gum arabic 90
– acacia gum 90
– Sudan gum 90

h

halogen lamps 154
halogen light bulb 359, 360
handicraft trades 293
heat-conducting adhesives 317
heat-emitting surface 36
heat-sealing adhesives 30
heterogeneous material 351
high-alloy steel 316
high energy radiation 59
– UV light 59
higher-functional resins 64
high-frequency (HF) energy 279

high-melt systems 50
high-molecular-weight polyethylene glycol
    (PEG) 277
high-molecular-weight resins 362
high-molecular-weight systems 289
high-molecular-weight thermoplastics 290
high-performance acrylic PSA tapes 41,
    298
high-performance adhesive 50, 56, 63
– REDUX 56
high-performance foamed adhesive tapes
    298
– advantages 298
high-performance joining technique,
    see adhesive technique
high-proof alcohol 90
high-quality acrylic adhesive 357
high-strength adhesives 122, 160
high-strength polymer matrix 328
high-strength polystyrene (HIPS) 305
high-strength removable adhesives 367
high-strength steel 229
high-tech fibers 322
high-tech filaments 322
high-temperature adhesives 36
– polybenzimidazoles 36
– polyimides 36
high-viscosity polymer systems 29
Holothuria forskali 352
hook-and-loop fasteners 300
hot-melt additives 51
hot-melt adhesive(s) 26, 30, 34, 35, 49, 51,
    90, 121, 264, 265, 272
– advantages 49
– disadvantages 35, 49
– systems 296
hot-melt antislipping products 268
hot-melt labeling adhesives 271
hot-melt packaging adhesives 279
hot-melt systems 282
hot-seal adhesives 51
hot-seal coatings 51
hot-setting adhesives 109
hot-setting epoxy resins 121
hot-setting phenolic resins 121
hybrid bonding techniques 233
hydraulic flight regulation systems 201
hydrocarbon-based products 43
hydrogen atom 18
hydrogen bridge bonds 18
hydrophilic polyesters 277
hydroxyl groups 96
hydroxyl polyurethanes 69
hydroxymethyl phenol 23

i

indoor plant   91
– *Ficus benjamina*   91
industrial adhesives   258
industrial prefabrication technology   254
infrared (IR) radiation   77
infrared spectroscopy   200
infusion systems   343
injection anchor systems   239
injection-bonded framework   235
inkjet printer   286
inorganic adhesives   360
inorganic air-drying adhesives   359
inorganic fillers   313
– chalk powder   313
– quartz filler   313
inorganic materials   2, 11, 13
– ceramics   11
– glass   2, 11, 13
– metals   2, 11, 13
– stone   11
integral construction method   214
interfacial forces   13
intermolecular forces   17
International Organization of Standardization (ISO)   125
iron-doped gas discharge lamp   79
iron-gallate ink   21
isocyanate vapors   282
iso-octyl acrylate   41, 47
isotropic adhesives   309

j

jet aircraft   210
jewelry industry   363
joining aids   29
– welding electrodes   29

k

Kaurit glue   9
Kenics helices   116, 118
kresol novolaks   64

l

labor-intensive dip process   328
labor-intensive production process   218
lamination process   276
Langmuir adsorption isotherm   22
Langmuir's investigation methods   23
lap-shear test   131
latex polymers   325
lead oxide   9
license plates   337
light-emitting diodes (LEDs)   120, 153, 301

linear thermal expansion   128
– coefficient   128
liquid crystal displays (LCDs)   309
liquid isoprene rubber systems   74
liquid matrix resin   204
liquid-solid products   57
load-bearing structure   299
lockin thermography   153, 156
long-chain polymers   59
long-term static tests   137
long wavelength   77
– microwave radiation   77
– radio waves   77
lottery tickets   286
low-alloy steel   316
low deformation rate   43
low-energy surfaces   42
– polyethylene   42
– polypropylene   42
low-melt adhesives   297
low molecular base resins   36
low-molecular organic systems   38
low-molecular/plastified substances   36
low-molecular substances   35, 38
low-molecular-weight phenolic resin   53, 251, 367
low-molecular-weight polyamides   59
low-molecular-weight prepolymers   282
low-molecular-weight silica   361
low-molecular-weight substances   46, 289, 291
– diols   291
– resins   291
low-priced production techniques   222
low-reactivity metal, *see* gold
low-rubber domains   93
low-viscosity dispersion adhesive   282
low-viscosity liquid   16, 72
low-viscosity state   33
low-viscosity systems   32
Lufthansa Boeing 747-400   339
luminescent phenomena   19

m

macromolecular substance   35
magnesium carbonate   9
magnesium (oxide) phases   179
magnetic techniques   146
marine bivalve mollusks   348
marine glue   9, 90
marine organisms   347, 354
material-joining technique   1, 213
matrix resin material   204
mechanical engineering   316

medical instruments 343
– cannulas 343
– catheters 343
– keyhole surgery instruments 343
– respiratory masks 343
– syringes 343
medium-wave infrared cameras (MWIR) 154
membrane bonding 274
metal/acid reaction 111
metal backings 302
metal-polymer interface 183
metal-polymer joints 182
metal rods 129
metal salts 69
– dibutyltin dilaurate 69
– tributyltin acetate 69
methacrylate adhesives 71, 344
methyl cellulose 9
methyl methacrylate (MMA) 74
– based adhesives 75
– bone cements 344
– homopolymers 74
microencapsulated adhesives 37
Micropearl F-30 368
microplasma burner 111
microwave-excited mercury lamps 121
microwaves radiation 77
modified silane (MS) polymers, *see* MS polyurethanes
moisture-activated adhesive 90
moisture-curing polyurethanes 69
moisture-setting polyurethane adhesives 69
moisture-setting thermoplastic polyurethane 69
mounting optical elements 312
– epoxy 312
– polysulfide 312
MS polyurethanes 49
mucoid substance 350
multi-part adhesives 36
– atmospheric plasma treatment 112
– corona treatment 112
– low-pressure plasma treatment 112
Mytilus edulis foot protein 345

n

nanometric adhesive 48
napkin ring specimen 132
National Materials Advisory Board (NMAB) 145
natural adhesives 88, 89
natural resins-based products 43

natural rubber-based PSAs 46
– typical formulation 46
NCO-terminated polyether 261
neoprene rubber 30
neurosecretory-like cells 353
neutron radiography 146, 156
nitric acid 179
nitrile-phenolic adhesives 56, 57
nitrile rubber 211
– lattices 326
nonconductive adhesives 234
nondestructive testing methods 145
nonhydraulic binders 237
– anhydrite 237
– gypsum 237
– magnesia 237
– Sorel's cement 237
nonmetallic plate 147
nonmetallic substance 1
non-nanostructured surfaces 27
non-oxidizing acids 111
– dilute sulfuric acid 111
– hydrochloric acid 111
nonpolar organic materials 13
– polyethylene 13
– polypropylene 13
nonpolar polymer materials 113
– PE 113
– PP 113
– PTFE 113
nozzle systems 290

o

one-part reactive pre-polymer systems 52
one-part silicone adhesives 37, 85
– systems 36, 119
optical cement 312, 314
optical industry 312
optical lockin thermography (OLT) 149, 150
optical systems 314
organic/inorganic components 39
organic material 11, 99
– plastics 11
– rubber 99
– textiles 11
– wood 11
organic polymers 89
original equipment manufacture (OEM) 207
– systems 234
ortho hydroxyl methyl group 57
oxidative enzyme 350
oxidizing acids 111
– chromic acid 111

– nitric acid 111
– phosphoric acid 111

p

PAA process 137
packaging adhesives 268
packaging industry 257, 291
– introduction 257
– marketing aspects 257
packaging materials 258
paddle-shaped cilia 350
pane-fitting technologies 225
paper industry 257
paper recycling 290
paperboard containers 258
paper-bonding applications 276
parahydrogen atoms 56
parasitic insects 354
pavement markings 335, 336
peel force-peel rate curve 45
peel ply fabric 202
peel-ply method 217
peel ply treatment 202, 203
peel test 134–137
peptide bonds 89
phenol granules 350
phenolic epoxy resin primers 212
phenolic material 351
phenolic resin, *see* Tegofilm
phenolic resin adhesives 1, 23, 39, 53, 56, 57, 95, 211, 212
– applications 57
– chemistry 53
– formulation 56
– primer 32, 209, 210
phenol novolaks 64
phosphoric acid 181, 217
photoelectrical scanning systems 305
physical bonds 18
– dispersion forces 18
– induced dipoles 18
– permanent dipoles 18
physical interactive forces 12
physical processes 33
– melt evaporation 33
– melt solidification 33
physical surface preparation techniques 111
– flame treatment 111
– microplasma treatment 111
pickling process 217
pick-up adhesives 277
plant kingdoms 89
plant tissues 356

plastic bottles 273
plastic cartridges 239
plastic films 286
plastic materials 22
– polypropylene 22
plastic rubbers 263
– styrene-butadiene-styrene 263
– styrene-isoprene-styrene 263
plastic surfaces 335
plastified epoxy adhesive 108
plastisol adhesives 34
Poisson's ratio 102, 103
polar groups 18
– alcohols 109
– epoxy resin 18
– esters 109
– ketones 109
– polar hydrocarbons 109
– polyvinyl chloride 18
polarization theory 17–20
polyaddition adhesives 116
– epoxy resins 116
polyaddition reaction 53
polyamide resins 50
polycarbonate (PC) roof 227
polychloroprene-derived solvent containing adhesives 333
polychloroprene rubber 332
polyester adhesives 240
polyester fibers 326
polyester polyols 70
polyether polyols 68
polyethylene adhesion 25
polyethylene glycol 345
polyethylene pieces 17
polyethylene residues 25
polyethylene-styrene (PES) 328
polyethylene surface 16
polyethylene terephthalate (PET) 270
– soft drink bottles 288, 291
polyhydroxy-benzoic acid 98, 99
– derivatives 99
polyimide adhesives 216
polymer chains 232
polymer components 33, 240
polymer dynamics 26
polymer materials 12, 128
– bonding 191
polymer matrix 99
polymer network 137
polymer substances 35
polymer-substrate interface 159
polymer systems 33
polymer-textile composites 320

polymethyl methacrylate (PMMA) 47
polyolefin copolymer-based hot melts 267
polyols 68
polyphenolic molecules 245
polypropylene adhesion 25
polysaccharide 89
– gum arabic 90
polyunsaturated monocarboxylic acids 362
polyurethane adhesives 38, 66, 68–70, 96, 280, 281
– polybutadiene polyols 70
– polyesters 70
– polyethers 70
– properties 69
– structure 69
polyurethane-based reactive hot melts 52
polyurethane films 280
polyurethane hot melts 35, 282, 284
– adhesives 280
polyurethane tapes 304
polyvinyl acetates 332
– derived adhesives 275
– dispersions 249
– homopolymer 286
polyvinyl alcohol 276
polyvinyl alkyl ethers 41
polyvinyl homopolymers 279
polyvinylchloride (PVC) plastisol adhesives 224, 368
– copolymers 30
– particles 51
– plastisol 226, 227
polyvinylidene chloride (PVDC) 261
postcuring dispersion adhesives 250
postcuring polyurethane hot melts 296
Power Strips™ 365
prepolymer chains 232
pressure-sensitive adhesives (PSAs) 2, 9, 26, 29, 30, 32, 40–47, 91, 115, 122, 159, 222, 311, 365, 372
– acrylic-based 310
– adhesives 303
– chemistry 40
– coated envelopes 287
– coatings 270, 287
– dynamic mechanical analysis (DMA) spectra 44
– film 19, 287, 300, 310, 343, 365
– foam tapes 298, 299
– foil 356
– formulations 46, 47
– introduction 40
– packaging tapes 264, 266
– peel resistance 45
– physical properties 43
– properties 44, 303
– tapes 299, 358
– viscoelastic behavior 44
pressure-sensitive films 305
pressure-sensitive microsphere-type adhesive 366
pressure-sensitive rubber 293
primary/secondary amines 59
primers 29, 39, 94–96
– chemistry 39, 94
– classification 29
– properties 39
– survey 29
print-finishing industry 278
process equipment construction 316
profile-reinforced integral aluminum plates 236
prolamin gel 345
proteins 89
– fibronectin 89
– keratin 89
Prussian Academy of Sciences 7
PTFE adhesive films 303
PTFE tapes 303
public transport advertising 338
pulse echo 156
– flow chart 148
– technique 148
pulse thermography 153
PU resin adhesives 281
– adhesive sets 223
PU resin hot melts 283
PU systems 228
PVF adhesives 57
PVF powder 57
pyrogenic silicic acid 65

r

radio frequency identification (RFID) system 293
– antennas 310
– technology 293
radiographic techniques 146, 147
rayon fibers 329
reaction adhesives 27
– adhesive behavior 27
reactive acrylate/methacrylate systems 72
reactive adhesive systems 117, 313
reactive hot melts (RHMs) 52, 250
reactive low-viscosity diluents 63
– hexanediol diglycidyl ether 63
reactive systems 249
ready-made adhesive films 57

redispersable polymers 277
refractive index 314
Reissner equations 131
relaxation test 137
removable adhesive systems 365
repair technology 221
resol condensation 56
resol-type phenolic resins 95
resol-type resins 54, 57
resorcinol formaldehyde latex (RFL) 325, 326
– dipped textiles 325–327, 330
– systems 326, 327, 330, 331
reusable products 343
– endoscopes 343
– surgery instruments 343
rheology additive 65
ribbon-inserting machines 284
ring-opening polymerization 59
risk priority number (RPN) 123
rotor blade 195, 196
– profile 194
RTV-1 silicones 86
RTV-2 silicones 86
rubber adhesives 227
rubber-based material 136
rubber-based PSAs 41, 269
rubber reinforcements 322
rubber-resin adhesives 304
rubber solutions 9
rubber-steel damping element (Schwingmetall®) 360
rubber-textile composites 331
rubber-toughened epoxies 228

s

saddle-back stitched brochures 286
safety-relevant components 222
– seat belts 222
salt spray tests 141
scanning electron microscopy (SEM) 24
– images 187
– microscopy 366
screen-printable adhesives 301
sealing wax 90
secondary neutral particle mass spectrometry (SNMS) 179
Se-gin 5
self-adhesive film 337, 339
self-adhesive gummings 94
self-adhesive labels 269
self-adhesive products, band-aids 343
self-adhesive reflective sheeting 334, 337
self-adhesive substances 8
self-bonding varnishes 52

semiconductive organic polymers 293
sensory-secretory complexes 352
shaft-hub connections 318
shear stress–shear strain curves 160
shear stress–shear strain diagrams 162
shelloic acid 21
shipbuilding industry 235
shoe industry 9
short-fibril fibers 325
– cotton yarns 325
short-term stress cycles 163
short wavelength 77
– cosmic radiation 77
– X-rays 77
shot-blast ceramic (SBC) 157, 185
shot blast glass (SBG) 157
– surface preparation 158
shot-blast steel (SBS) 157, 185
– block copolymers 332
side gluing 283
silane adhesion promoters 188
silicone adhesives 85, 293, 316
– RTV-1 adhesives 85
silicon resin systems 37
silver-paste printed polyester circuit modules 309
single lap shear specimen 130
SIS block copolymers 332
small-scale industries 293
solar modules 299
solidified glue broth 92
solid-state material 13, 15, 19
solid-state polymers 20
solid rubber polymer 325
solvent-based adhesives 48, 290, 298
solvent-based systems 332
solvent-containing acrylate adhesives 71
solvent-containing contact adhesive 33, 34, 38, 47, 48
solvent-containing isocyanate systems 328
solvent-containing PU resins 261
solvent-containing synthetic resin adhesives 238
solvent-containing systems 71
solvent-free adhesive systems 71
solvent-free liquid phase 261
solvent-free systems 262, 263
solvent-free thermoplastic hot-melt resins 296
sophisticated test cabinets 143
– weatherometer 143
– xenotest 143
spot-welded adhesive-bonded joint 373
spray adhesives 294

spring-loaded transducer 147
stainless steel 27, 48, 319
– structures 319
starch-based adhesives 263
– paper bags 263
static mixer 118
steel plates 241
steel surfaces 27
stress-strain analysis 126
stress-strain curve 132
stretch-release tapes 372
structural adhesives 294
structural bonding 221
structural sealant glazing systems (SSGS) 242
styrene butadiene rubber (SBR) 295
– lattices 326
sulfur-free systems 329
sulfur vulcanization 359
Sulzer static mixer 118
surface-active substances 48
surface bonding 123
surface preparation methods 169, 174
surface tension quantities 17
surface-to-surface adhesive joint 229
surface-to-surface technique 213
surface treatment methods 114
– schematic presentation 114
synthetic fibers 321
synthetic polymers 270
synthetic resin-based adhesives 286
synthetic thermoplastics 289
– rubber 93, 283
synthetic wood adhesives 249

t
tailor-made polyurethane prepolymers 280
tail rotor blade 199
tail tie adhesives 277, 278
tamper-evident bottle closures 273
tamper-indicating labels 341, 342
tanning process 353
tapered double cantilever beam (TDCB) 133
– specimen 133
teflon/silver 17
Tegofilm 9
temple of Artemis 5
tenside-based dry lubricant coat 171
tensile lap shear specimens 162, 168
tensile shear tests 128, 129, 160
tertiary amines 69
– *N*-methyl-morpholine 69
– *N*-triethylene diamine 69

tetrahydrofuryl methacrylate 84
textile cleaning products 321
textile industry 50, 320
– history 320
textile-reinforced rubber components 323
thermal expansion coefficient 102
thermal joining technique 229, 230, 233, 365
– bonding 230
– clinching 229, 230
– laser welding 230
– riveting 229, 230
– soldering 229
– spot-welding 230
– welding 229
thermally conductive adhesives 307
thermal paint-curing processes 115
thermal wave 149
thermoplastic components 220
thermoplastic elastomers (TPE) 294
thermoplastic hot melts 50, 52
thermoplastic materials 71
– PVC 71
thermoplastic methyl methacrylate copolymers 71
thermoplastic polymers 279
thermoplastic polyurethane elastomer film 357
thermoplastic state 33
thermoplastic substances 290
thick adherent specimen (TAS) 131
– geometry 131
thixotroping agents 65
three-dimensional airship girders 7
tissue adhesives 344
tissue paper 275
– products 274
tomb of Tutankhamun 5
torture tests 141
traffic signs 335
transportation systems 207
two-part adhesives 35, 36, 115
– acrylic adhesives 35
– cold-setting epoxy resins 35
– polyester 35
– polyurethanes 35
two-part epoxy resins 63
– adhesives 63, 222
– resin systems 62
two-part polyurethane adhesive 70
two-part silicone adhesives, *see* RTV-2 silicones
two-shot system 282
two-stage reaction 66

u

ultra-high molecular weight (UHMW) PE 303
ultrasonic bath 157
ultrasonic lockin thermography 149, 151, 156
ultrasound burst phase thermography 151, 152
ultrasound equipment 315
ultrasound pulse 149
ultrasound thermography 157
urea-based adhesives 253
– curable acrylic 47
– reactive hot melt 53
– setting acrylate adhesives 77
UVA/UVB radiation 80
UV curing process 79, 81
– characteristic parameters 81
UV light 32, 37, 282, 314, 330, 358
– radiation 77, 120, 121, 299
– sensors 121
– source 77
– stabilizers 47, 79
UV-setting acrylics adhesives 71–73, 77, 80–84, 313, 315, 316
– formulations 84
UV-setting polyacrylates 53
UV-setting systems 73
UV/visible band 77

v

VDA alternating climate test 178
VDA-KKT test 168, 170
ventilation systems 48
vinyl pyridine (VP) lattices 326
viscoelastic-plastic deformation 137
visual inspection systems 232
visual/optical techniques 146
volatile organic compound (VOC) 126, 332
– emission 126, 144, 238
vulcanization accelerators 56
vulcanization agent 56
– aid 56
– sulfur 56
– zinc oxide 56

w

water-activated adhesive 264
– strips 264
– tapes 266
water-based products 276
water boil tests 141
waterborne adhesives 271, 273, 279, 289, 291
– labeling 270
waterborne products 268
waterborne system 291, 328
water jets 149
water-reactivated solid adhesives 91
water-resistant adhesive forces 57
water-resistant chelate compounds 57
water-resistant plywood 9
water-soluble hot melt 53
water-soluble polymers 277, 278
water-vapor pressure 36
Wedge impact method 138
Wedge test 133, 134
weight-to-payload ratio 206
well equipped motor 222
Wenzel's roughness factor 16
Western European adhesives market 372
wetting process 16
window-washing platforms 341
wood 245, 246
– adhesives 55, 248
– based materials 243, 247
– constructions 243, 257
– macroscopic structure 246
– molecular structure 245
– prospects 257
wood matrix 243
– cellulose 243
– lignin 243
wood working glue, Xylokolla 6

x

X-ray photoelectron spectroscopy (XPS) 23, 180

y

Young Dupre equation 112, 113